HEALTHCARE TECHNOLOGIES SERIES 3

Portable Biosensors and Point-of-Care Systems

Other volumes in this series:

Volume 1 ***Nanobiosensors for Personalized and Onsite Biomedical Diagnosis*** Prof. David A. Clifton (Editor)
Volume 2 ***Machine Learning for Healthcare Technologies*** Prof. David A. Clifton (Editor)
Volume 4 ***Biomedical Nanomaterials: From Design To Implementation*** Dr Thomas J Webster and Dr Hilal Yazici (Editors)

Portable Biosensors and Point-of-Care Systems

Edited by
Spyridon E. Kintzios

The Institution of Engineering and Technology

Published by The Institution of Engineering and Technology, London, United Kingdom

The Institution of Engineering and Technology is registered as a Charity in England & Wales (no. 211014) and Scotland (no. SC038698).

First published 2017

The Institution of Engineering and Technology
Michael Faraday House
Six Hills Way, Stevenage
Herts, SG1 2AY, United Kingdom

www.theiet.org

British Library Cataloguing in Publication Data
A catalogue record for this product is available from the British Library

ISBN 978-1-84919-962-9 (hardback)
ISBN 978-1-84919-963-6 (PDF)

Typeset in India by MPS Limited
Printed in the UK by CPI Group (UK) Ltd, Croydon

For Angela-Lu

Contents

List of contributors

Leonidas Alexopoulos
Assistant Professor
Systems Bioengineering Group
Department of Mechanical Engineering
National Technical University
of Athens
Athens
Greece

Pratik Banerjee
Assistant Professor
Division of Epidemiology,
Biostatistics, and Environmental
Health Science,
School of Public Health,
University of Memphis
Robison Hall,
Memphis, TN
United States of America

Joshua Balsam
Division of Chemistry and Toxicology
Devices,
Office of In Vitro Diagnostics and
Radiological Health, FDA,
Silver Spring,
Maryland
United States of America

Hugh Alan Bruck
Department of Mechanical Engineering,
University of Maryland College Park
(UMCP),
College Park,
Maryland
United States of America

Nikos Chaniotakis
Professor
Department of Chemistry
University of Crete
Heraklion, Crete
Greece

Bryan Allen Chin
Breeden Professor
Director, Materials Research
and Education Center
Director, Auburn University
Detection and Food Safety Center
Auburn, Alabama
United States of America

Michel Drancourt
Professor
Aix Marseille University
INSERM, CNRS, IRD, URMITE
Marseille
France

Brendan O'Farrell
President
DCN
Carlsbad, California
United States of America

Konstantinos P. Ferentinos
Informatics Laboratory
Department of Agricultural
Economics and Development
Agricultural University of Athens
Athens
Greece

Sotirios Kakabakos
Research Director
Institute of Microelectronics
National Center for Scientific Research
"Demokritos" Ag. Paraskevi,
Athens
Greece

George Kanakaris
Systems Bioengineering Group
Department of Mechanical Engineering
National Technical University of Athens
Athens
Greece

Spyridon E. Kintzios
Professor
Dean, School of Food, Biotechnology
and Development Director,
Laboratory of Cell Technology
Department of Biotechnology
Agricultural University of Athens
Athens
Greece

Konstantin N. Konstantinov
Professor
Department of Molecular Genetics
and Microbiology
University of New Mexico,
Health Sciences Center
Albuquerque, New Mexico
United States of America

Eleni Makarona
Institute of Microelectronics
National Center for Scientific Research
"Demokritos" Ag. Paraskevi, Athens
Greece

Andres W. Martinez
Assistant Professor
Department of Chemistry and
Biochemistry,
California Polytechnic State University,
San Luis Obispo, CA
United States of America

Konstantinos Misiakos
Institute of Microelectronics
National Center for Scientific
Research "Demokritos" Ag. Paraskevi,
Athens
Greece

Georgia Moschopoulou
Laboratory of Cell Technology
Department of Biotechnology
Agricultural University of Athens
Athens
Greece

Isabelle C. Noxon
Department of Chemistry and
Biochemistry,
California Polytechnic State University,
San Luis Obispo, CA
United States of America

Avraham Rasooly
Office of Cancer Complementary and
Alternative Medicine,
National Cancer Institute,
Rockville, Maryland
United States of America

Reuven Rasooly
Western Regional Research Center,
Agricultural Research Service,
U.S. Department of Agriculture,
Albany, California
United States of America

Shubham Subrot Panigrahi
Department of Food Process
Engineering,
National Institute of Technology
Rourkela, Rourkela, Odisha
India

Panagiota Petrou
Institute of Microelectronics
National Center for Scientific Research
"Demokritos" Ag. Paraskevi, Athens
Greece

Ioannis Raptis
Institute of Microelectronics
National Center for Scientific Research
"Demokritos" Ag. Paraskevi,
Athens
Greece

George Roussos
Department of Chemistry
University of Crete
Heraklion, Crete
Greece

Emily Roy
Division of Epidemiology, Biostatistics,
and Environmental Health Science,
School of Public Health,
University of Memphis
Robison Hall, Memphis, TN
United States of America

Robert L. Rubin
Professor
Department of Molecular Genetics
and Microbiology
University of New Mexico,
Health Sciences Center
Albuquerque, New Mexico
United States of America

Preetam Sarkar
Professor
Department of Food Process
Engineering,
National Institute of Technology
Rourkela, Rourkela,
Odisha
India

Spencer A. Schultz
Department of Chemistry and
Biochemistry,
California Polytechnic State University,
San Luis Obispo, CA
United States of America

Tyler A. Sisley
Department of Chemistry and
Biochemistry,
California Polytechnic State University,
San Luis Obispo, CA
United States of America

Petr Skládal
Professor
Department of Biochemistry and
CEITEC RG Nanobio
Masaryk University
Brno
Czech Republic

Christos Sotiropoulos
Professor
Hematology Division, Department of
Internal Medicine
University of Patras Medical School
Rion of Patras
Greece

Alexandre Tsoukalis
CEO
Micrel Medical Devices S.A.
Gerakas
Greece

Constantinos P. Yialouris
Professor
Informatics Laboratory
Department of Agricultural
Economics and Development
Agricultural University of Athens
Athens
Greece

Howard C. Wikle, III
Department of Research
Associate Professor
Materials Engineering
Auburn University
Auburn, Alabama
United States of America

Preface
Beyond the tricorder

We currently experience a scientific and technological revolution in the next generation of analytical and diagnostic devices, which are characterized by their very small size, ease of use, multi-analytical capacities and speed. A novel terminology, that was essentially unheard of 10 to 15 years ago, is now populating the jargon of scientific publications. Expressions like point-of-care (POC), point-of-care testing (POCT), point-of-test (POT) and point-of-interest//point-of-incidence (POI) are becoming increasingly familiar and popular among scientists and technologists.

How could one define a true portable diagnostic or analytic system? For a start, it has to be really small, ideally handheld or even smaller, as in the case of wearable and implantable sensors. This is not always feasible, but most POC devices nowadays are fairly miniaturized to the size of a small instrument easily transportable in a suitcase. Second, it should be robust enough to operate under resource-limited conditions, for example during travel by sea or air, or at home or in underdeveloped areas, using as less reagents as feasible (ideally, none at all) and providing a result in the fastest possible way. Third, they must be easily operated by a lay user, for example an ordinary person desiring to run a health scan but not having the background of a physician or a medical lab expert. Finally, multipurpose analytical capabilities and low cost are frequently listed among the top specifications of a POC device.

Like many other scientific and technological achievements, aspirations and expectations for an 'absolute' portable, all-in-one device have been fuelled by art and literature. The dominant influence in the field of POCT is righteously attributed to the sci-fi TV series *Star Trek* with the introduction, in 1966, of the Tricorder concept, a handheld gadget able to remotely scan the health status of an individual (human or alien, let us not forget this detail) and provide a diagnosis within a few seconds. So strong was the influence of the Tricorder idea that some early portable devices started were named after it (e.g., the Berkeley Tricorder, capable of monitoring a wide range of health-related signals and wireless transmitting them). More critical, however, was the launch, in 2012, of the *Qualcomm Tricorder XPrize* Challenge (http://tricorder.xprize.org/), with the aim to stipulate the development of a portable, user-friendly medical scanner, able to diagnose 13 predetermined pathological conditions (including stroke, tuberculosis, hepatitis A and pneumonia) and three other additional ailments to be decided by the individual contesters' teams. Although the competition is still running (with the prototypes of the seven finalist teams now in the qualification stage), it has been already successful in stirring the interest for a concerted effort towards the realization of a Tricorder-like device, as documented by the preliminary participation of 312 groups from 38 countries. Other similar initiatives have preceded or followed the *Tricorder XPrize*, including the *Nokia X-Prize Challenge* (indicating the anticipated core role of smartphones in POCT) and the recent European Commission *FoodScanner* initiative.

In spite of the publicity of these and other initiatives, criticism regarding the limits of POCT abound. In particular, there is concern whether measuring a limited set of diagnostic parameters (e.g. vital medical signals such as blood pressure, temperature and heart rate) or analytical targets (i.e. caloric value and percentage of protein and fat in food) suffices for the interpretation of the condition of the assayed sample, this being for example the health status of a patient or the safety and quality of a ready-to-eat food item. Furthermore, requiring a fast response time, combined with a simplification of sample preparation (vs the often tedious and lengthy standard process of sample purification and concentration) in a multidimensional, multipurpose analytical setting should also require inevitably the sacrifice of part of the retrievable information. For example, dedicated biosensors often fail to provide an analytical spectrum of a sample's composition. Similarly, lateral-flow assays, such as pregnancy tests, provide only qualitative results based on a cut-off value. Last, but not least, in order to achieve a minimally invasive diagnostic process, assays should be preferably done in air samples or using sample irradiation (e.g. by means of near-infrared spectroscopy) or by using easy-to-obtain biological samples, such as saliva and urine. Needless to say, such approaches may be ideal for POCT but bear the risk of retrieving non-representative information from the original target. Therefore, further scientific and technological development is due in order to allow for current portable devices and principles to reach their full potential. Fortunately, as it will be demonstrated in detail in the present book, this is not only feasible but, in several cases, a part of real, day-to-day analytical and diagnostic practice. Current and emerging systems and technologies surpass by far the perspective of the mythical Tricorder, having actually obtained the status of miniaturized multi-analytical platforms, that are directly comparable, in terms of capacity, to their conventional, bulky counterparts.

A common feature of the majority of portable detection systems is the fact that they are based on multidimensional (also known as hyphenated) analytical techniques rather than a simple working principle. Even in the latter case, miniaturization of the measuring unit and, often enough, the read-out device requires the flawless integration of several subcomponents, including the sensing biorecognition element, the signal transducer, the detection unit data processing interface, data classification and interpretation and finally the display of results to the end-user. In other words, the development of a working POC/POT system represents an interdisciplinary feat, in which successful integration of various and scientifically distant subcomponents – from cell biology and biochemistry to advanced optics to electronics and software engineering – is the indispensable goal.

Portable biosensors and POC systems are the first comprehensive, in-depth review of this fascinating field. More specifically, the book reflects the convergence of several scientific and technological disciplines – from nanotechnology to life sciences – in the development of fully operational systems, integrating the latest research results from a vast array of different sectors in such way to guarantee their flawless compatibility in a single platform.

The book is divided into four parts. *Part I (Chapters 1–6)* is an in-depth analysis of the various technologies upon which portable diagnostic devices and biosensors are built. As a homage to the – very close to the end-user – nature of optical systems, *Chapter 1* (by Avi Rasooly and co-authors) is dedicated to optical detection, as the prevalent approach for biomedical testing. In addition to reviewing current POCT technologies, the combination of low-cost portable imaging devices, simple optical configuration, and new computational and optical signal enhancement techniques is demonstrated to enable high-sensitivity florescence detection for POCT in resource-poor settings for various applications, such as the microbial toxins activity analysis of botulinum neurotoxin A and Shiga toxin 2. Actual examples of novel approaches for increasing the sensitivity and performance of low-cost devices are given, including the employment of image stacking-based computational signal enhancement. *Chapter 2* (by Andres Martinez and co-authors) is an illustrative display of the historical evolution of paper-based diagnostic devices. Currently, commercially available paper-based diagnostic devices (such as dipstick assays and lateral-flow assays) are analysed as the portable testing paradigm due to their low cost and simplicity, at the same time presenting a very low entrance barrier, both in terms of the cost of equipment and the required technical expertise. For example, microPADs are now being developed to detect a wide range of analytes, including environmentally concerning substances (e.g. volatile pollutants), defence-associated compounds (such as explosives), biomarkers and drugs. In addition, directions for future developments and features of the next generation of paper-based devices are presented. It follows that *Chapter 3* (by Brendan O'Farrell) is an extensive review of advanced lateral-flow technologies, which represent the most mature and widely applicable diagnostic commercial biosensors currently in use. They provide unique advantages to field testing, including the ability to deliver accurate, sensitive and often quantitative results, in parallel offering a high degree of operability by the end-user. Different lateral-flow architecture, formats and device designs are presented, along with their advantages and disadvantages. Performance considerations and conditions for optimization are described, nicely framed with actual application cases. Strategic considerations regarding future developments in the market place are also presented, which call for a broader focus on issues such as the selection of the right sample handling methods, reader technologies and device design. *Chapter 4* (by Robert Rubin and Konstantin Konstantinov) takes us to a further journey into immunosensor technology, particularly utilizing electrochemical methodology to measure antibody/antigen interactions. Because electrochemistry is inherently flexible, adaptable, and relatively amenable to low-cost fabrication and miniaturization, electrochemical immunosensors have particular promise for portable POC/POT applications. Different types of electrochemical immunosensors (e.g. amperometric, coulometric and impedimetric) are described, along with performance parameters such as antibody origin and specificity, assay duration, component reusability and the possibility of label-free application. These and other factors as well as desirable features are considered for the challenging conversion of potential POC electrochemical immunosensing devices developed in the laboratory to actual field-based settings. In *Chapter 5* (by Howard Wikle III and Bryan Chin), the fascinating technological field of magnetoelastic biosensors is presented. In this particular type of acoustic wave gravimetric sensors resonating platform (resonator), the biorecognition platform is composed of an amorphous magnetostrictive ferromagnetic alloy, which displays a strong coupling between the elastic and magnetic fields. Magnetoelastic biosensors have unique advantages for portable biodetection applications, due to the wireless design, very high mass sensitivity, low mechanical damping of the resonator platform and the rapid capture and binding of target species by the biorecognition layer, leading to very short assay duration (in the range of minutes). The origin of magnetostriction and magnetoelastic coupling, as well as the fabrication processes for the magnetoelastic resonator platform and the biorecognition element are reviewed, together with case applications. *Part I* concludes with a review, in *Chapter 6* (by Spyridon Kintzios) of cell-based biosensors currently in use with POC/POT systems. Even though the use of cells as biosensing elements is generally associated with relatively low cost, high sensitivity and speed of assay and relative – depending on the target application and sample matrix – resistance to interfering compounds, the commercialization of cell-based systems is still hampered by the limited cell storability and detection selectivity. This is about to change with the advent of the so-called designer cells, lying on the border between cells of natural origin or genetically engineered ones. Two major technologies for constructing 'designer' cells are presented, namely synthetic gene circuitry and membrane engineering. Moreover, commercial cell-based biosensors are prevalent in field applications for toxicity testing, having become popular during the last 15 years as reliable, cost-efficient and high throughput tools for a variety of toxicology tests, including acute toxicity testing, toxicokinetics, genotoxicity and mutagenicity. A vital discrimination in the design of a functional cell-based POC/POT system should be based on the prokaryotic or eukaryotic (in particular mammalian) origin of the cell, since this is a definite factor for the system's potential but also its limitations. This aspect is also assessed, along with a full update of commercial systems.

In *Part II*, advances in the design and optimization of special components of biosensor systems and handheld devices are presented. *Chapter 7* (by Nikos Chaniotakis and George Roussos) deals with nanocomposite materials used, currently or potentially, in biosensing technology, with the advantage of their high controllability and customizable fabrication, offering an ideal solution to the system miniaturization process. Miniaturization itself is an engineering challenge, presenting the researcher with a series of problems which are reviewed in this chapter. A detailed analysis is given of the dramatic changes of a material's chemical and physical properties due to size reduction below a critical level. Carbon nanocomposites are especially reviewed in-depth. Miniaturization is also a prerequisite for the fabrication of implantable biosensors, which undoubtedly represent the ultimate form of a portable diagnostic system. In *Chapter 8*, Panagiota Petrou and her co-authors handle meticulously the topic of monolithically integrated optoelectronic biosensors, their operating principle lying at the core of evanescent field optics, that is the study of the part of the electromagnetic field of the wave-guided light which decays exponentially in the medium above the sensors surface (with a depth of tens to hundreds of nanometres), depending on the changes in the refractive index taking place on the transducer surface, as a result of biorecognition interactions. Integrated optics technology is one of the most promising solutions towards the development of truly portable lab-on-a-chip platforms. The different categories of integrated optical sensors, such as grating-coupled waveguide sensors, microring resonators, photonic crystal waveguides, integrated interferometers, silicon nanowires and slot waveguides are presented and conceptually analysed, including examples of applications with performance details. A portable biosensor or other POC system is meaningless without a suitable, efficient signal-processing system that reads the measurement signal, providing the user with proper information or advice in a clear and explicit way. In *Chapter 9* (by Constantinos Yialouris and Konstantinos Ferentinos), techniques concerning the processing of time-series produced by portable biosensors are presented, including prediction and pattern recognition. Currently popular and successful computational intelligence techniques in time-series analysis, such as artificial neural networks, support vector machines and genetic algorithms, are explained along with case applications. Moreover, emphasis is given in three techniques that can be used in a final product, by developing an adequate app for a mobile POC device.

Having covered the technological background, its practical implementation comes next. In *Part III*, a wide scope of applications of portable biosensors and handheld POC devices is presented, ranging from the support of primary healthcare to food and environmental safety screening. Diverse topics are covered, including counterterrorism, travel medicine and drug development. Food safety control is a global daily concern, especially in view of environmental shifts caused by the climatic change, among other challenges. In *Chapter 10*, Pratik Banerjee and his co-authors tackle the integration of nanotechnology in food quality biosensing. Various groups of nanosensors are reviewed, including optical biosensors, such as fibre optic sensors, PEBBLE (probes encapsulated by biologically localised embedding) nanosensors and surface enhanced Raman scattering, as well as chemical and biological nanosensors. Exemplary applications include the detection of bacteria such as *Escherichia coli O157:H7*, *Listeria monocytogenes* and *Salmonella enteritidis* and associated toxins, fungal pathogens and organic biohazards, such as pesticide residues. Currently, the need for efficient early warning systems against terrorist attacks is more pressing than ever. The urgency of the availability of on-site and near real-time bioanalytical procedures for monitoring biowarfare and/or potential pandemic infectious disease agents is reflected on the emergence of the application-specific term point-of-incidence (POI). The quiver of available POI technologies is exhaustively reviewed by Petr Skládal in *Chapter 11*, from paper and lateral flow based assays and lab-on-chip/lab-on-disk platforms, to electrochemical detectors and surface acoustic wave immunosensors. Customized approaches for a large number of individual detection targets are described in detail, for example uropathogens, *Clostridium difficile*, human enteroviruses,

Ebola, influenza and botulinum neurotoxin type E – to name just a few of the applications displayed in this chapter. Travellers' safety is by default a dominant area for the employment of POCT. In particular, travelling by air or sea excludes the possibility of rapid access to fully or at least adequately equipped medical laboratories, hence the need for portable medical screeners – namely, a tricorder-like device. In *Chapter 12* (by Michel Drancourt) POC approaches in so-called medical deserts (prominently ships and commercial aircrafts) are reviewed in quite informative manner. The surveillance of travel safety incidences or their probability requires a strategic prioritization of assessment targets. For example, remote detection of fever, an otherwise non-specific clinical sign of infection, is critical to the selection of individuals presenting an increased risk of infectious disease. The internationally established rules for documenting infection risk during air and sea travel are detailed, including guidelines by the World Health Organization. Also suggested in this chapter are actions and measures to be set in arrival points after an initial confirmation of a positive signal, ultimately utilizing a syndromic POC kit. Not surprising at all, POCT in veterinary medicine is a rapidly expanding field, surpassing in terms of different applications even the human medical sector. This phenomenal development is due both to the economic value of farm animal products, the expanding pet ownership and the increased risk for human infection presented by zoonotic disease agents, like influenza A virus subtype (H1N1). *Chapter 13* (by Georgia Moschopoulou) provides a state-of-the-art assessment of portable biosensors for veterinary use. The review is classified in two major groups, farm animals and pets (mainly cats and dogs). Key applications include infectious disease (such as mastitis, influenza, porcine circovirus, classical swine fever virus, rabies and many others) as well as non-infectious pathological conditions like diabetes mellitus in pets (a prominent field for use of POC devices such as portable blood glucose meters), clinical and subclinical ketosis, hyperlactemia and fertility problems.

Finally, *Part IV* of the book is dedicated to the presentation of commercially available products, excluding these that have been already mentioned in previous chapters. In *Chapter 14*, George Kanakaris and his co-authors provide a literal gigantic review of 104 POC in-vitro-diagnostics companies and their products, their technical aspects and references to their clinical performance. The chapter is nicely complemented with a comprehensive table containing the majority of target biomarkers currently being the object of POC systems, along with the respective technologies utilized for each, and the companies offering a relevant product, as well as a discussion on the common attributes of commercialized point of care systems. With a detailed presentation of more than 150 applications, this chapter is certainly a definite source of reference for the description of the state of the art of the biosensor industry. In *Chapter 15* (by Spyridon Kintzios), technologies are reviewed having currently achieved a technology readiness level value of 7 or higher, therefore being promising for future commercialization and especially addressing the need for end-user/consumer-based operation. Emphasis is also given in advances in conceptual approaches and operating principles (e.g. breath analysis, ultra-miniaturized and endoscopic biosensors), novel applications of existing commercial portable biosensors (in particular food safety screeners and wearable sensors) and novel biomarkers. The concluding *Chapter 16* (by Alexandre Tsoukalis) is a special market case study regarding the POC infusion systems combined with intelligent patient monitoring. This is a critical area of modern medical technology that enables the sophisticated treatment both at hospital setting and at home. The particular features and applicability of different infusion pumps (e.g. large volumetric pumps, syringe pumps, ambulatory pumps and home care elastomeric pumps) are reviewed along with an analysis of problems related to the POC. The design principles of 'smart' infusion pumps are outlined in order to ensure large-scale application without compromising treatment safety. Last, a novel fully portable smart pump, readily connectable to an internet of things network is presented as a technological example of future directions in the field of intelligent POC treatment.

The structure and content of *portable biosensors and POC systems* is designed for maximum reception and impact by a broad audience of researchers and experts in the diagnostic and analytical industry. We envisage that people working in medical diagnostics, biosensors, environmental and food safety monitoring, electronics and material engineers, defence/homeland security experts and professors/post-graduate students in these sectors will benefit greatly by reading this book.

This book would never have been realized without the dedicated efforts of all contributors, who sacrificed considerable part of their valuable time to deliver their chapters in due time. It has been both a pleasure and a privilege to work together with colleagues of the highest calibre in their respective field of expertise. In particular, I would like to thank Dr Georgia Moschopoulou, my close associate and also a chapter contributor, for her critical help in editing incoming manuscripts.

Finally, I would like to express my gratitude to the Institution of Engineering and Technology for entrusting me with this challenging project. The personal attention and support provided by Jennifer Grace, Assistant Editor, throughout this demanding, but very rewarding journey, was the key element to the successful completion of the book at hand.

Spyros Kintzios
Athens, Greece
August 2016

Part I

Background science and technology

Chapter 1
Portable optical detectors for point-of-care diagnostics

Reuven Rasooly[1], Hugh Alan Bruck[2], Joshua Balsam[3] and Avraham Rasooly[4]

1.1 Introduction

Point-of-care testing (POCT), which is diagnostic testing performed on site, has the potential to improve healthcare and healthcare delivery. The motivation for POCT is to bring medical testing conveniently and immediately to the patient. Samples can be obtained and tested, and results are analyzed immediately at or near the location of the patient, thus enabling more rapid medical diagnostics and treatment. Early POCT research included the work of Clark and Lyons in the 1960s, which resulted in the "enzyme electrode" for glucose measurement using the enzyme glucose oxidase (GOD) [1]. This work was the first demonstration of a biosensor, as well as the first glucose monitor. More comprehensive portable POCT technology was developed in the early 1990s, with a portable simultaneous multiple analyte whole-blood analyzer for optical monitoring of chemical reactions at nine wavelengths [2]. However, the most significant advance in POCT technology has been the emergence of modern consumer handheld devices, such as the smartphone, which enables POCT devices to be portable, and handheld instruments, such as lab-on-a-chip (LOC), leading to the proliferation of POCT, especially in resource-poor settings.

1.1.1 Medical applications for POCT

Most basic medical diagnostics performed as POCT including EEG, measuring O_2 saturation, blood pressure, heart rate and lung and heart sound, ultrasound, measuring reflex, electromyography, ophthalmoscopy, and others. Many of these tests

[1]Western Regional Research Center, Agricultural Research Service, U.S. Department of Agriculture, Albany, California 94710

[2]Department of Mechanical Engineering, University of Maryland College Park (UMCP), College Park MD 20742

[3]Division of Chemistry and Toxicology Devices, Office of In Vitro Diagnostics and Radiological Health, FDA, Silver Spring, MD 20993

[4]Office of Cancer Complementary and Alternative Medicine, National Cancer Institute, Rockville, MD 20850

are carried out with simple low-cost devices, such as acoustic stethoscope, reflex hammer, ophthalmoscope, retinoscope, or otoscope. New consumer electronics technologies have expanded POCT to diagnostic techniques that can be performed by medical staff and patients, such as biological assays or microbial analysis, which had previously been conducted only in a laboratory environment by trained medical professionals. Although blood glucose testing originated in the 1960s, it is still the most common POCT application; many new POCT techniques have since been developed, including blood gas and electrolytes analysis, rapid coagulation testing, rapid cardiac markers diagnostics, drugs of abuse screening, urine strips testing (leukocytes, nitrite, urobilinogen, protein, pH, hemoglobin, specific gravity, ketone, bilirubin, and glucose), skin cancer, food pathogens screening and infectious disease testing, and cholesterol screening. Various test kits, such as the pregnancy test, HIV salivary assay, urinary tract infection, urine test strip, and fecal occult blood analysis, have also expanded the use of POCT.

1.1.2 Portable technologies for POCT

In recent years, many cheaper, faster, and more sophisticated POCT devices have been introduced. Many of these devices are based on consumer electronics technologies, such as the smartphone or tablet, which are ubiquitous, low-cost devices with computing capabilities. Portable telecommunication devices are also being integrated into medical devices of physical measurements to provide new capabilities (e.g., data recording or analysis tools), which enable them to be converted into telemedicine devices (e.g., smartphone attachment for stethoscope, smartphone-based ambulatory blood pressure monitoring, and pocket mobile smartphone system for the POC submandibular ultrasonography). In addition, the integration of new technologies, such as microfluidics, lateral flow, and LOC, into physical measurement devices have enabled the development of highly sensitive, low-cost biological assays that, in some cases, even outperform "conventional" detection technologies. In particular, LOC technologies [3] have enabled the performance of various chemical and biological assays outside of laboratory environments and are a promising approach for POCT in many settings [4,5]. Several examples of POCT technologies for biological assays include microchip enzyme-linked immunosorbent assay (ELISA)-based detection of ovarian cancer human epididymis protein 4 (HE4) biomarker in urine [6], detection systems for melanoma or skin lesions [7–9], loop-mediated isothermal amplification genetic testing device [10], acoustic wave enhanced immunoassay [11], and colorimetric reader [12].

1.1.3 Optical detection and analysis

Optical detection and analysis is probably the most common detection approach used in biologic research for both clinical and industrial laboratory analyses. Optical detection is based on one of several optical modalities, including light absorbance, fluorescence, polarization colorimetry, spectrometry, or luminescence. Several types of optical detectors used for biodetection employ photodiodes [13–16], photomultipliers [17–22], or avalanche photodiodes, which are inherently spot detectors designed to analyze very small areas with exceptionally high sensitivity. In contrast to these optical detectors, charge-coupled device (CCD) or complementary metal–oxide–semiconductor (CMOS) is a relatively simple, low-cost, sensitive imaging device that can be used for optical analysis of large areas. Such imaging devices have already been employed in several array assays [23–26]. Their main advantage over photodiodes, photomultipliers, or avalanche photodiodes, which are only capable of "serial" analysis of samples, is that they can be used for analyzing light from a large enough area that it can cover the entire surface of a multichannel device, such as a LOC or microarray [27–29], enabling "parallel" multi-sample analysis. This has made CCD- or CMOS-based detectors an ideal choice for multichannel detection, enabling a large number of samples to be analyzed simultaneously.

Many consumer electronics and wireless telecommunications, such as smartphones, tablets, webcams, flatbed scanners, and even computer-gaming consoles, are essentially powerful portable computers with imaging capabilities that can be easily converted into low-cost, portable optical imaging platforms for POCT. As many biological assays and clinical tests (e.g., ELISA, microarray analysis, and DNA sequencing) utilize optical detectors, there are many potential applications, reagents, and assays for optical detection. Therefore, combining consumer electronics devices possessing optical capabilities with biological assays and clinical testing utilizing optical detection have led to many POCT devices, including a reader for lateral flow immunochromatographic assays [30], fluorescence detector [31–33], wide-field fluorescent microscopy [34], capillary array for immunodetection for *Escherichia coli* [35], lens-free microscopy [36], or fluorescent imaging cytometry [37]. Such technologies are especially useful in resource-poor settings [4,5].

1.1.4 Examples for the broader use of optical detection in medicine

There are several examples in which optical detection has been utilized broadly in medical applications that are relevant to POCT.

1.1.4.1 Ophthalmology

One example of the utilization of optical detection in POCT for imaging is ophthalmology, in which a digital retinoscope has been developed by combining a mobile smartphone camera and a retinoscope [38]. In smartphone retinal (fundus) diagnostics [39,40], the smartphone-based retinal camera is used as a portable wide-field imaging device for acquisition, storage, and analysis of fundus images, which can be directly transmitted from the phone via the wireless telecommunication system for remote evaluation [41].

1.1.4.2 Endoscopy

One application of microscopy in POCT is endoscopy, in which a low-cost microendoscope based on a fiber-optic, fluorescence digital microscope with subcellular resolution has been used for the detection of cervical neoplasia [42], gastrointestinal neoplasia [43], head-and-neck cancer [44], esophageal squamous cell carcinoma [45], and middle-ear cholesteatoma [46]. In addition to endoscopy, a mobile fiber-optic sensor has been used for detection of oral and cervical cancer [47].

1.1.4.3 Flow cytometry

Portable cytometers have many clinical applications for POCT. Recently, optofluidic fluorescent imaging cytometry on a smartphone with a spatial resolution of $\sim$2 μm was described [37,48]. While very mobile and versatile, the flow rate of this system is $\sim$1 μL min^{-1}, which limits analysis to small volumes. New high flow rate imaging cytometry for rare cells was recently reported [49–51]. Smartphones have also been used effectively for microscopy [34,35,37,48,52–56]; however, the cameras in these phones are often limited in their sampling rates (e.g., many phones are limited for 30 fps) and are less versatile with their optical systems (e.g., inability to change lenses) than devices such as webcams for flow cytometry applications.

1.1.4.4 Wide-field imaging

Wide-field imaging can be used for many clinical applications, such as cancer detection, with diagnostics that include the use of smartphones for cervical cancer visual inspection of the cervix through the application of visual inspection with acetic acid screening [57], as well as Melanoma [58,59]. However, the accuracy of such applications has been questionable [60,61].

1.1.5 Smartphone and webcam-based POCT

Smartphones are an exceptional platform for optical detection due to their ubiquity and sophisticated electronics, which has attracted the development of many new POCT devices. Practically, all smartphones are equipped with one or more optical imaging sensors. Newer smartphones are equipped with large CMOS (e.g., 8–12-MP CMOS) sensors, which enable very high-resolution imaging. An inherent feature of smartphone-based optical detection is their connectivity, which enables seamless telemedicine, in which data from LOC assays and other clinical imaging data can be analyzed globally allowing for clinical expertise to be brought to remote areas and to low-income countries where such clinical expertise does not exist.

The main applications utilize the imaging capabilities of mobile devices including digital microscopy and its application for wide-field imaging, endoscopy, assay analysis and flow cytometry, and macroscopy for biochemical and immunological assays analysis. Examples for such devices include portable ELISA [62] and a smartphone-based device [63] that enables POCT of laboratory-quality immunoassays for diagnosis of infectious diseases. Other applications include wide-field computational imaging of pathology slides using lens-free on-chip microscopy [64], phone-assisted microarray reader for mutation detection [65], mobile phone camera for DNA detection [66], detection technologies for rapid antimicrobial susceptibility test based on single-cell morphological analysis [67], microfluidic $CD4^+$ and $CD8^+$ T-lymphocyte counters for point-of-care HIV diagnostics using whole blood [68], and paper-based multiplexed transaminase test for low-cost, point-of-care liver function testing [69].

In addition to smartphones, webcams [49,51,70–73] can also be used for optical detection. Similar to smartphones, webcams utilize CMOS detectors for real-time feed of video images into computers. The low cost (often less than $10) and availability of webcams make them attractive as low-cost optical detectors for clinical assays and microscopy applications in resource-poor settings. Unlike the smartphone cameras that are integrated with the phone, webcams are far more flexible to install in a diagnostic device, simplifying the device design and fabrication. In addition, the lenses can be easily changed, filters can be added, and they can also potentially reduce the cost of the device. The use of webcams has been demonstrated in several medical diagnostic techniques, including fluorescence plate readers [5,74–76], detectors for LOC [4,27,29,71,72,77], flow cytometer [49,50], and fluorescence microscopes [78].

However, unlike highly sensitive photomultipliers or avalanche photodiodes, a fundamental question that needs to be addressed to utilize a webcam or smartphone for POCT is whether an approach to improving the low sensitivity of the CMOS webcam can be developed to realize the same sensitivity as significantly more expensive cooled CCDs for optical detection, which would be suitable for more expensive laboratory systems currently utilizing photomultipliers. In the following sections, we will discuss several approaches that have been developed to address this challenge to enable portable image-based optical detection technologies for POCT.

1.2 Portable CMOS and CCD imaging-based detection technologies

POCT will involve various biochemical and immunological assays, many of which are based on optical detection modalities (e.g., densitometry, colorimetry, and fluorescence). One of the most commonly used immunological assays for medical diagnostics is the ELISA [79,80]. ELISA was designed as a laboratory test; however, many portable immunology–based diagnostics devices based on CCD detection systems combined with LOC [4,24,26,28,74,81–83] have enabled sensitivity comparable to laboratory ELISA plate readers for many applications, including microchip ELISA-based detection of ovarian cancer HE4 biomarker in urine [6], and a portable ELISA detector utilizing a phone camera as a spectrometer for detection of interleukin-6, a protein used diagnostically for several types of cancer, and Ara h1, one of the principle peanut allergens [62]. A smartphone dongle combined with microfluidics was developed for immunodetection of infectious diseases at point-of-care [84]. A simple CCD-based detector combined with LOC was developed for immunological detection of microbial toxins, including the detection of staphylococcal enterotoxins utilizing densitometry [78,85]. The system was also automated for high-throughput toxin detection [86]. In addition to densitometry-based detection, similar optical configurations have been used for fluorescent detection of toxins activity, including botulinum neurotoxin a (BoNT-A) activity [4,28,29] and analysis of Shiga toxin 2 (Stx2) activity [87].

1.2.1 POCT bioassay for foodborne toxins

Stx, or Shiga toxins, are foodborne toxins that inhibit protein synthesis within target cells. They are produced by the bacterium *Shigella dysenteriae* and shigatoxigenic group of *E. coli*, which includes serotypes O157:H7, O104:H4, and other enterohemorrhagic *E. coli*. It is considered one of the main causes of food poisoning.

A POCT system was developed for label-free fluorescence analysis of Stx2 activity [87]. An immunoassay can detect the presence of the toxin, but it cannot distinguish between the active form of the toxin, which poses a threat to life, and the inactive form, which can bind to antibodies but show no toxicity. The toxin activity cell assay is based on Stx inhibition of green fluorescent protein (GFP) production in Vero cells exposed to the toxin. The GFP expression is inhibited by the toxin, which reduces the light emission measured by the CCD detector. This simple bioassay does not require cell labeling or complex instrumentation.

1.2.2 Webcam-based fluorescence plate reader for POCT of foodborne toxins

Plate readers are commonly used optical detectors for immunodetection (e.g., ELISA) or chemical assays for multi-analyte detection (e.g., 96-well plates). Several optical modalities used in plate readers including absorbance, fluorescence, or luminescence detected using a sensitive optical detector (e.g., photomultipliers). So the application of low-sensitivity, high-noise imagers for plate readers requires developing new approaches to improve sensitivity and reduce noise to acceptable levels to detect faint optical signals. The configuration of a simple imaging fluorescence plate reader with four modules used in our previous work for many applications [5,29,70,72,75–77,86–88] is shown schematically in Figure 1.1(a) and a photo in Figure 1.1(b).

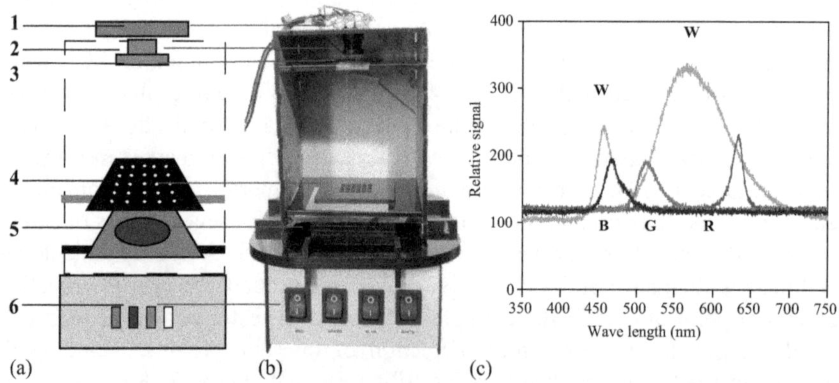

(a) (b) (c)

Figure 1.1 Webcam-based fluorescence plate reader. (a) A schematic configuration of the webcam-based plate reader with the main system components highlighted in the schematic: (1) a webcam camera mounted in a homemade acrylic box; (2) interchangeable lens (with a green band-pass emission filter (3) mounted on the end of the lens); black acrylic sample chip (4); blue band-pass excitation filter (5); and multiwavelength LED (6). (b) A photo of a webcam-based plate reader. (c) The excitation spectra (measured by a spectrometer) of the multiwavelength LED for the white, W, blue, B, green, G, and red, R

The main components of the simple CCD detection system are (1) a camera, (2) lens, (3) emission filter mounted on the end of the lens, (4) assay plate, (5) excitation filter, and (6) multiwavelength light-emitting diode (LED). An image of the detector is shown in Figure 1.1(b). In this configuration, the excitation source is directly in line with the detector, so good quality excitation and emission filters are essential for blocking excitation light reaching the detector while permitting the fluorescence emission to be captured. In this setup, to increase sensitivity, a Point Gray Research *Chameleon* camera equipped with a C-mount CCTV lens (Pentax 12 mm *f*/1.2) can be used as the photodetector. The advantage of this camera is that it has a high sensitivity and a range of exposure times. This type of cameras overcome some of the limitations of conventional low-cost webcams in which the exposure time is determined automatically on the basis of the light exposure of the CMOS, and a low-cost 8-bit CMOS is operated at 30 fps in which the framing rate is fixed at 1/30 s and cannot be controlled by the user. In addition, the typical lens of many low-cost webcams has a fixed aperture and fixed focal length with just a variable focus, whereas a more advanced camera enables the use of interchangeable lenses with better controls and image quality.

1.2.2.1 LED illumination module

For optical detection, illumination is a critical component. A versatile illuminator has been described in previous work [29], which consists of a multiwavelength LED illumination box equipped with white and RBG LED illuminating in the red, green, blue, or white spectra (Figure 1.1(c)). It can excite multiple fluorophores in excitation ranges of 450–650 nm (red 610–650 nm, green 512–550 nm, and blue 450–465 nm). This broad wavelength enables the use of many fluorophores and many imaging applications [5,29,70,72,75–77,86–88], including fluorescent microscope illumination [78]. The excitation and emission filters are very critical for image quality. For excitation, a 20-nm band-pass filter with a 486-nm center wavelength was used (D486/20×), and for emission, a 50-nm band-pass filter with a center wavelength of 535 nm (HQ535/50M filters, both from Chroma Technology Corp., Rockingham, VT) was used.

1.2.2.2 Assay plate

A laminated assay plate for small volumes ($\sim$30 μL) can be fabricated as described in previous work [28,74], using a rigid polymer core (e.g., 3.2-mm black poly(methyl methacrylate)) with laser-machined wells. The bottom of the sample wells was laminated with thin polymer (e.g., polycarbonate (PC) film) bonded with double-side adhesive. The thick core provides rigidity to the plate, and the black color minimizes optical reflection and crosstalk between the wells.

1.2.3 Fluorescence detection of Stx2 activity

To evaluate Stx2 activity, the inhibition of GFP fluorescence protein synthesis was measured using the fluorescence detector [87]. In this work [87], transduced Vero cells with GFP fluorescence construct were cultured for 24 h and treated with various amounts of Stx $\left(100 \text{ ng mL}^{-1} \text{ to } 0.01 \text{ pg mL}^{-1}\right)$. The cells were imaged by

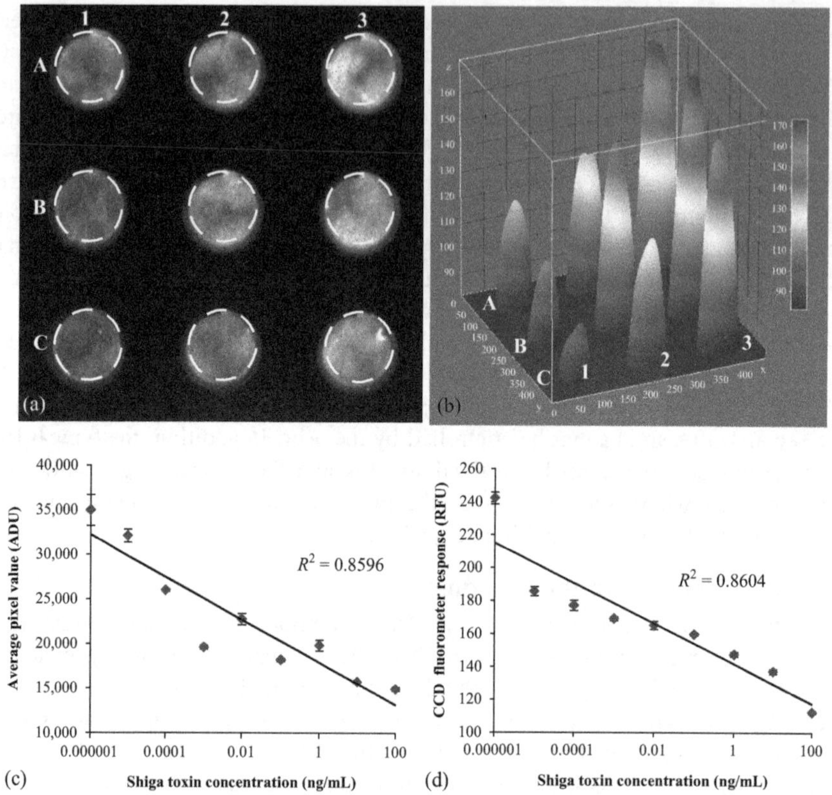

Figure 1.2 Shiga toxin (Stx) activity analysis. Vero cells expressing GFP in response to various concentrations of Stx. The Vero cells were treated with various amounts of Shiga toxin. The plate was illuminated by a blue LED equipped with a blue excitation filter, and the plate analyzed with a CCD detector equipped with a green emission filter in a still single frame mode. (a) image of a nine-well plate of Stx analysis with the amount of toxin used: well 3-A control no toxin, well 3-B 0.01 pg mL^{-1}, well 3-C 0.1 pg mL^{-1}, well 2-A 1 pg mL^{-1}, well 2-B 10 pg mL^{-1}, 2-C 100 pg mL^{-1}, well 1-A 1 ng mL^{-1}, well 1-B 10 ng mL^{-1}, and well 1-C 100 ng mL^{-1}. (b) The corresponding ImageJ 3D image. (c) Average signal brightness was plotted against the various Stx concentrations. (d) The same toxin concentrations were analyzed with a fluorometer

the detector (Figure 1.2(a)), and the signal was quantified using the National Institutes of Health open-source image processing software ImageJ (Figure 1.2(b)). In the control with no toxin, there is no signal inhibition, and the signal is high (Figure 1.2(b) well A3 in). The signal decreases proportionally to the level of toxin at the highest level of toxin $\left(100\ \mathrm{ng\ mL^{-1}}\right)$ shown in Figure 1.2(b) well C1 with the lowest signal. The average luminous intensity value was plotted against Stx concentration (Figure 1.2(c)). There is a high negative correlation between the fluorescence intensity signal and the concentration of the toxin. An increase in Stx2 concentration results in decreased GFP fluorescence intensity ($R^2 = 0.85$, p-value of 0.0037). The samples with the same toxin concentrations were also analyzed with a commercial fluorometer (Figure 1.2(d)), which indicated a very strong correlation between CCD measurements and the fluorometer ($R^2 = 0.86$, p-value of 0.0030). The limit of detection (LOD) of the CCD camera measurement was 0.1 pg of Stx calculated as the mean pixel intensity value represented by a digital number in three control samples (cells with no toxin) minus three times the standard deviation of those samples. These data suggest that a simple low-cost CCD-based detector can be used as a fluorescent detector for cytotoxicity assays [87].

1.2.4 Application of the webcam-based fluorescence plate reader to other food bone toxins

Similar fluorescent detectors were also used for BoNT-A activity analysis [4,28,29]. A fluorescence resonance energy transfer assay was used that measures cleavage of a fluorophore-tagged peptide substrate specific for BoNT-A (SNAP-25) by the toxin light chain with a sensitivity of 0.5 nM, which is the reported sensitivity of the SNAP-25 in vitro cleavage assays. Similar detectors were also used for food safety immunological assays in detection of staphylococcal enterotoxin B (SEB) using enhanced chemiluminescence [85] with an LOD of $0.01\ \mathrm{ng\ mL^{-1}}$, which is approximately ten times more sensitive than traditional ELISA [83], for SEB detection utilizing colorimetric assays [74] or densitometry detection with a LOD of $0.5\ \mathrm{ng\ mL^{-1}}$ [75].

1.3 Computational enhancement of the sensitivity of webcam-based detectors

Sensitive optical readers utilize high-cost, high-quality, high-resolution, low-noise, and sensitive detectors (e.g., photomultipliers, avalanche photodiodes, or cooled CCDs). The application of low-cost detectors (e.g., ~\$10 webcam) is challenging because of their low sensitivity, low resolution, and high noise. Webcams for plate readers require improved sensitivity and reduced noise. The computational method of image stacking [71–73] was used for sensitivity enhancement of digital images consisting of a large array of pixels containing signals along with random noise and compression artifacts, which reduce signal detection sensitivity. The quality of the signal can be measured using the signal-to-noise ratio (SNR).

1.3.1 Image stacking-based computational signal enhancement

In video mode, many images (e.g., 30 fps) of the same object are captured. Image-stacking software enables calculation of the average value of each pixel of each frame to generate an "averaged" single image from all the frames. In this

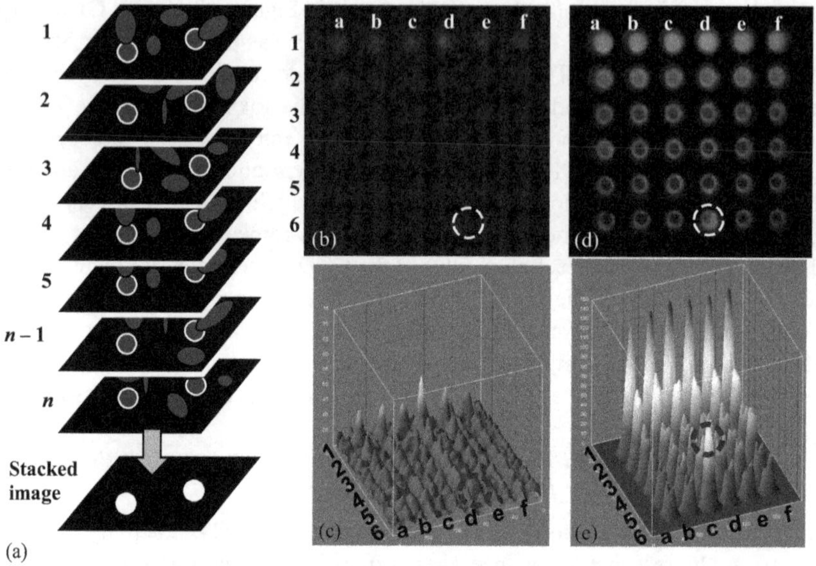

Figure 1.3 *Image-stacking enhancement. (a) A schematic of image stacking for enhanced imaging (the webcam in video mode captures n individual frames). In each frame, there is an underlying signal of interest (marked with white circles) and interfering noise. If many individual frames are averaged together, the standard deviation of this background noise is greatly reduced, and its mean value can be subtracted from each pixel, resulting in an enhanced SNR. (b) A 36-well plate with six concentrations of fluorescein (rows 1–6) each in six replicas (columns a–f). The plate was illuminated by a blue LED equipped with a blue excitation filter, and the plate was measured with a green emission filter. The signals of the wells were detected by a CMOS webcam equipped with the original f3.8 5-mm lens operating in a still single-frame mode. (c) The corresponding ImageJ 3D image. (d) image stacking enhancement, the same plate was analyzed in video mode image enhanced by image stacking, the fluorescein concentrations used: row 1: 500 nM, row 2: 250 nm, row 3: 125 nM, row 4: 60 nM, row 5: 30 nM, and row 6: control (water). Row 5 column d (marked with a circle) is a reference point which can be used to orient the plate. (e) the corresponding 3D image of (d) using ImageJ*

"averaged image," the random noise of each frame will be averaged to zero after image stacking, whereas the signal present in all frames will be a constant value in the averaged single image. So, image stacking enables reducing the random noise and compression artifacts in video images when creating a still "averaged image" with improved SNR [73].

A schematic of stacking is shown in Figure 1.3(a). In video mode, the webcam captures *n* individual frames. In each frame, pixels with a signal (marked with white circle) and pixels with random noise are captured, and in each individual frame, the signal and the noise are not distinguishable. In the stacked image containing the average value of each pixel for all frames, the random noise average is low, whereas the signal is constant with an increasing S/N ratio.

To demonstrate the improved sensitivity, a generic CMOS-based webcam plate reader equipped with an original 5 mm *f*3.8 lens was used for fluorescein (a common dye used in biological assays) samples in the range of $0-1$ mg mL^{-1} analysis. In Figure 1.3(b), an emission from a single frame of a 36-well plate (rows 1–6 in Figure 1.3(a)) in six replicas (columns a–f) loaded with six concentrations of fluorescein (0 μM (i.e., water) to 500 μM) imaged with the webcam and analyzed with ImageJ software to quantify the intensity of user-specified areas of the image. While in the still single frame (Figure 1.3(b)), the only signal is detected in row 1 at 500 μM, and there is no visible signal in the control (water, row 6) except row 6, column e (marked with a circle), a reference point that can be used to orient the plate. In a three-dimensional (3D) view from ImageJ (Figure 1.3(c)), the signal level for each well corresponds to the concentration of the fluorescein, suggesting that there is no strong signal except from the 500-μM samples (row 1) with an LOD (calculated on the basis of the control (water, row 6)) of 1000 μM.

Since image stacking was used to improve CMOS sensitivity, the 36-well plate was also detected by a CMOS webcam, operating in a video mode enhanced by image stacking (Figure 1.3(d)) with the corresponding ImageJ image (Figure 1.3(e)). It indicated a very good signal with a LOD of 60 μM, comparable to a conventional plate reader.

These results suggest that image stacking enables an increase in sensitivity of inexpensive (e.g., $10) webcams, and that it may be practical to develop a low-cost florescence plate reader (e.g., for ~$100) with the sensitivity and capacity to detect multiple analytes. Such a webcam-based plate reader can be used for telemedicine, enabling the transfer of data between distant locations to provide medical diagnostics and remote sites. The plate reader can also be used for other optical modalities, including absorbance, densitometry, and luminescence.

1.4 Capillary arrays as waveguides for enhancing the sensitivity of optical detectors

In addition to computational image enhancement through image stacking, the use of capillaries has enabled an $\sim 100\times$ increase in detection sensitivity [31,33,89] for image-based detection. In this array, light-wave energy propagating through the capillary walls interacts directly with and excites fluorophore molecules via evanescent waves (i.e., acts as a "waveguide"), which in turn emit light that can be detected at the end of the capillary by the imaging detector. These capillary array waveguides are 3D detection systems (Figure 1.4), in which *X* columns and *Y* rows of capillaries are arrayed in two dimensions and light propagates via the capillary walls to provide a third *Z* dimension for illumination along the axes of the capillaries oriented normal to the array distribution. Combining computational image

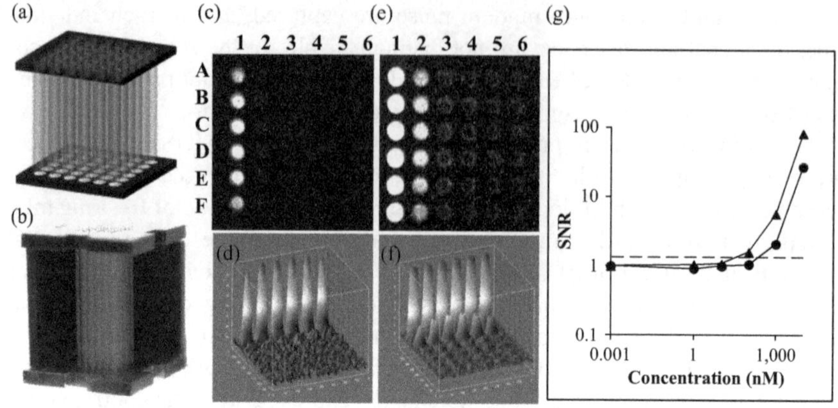

Figure 1.4 Capillary array fluorescence detection combined with image-stacking enhancement. (a) Schematic of the capillary array, (b) photo of the capillary array, (c) capillary array detection of fluorescein. A 36-capillary array was loaded with six concentrations of fluorescein (columns 1–6) each in six replicas (rows A–F). The plate was illuminated by a blue LED equipped with a blue excitation filter, and the plate was measured with a green emission filter. The signals from the array were detected by the CMOS Webcam operating in a still single frame mode. (c) The corresponding 3D image from ImageJ. (d) The same plate analyzed in video mode and image enhanced by image stacking (e) with the corresponding 3D image from ImageJ in (f). The signal-to-noise ratios (SNR) for both modes were plotted against the various fluorescein concentrations (g). Triangles are data points for video-captured stacked images, and circles are single image mode with no stacking. The LOD (the mean plus three times standard deviation of a control (water)) is marked as a dashed line. The fluorescein concentrations used were (columns 1–6) 10,000 nM, 1,000 nM, 100 nM, 10 nM, 1 nM and control (water)

enhancement with capillary array waveguide amplification resulted in an ~1,000× increase in detection sensitivity [89].

1.4.1 Webcam detector with capillary array

A 6×6 array of glass capillaries (Figure 1.4) was used for fluorescence detection. To orient all 36 capillary channels toward the camera image sensor simultaneously, two laser machined six-by-six arrays of holes in two 3.2-mm thick plates of black acrylic were fabricated to hold the capillaries in a parallel configuration. The capillary array [31,33,89] is shown schematically in Figure 1.4(a) (actual photo shown in Figure 1.4(b)) was used for fluorescence amplification. The length of the capillaries used is 32 mm with an inner diameter of 0.8 mm. For the fluorescence detection used in this work, the array was placed on the excitation filter (Figure 1.1(a5)) at the same position used for the plate assay (Figure 1.1(a4)). The light is emitted by the multiwavelength LED illuminator passed through the excitation filter, carried through the capillaries to the emission filter mounted on the end of the lens and captured by a webcam.

1.4.2 Fluorescein detection using capillary array

To further demonstrate the capabilities of the technology, a 36-capillary array was loaded with six concentrations of fluorescein (Figure 1.4(c), columns 1–6) each in six replicas (rows A–F). The plate was illuminated by a blue LED, and the signals were detected by the CMOS Webcam operating in a still single frame mode, the corresponding ImageJ 3D image shown in Figure 1.4(d). The same array analyzed in video mode and image enhanced by image stacking (Figure 1.4(e)) with the corresponding ImageJ 3D image are shown in Figure 4(f). The SNRs for both modes were plotted against the various fluorescein concentrations (Figure 1.4(g)). Triangles are data points for video-captured stacked images and circles are single image mode with no stacking. The LOD (the mean plus three times standard deviation of a control (water)) is marked as a dashed line. The fluorescein concentrations used were (columns 1–6): 10,000 nM, 1,000 nM, 100 nM, 10 nM, 1 nM and control (water). The plot in Figure 1.4(g) clearly shows the signal amplification of image stacking.

Image stacking can be combined with capillary array to increase sensitivity.

1.5 Smartphone-based fluorescence detection system using capillary array

In addition to the multipurpose fluorescence detector, which can be used for plates and capillary array shown in Figure 1.1, a compact and more mobile fluorescence detector based on a smartphone was developed [90] (Figure 1.5). For effective detection using the capillary array, the optical configuration must allow for uniform measurement between channels from a close distance when using such a capillary array sensor. This is a challenge due to the parallax inherent in planar imaging of long parallel capillary tubes with typical lens configurations. To enable effective detection from a short distance, orthographic projection system that forms parallel light projection images from the capillaries using an object-space telecentric lens configuration was developed [90]. This optical configuration results in a significantly higher degree of uniformity in measurement between channels, as well as a significantly reduced focal distance, which enables a more compact sensor.

1.5.1 Smartphone-based capillary array fluorescence detector

In this detection system [90], a low-cost LED flashlight (~$1) equipped with five white LED and a diffusion panel is used for the illumination system to enable wide spectra fluorescent excitation in the range of 450–740 nm. The light path includes mirrors to focus the light on the capillary array (Figure 1.5(a)). The main elements of the orthographic projection fluorescent sensor are shown in

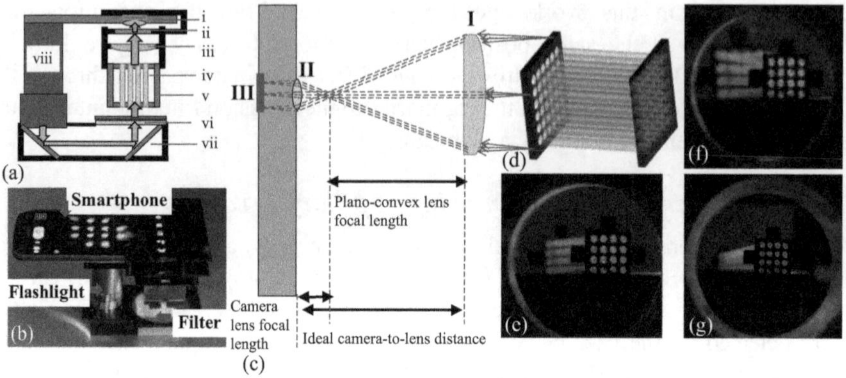

Figure 1.5 Configuration of smartphone-based capillary array fluorescence detector. (a) A schematic of the fluorescent sensor. The main components are (i) camera phone, (ii) emission filter, (iii) secondary lens, (iv) alignment fixture, (v) capillary tube array, (vi) excitation filters and a diffuser, (vii) light path with two mirrors, (viii) LED flashlight. (b) A photograph of the flashlight based device. (c) The optics used for the smartphone-based detector imaging with the capillary array utilizing the orthographic projection configuration. To demonstrate the effect of orthographic projection optics, a 16-tube capillary array (12.5 cm in length) imaged with a typical smartphone camera through a plano-convex lens (4.5 cm diameter, 15 cm focal length). (c) A schematic of the object-space telecentric optics with (i) the plano-convex secondary lens and the principal rays imaged by the lens of (ii) the phone camera that focuses the image onto (iii) the camera CMOS image sensor. (d) A schematic of the capillary array. The capillary array was imaged at three different camera-to-lens distances: (e) distance of 14.5 cm, (f) distance of 30 cm, and (g) 5 cm. The object-space telecentric condition is satisfied at (c) the ideal camera-to-lens distance, so that all capillaries of the array appear parallel. (d) Positioned too far for the camera-to-lens distance, and (e) positioned too close for the camera-to-lens distance. Configurations (d) and (e) both result in uneven images of the capillary array with the center capillaries appearing brighter than those on the edges during fluorescence measurement

Figure 1.5(a): (i) smartphone, (ii) emission filter, (iii) secondary lens, (iv) alignment fixture, (v) capillary tube array, (vi) excitation filters and a diffuser, (vii) light path with two mirrors, and (viii) LED flashlight. In the configuration shown in Figure 1.5, two excitation filters (Figure 1.5(a-vi)) are used to reduce the noise from the white LED. This optical configuration (Figure 1.5(a)) is very compact, with an optimal optical distance of 32 mm using a secondary lens, as compared to previous designs (see Figure 1.1) that utilized conventional lens systems (such as the stock lens that came with the camera). Without a secondary lens, distances as large as 140 mm were required between the imaging device and the capillaries to obtain images with uniform brightness across the capillary array.

1.5.2 Orthographic optical configuration

To image the 3D capillary array, orthographic projection optics using a telecentric lens to represent the 3D array in two dimensions was developed [90]. The basic optical configuration of the object-space telecentric optics is shown schematically in Figure 1.5(c), with a plano-convex lens (Figure 1.5(c-i)) and the principal rays imaged by the lens of a phone camera (Figure 1.5(c-ii)) focusing the image onto the camera's CMOS detector (Figure 1.5(c-iii)). The capillary array (Figure 1.5(d)) is imaged with the added plano-convex lens (4.5-cm diameter, 15-cm focal length) in three different positions. In Figure 1.5(c), the distance between the capillaries and the camera lens, which was minimized by proper secondary lens selection, is 32 mm to achieve both focus and uniform capillary illumination. Without a secondary lens, the minimum distance is 65 mm to achieve focus and approximately 140 mm to achieve uniform capillary measurements. As shown in Figure 1.5(e), at the ideal camera-to-lens distance in which the focal points of the camera phone lens and the secondary lens are aligned, the object-space telecentric condition is achieved, and a parallel projection of the 3D capillary array is seen in Figure 1.5(e). All the projection lines (i.e., light rays imaged by the telecentric lens) from the 3D capillary array are orthogonal to the projection plane, resulting in every plane of the captured image retaining straight lines and parallelism (i.e., an affine transformation) on the sensor. In Figure 1.1(f), the image is taken with the secondary lens at a distance too far from the camera, and in Figure 1.1(g) the distance is too close.

1.6 Summary of factors contributing to the sensitivity of low-cost optical detectors

The smartphone-based detector was used to measure serial dilutions of fluorescein [90] with an LOD of 1 nM with image stacking and 10 nM without image stacking, similar to the LOD obtained with a commercial plate reader. Moreover, the capillary array required a sample volume of less than 10 μL, which is an order of magnitude less than the 100 μL required for the plate reader, whereas the number of capillaries that can be imaged in the smartphone configuration is lower (only eight) than regular configuration (Figure 1.5(b)). In previous work [90], we described some of the factors contributing to improvements in the sensitivity of low-cost optical detectors. They include:

- The use of video mode combines with image stacking.
- The quality of filters is very critical. Using high-quality narrow-band filters at the emission/excitation wavelengths will reduce noise and improve detection sensitivity.
- Although camera phone lenses are not interchangeable, many webcams have the capability for changing the lens (e.g., we used an *f*/1.2 lens to maximize the amount of light transmitted to the sensor).

- Increasing the intensity of the LED illumination (the use of more LEDs) will increase fluorescent signal.
- The use of low-cost lasers equipped with line generator may increase light intensity and provide narrow wavelength illumination, making emission filters unnecessary.
- For single-frame imaging, some webcams allow for long exposure times (>1 s).

1.7 Conclusions

Optical detection is important for biological analysis in POCT. Examples have been provided of several low-cost optical technologies that demonstrate the potential for medical diagnostics utilizing readily available, low-cost consumer electronics capable of interconnectivity and telemedicine. However, to achieve high enough sensitivity to detect biological analytes at extremely low concentrations, new computational techniques have to be employed, such as video imaging mode combined with computational image-stacking algorithms to remove noise and increase sensitivity by nearly 10-fold, resulting in sensitivities similar to conventional plate readers.

An important element in biological analysis is LOC technologies that have the capability to perform various assays without the need of dedicated laboratories and complex equipment or expensive reagents as demonstrated for LOC-ELISA. The use of readily available polymers, such as acrylic and PC, for LOC-ELISA fabrication is compatible with mass production technologies used in various industries in low-income countries, as opposed to the specialized manufacturing technologies need for silicon or polydimethylsiloxane; they can substantially improve access to medical diagnostic technologies in low-resource settings.

The optical detection technologies described in this manuscript are simpler to fabricate and utilize lower cost components, which are more readily available in low-income countries. Since numerous medical diagnostics assays rely on optical detection, these technologies may enable the performance of many medical diagnostics assays in a clinic instead of laboratory environment. These technologies can be used for numerous biomedical applications, ranging from cardiovascular and cancer diagnostics to food and water safety. The simplicity and the low cost of the technologies can facilitate bringing modern health and medicine to low-resource settings, in which the majority of the world's population lives and which do not currently have access to such critical medical diagnostics.

Acknowledgment

This work was supported by the US Department of Agriculture's Western Regional Research Center, the FDA's Center for Devices and Radiological Health, Division of Biology, and the National Cancer Institute. The views expressed are those of the authors and do not represent those of the US government.

References

[1] Clark L.C. Jr., Lyons C. 'Electrode systems for continuous monitoring in cardiovascular surgery'. *Annals of the New York Academy of Sciences*. 1962;**102**:29–45.

[2] Schembri C.T., Ostoich V., Lingane P.J., Burd T.L., Buhl S.N. 'Portable simultaneous multiple analyte whole-blood analyzer for point-of-care testing'. *Clinical Chemistry*. 1992;**38**(9):1665–70.

[3] Herold K.E., Rasooly A., eds. *Lab on a Chip Technology*. 2009/01/20 ed. Norfolk, UK: Caister Academic Press; 2009.

[4] Sun S., Ossandon M., Kostov Y., Rasooly A. 'Lab-on-a-chip for botulinum neurotoxin a (BoNT-A) activity analysis'. *Lab on a Chip*. 2009;**9** (22):3275–81.

[5] Sun S., Yang M., Kostov Y., Rasooly A. 'ELISA-LOC: lab-on-a-chip for enzyme-linked immunodetection'. *Lab on a Chip*. 2010;**10**(16):2093–100.

[6] Wang S., Zhao X., Khimji I., *et al*. 'Integration of cell phone imaging with microchip ELISA to detect ovarian cancer HE4 biomarker in urine at the point-of-care'. *Lab on a Chip*. 2011;**11**(20):3411–8.

[7] Rosado L., Castro R., Ferreira L., Ferreira M. 'Extraction of ABCD rule features from skin lesions images with smartphone'. *Studies in Health Technology and Informatics*. 2012;**177**:242–7.

[8] Wadhawan T., Situ N., Rui H., Lancaster K., Yuan X., Zouridakis G. 'Implementation of the 7-point checklist for melanoma detection on smart handheld devices'. *Conference Proceedings: Annual International Conference of the IEEE Engineering in Medicine and Biology Society IEEE Engineering in Medicine and Biology Society Conference* 2011;**2011**: 3180–3.

[9] Boyce Z., Gilmore S., Xu C., Soyer H.P. 'The remote assessment of melanocytic skin lesions: a viable alternative to face-to-face consultation'. *Dermatology*. 2011;**223**(3):244–50.

[10] Stedtfeld R.D., Tourlousse D.M., Seyrig G., *et al*. 'Gene-Z: a device for point of care genetic testing using a smartphone'. *Lab on a Chip*. 2012;**12** (8):1454–62.

[11] Bourquin Y., Reboud J., Wilson R., Zhang Y., Cooper J.M. 'Integrated immunoassay using tuneable surface acoustic waves and lensfree detection'. *Lab on a Chip*. 2011;**11**(16):2725–30.

[12] Lee D.S., Jeon B.G., Ihm C., Park J.K., Jung, M.Y. 'A simple and smart telemedicine device for developing regions: a pocket-sized colorimetric reader'. *Lab on a Chip*. 2011;**11**(1):120–6.

[13] Capitan-Vallvey L.F., Asensio L.J., Lopez-Gonzalez J., Fernandez-Ramos M.D., Palma A.J. 'Oxygen-sensing film coated photodetectors for portable instrumentation'. *Analytica Chimica Acta*. 2007;**583**(1):166–73.

[14] Mac Sweeney M.M., Bertolino C., Berney H., Sheehan M. 'Characterization and optimization of an optical DNA hybridization sensor for the detection of

multi-drug resistant tuberculosis'. *IEEE Engineering in Medicine and Biology Society*. 2004;**3**:1960–3.

[15] Claycomb R.W., Delwiche M.J. 'Biosensor for on-line measurement of bovine progesterone during milking'. *Biosensors and Bioelectronics*. 1998;**13**(11):1173–80.

[16] Bruno A.E., Barnard S., Rouilly M., Waldner A., Berger J., Ehrat M. 'All-solid-state miniaturized fluorescence sensor array for the determination of critical gases and electrolytes in blood'. *Analytical Chemistry*. 1997;**69**(3):507–13.

[17] Moehrs S., Del Guerra A., Herbert D.J., Mandelkern M.A. 'A detector head design for small-animal PET with silicon photomultipliers (SiPM)'. *Physics in Medicine and Biology*. 2006;**51**(5):1113–27.

[18] Takei M., Kida T., Suzuki K. 'Sensitive measurement of positron emitters eluted from HPLC'. *Applied Radiation and Isotopes*. 2001;**55**(2):229–34.

[19] Ruiz-Martinez M.C., Berka J., Belenkii A., Foret F., Miller A.W., Karger B.L. 'DNA sequencing by capillary electrophoresis with replaceable linear polyacrylamide and laser-induced fluorescence detection'. *Analytical Chemistry*. 1993;**65**(20):2851–8.

[20] Tibbe A.G., de Grooth B.G., Greve J., Liberti P.A., Dolan G.J., Terstappen L.W. 'Cell analysis system based on immunomagnetic cell selection and alignment followed by immunofluorescent analysis using compact disk technologies'. *Cytometry*. 2001;**43**(1):31–7.

[21] Tsukagoshi K., Jinno N., Nakajima R. 'Development of a micro total analysis system incorporating chemiluminescence detection and application to detection of cancer markers'. *Analytical Chemistry*. 2005;**77**(6):1684–8.

[22] Roda A., Manetta A.C., Portanti O., *et al.* 'A rapid and sensitive 384-well microtitre format chemiluminescent enzyme immunoassay for 19-nortestosterone'. *Luminescence*. 2003;**18**(2):72–8.

[23] Taitt C.R., Anderson G.P., Ligler F.S. 'Evanescent wave fluorescence biosensors'. *Biosensors and Bioelectronics*. 2005;**20**(12):2470–87.

[24] Ngundi M.M., Qadri S.A., Wallace E.V., *et al.* 'Detection of deoxynivalenol in foods and indoor air using an array biosensor'. *Environmental Science and Technology*. 2006;**40**(7):2352–6.

[25] Moreno-Bondi M.C., Taitt C.R., Shriver-Lake L.C., Ligler F.S. 'Multiplexed measurement of serum antibodies using an array biosensor'. *Biosensors and Bioelectronics*. 2006;**21**(10):1880–6.

[26] Ligler F.S., Sapsford K.E., Golden J.P., *et al.* 'The array biosensor: portable, automated systems'. *Analytical Sciences*. 2007;**23**(1):5–10.

[27] Kostov Y., Sergeev N., Wilson S., Herold K.E., Rasooly A. 'A simple portable electroluminescence illumination-based CCD detector'. *Methods in Molecular Biology*. 2009;**503**:259–72.

[28] Sapsford K.E., Sun S., Francis J., Sharma S., Kostov Y., Rasooly A. 'A fluorescence detection platform using spatial electroluminescent excitation for measuring botulinum neurotoxin A activity'. *Biosensors and Bioelectronics*. 2008;**24**(4):618–25.

[29] Sun S., Francis J., Sapsford K.E., Kostov Y., Rasooly A. 'Multi-wavelength spatial LED illumination based detector for in vitro detection of botulinum neurotoxin A activity'. *Sensors and actuators B: Chemical*. 2010;**146** (1–8):297–306.

[30] Mudanyali O., Dimitrov S., Sikora U., Padmanabhan S., Navruz I., Ozcan A. 'Integrated rapid-diagnostic-test reader platform on a cellphone'. *Lab on a Chip*. 2012;**12**(15):2678–86.

[31] Balsam J., Bruck H.A., Rasooly A. 'Orthographic projection capillary array fluorescent sensor for mHealth'. *Methods*. 2013;**63**(3):276–81.

[32] Balsam J., Rasooly R., Bruck H.A., Rasooly A. 'Thousand-fold fluorescent signal amplification for mHealth diagnostics'. *Biosensors and Bioelectronics*. 2013;**51C**:1–7.

[33] Balsam J., Bruck H.A., Rasooly A. 'Capillary array waveguide amplified fluorescence detector for mHealth'. *Sensors and Actuators B: Chemical*. 2013;**186**:711–7.

[34] Zhu H., Yaglidere O., Su T.W., Tseng D., Ozcan A. 'Wide-field fluorescent microscopy on a cell-phone'. *Conference proceedings: Annual International Conference of the IEEE Engineering in Medicine and Biology Society IEEE Engineering in Medicine and Biology Society Conference* 2011;2011:6801–4.

[35] Zhu H., Sikora U., Ozcan A. 'Quantum dot enabled detection of *Escherichia coli* using a cell-phone'. *Analyst*. 2012;**137**(11):2541–4.

[36] Tseng D., Mudanyali O., Oztoprak C., *et al*. 'Lensfree microscopy on a cellphone'. *Lab on a Chip*. 2010;**10**(14):1787–92.

[37] Zhu H., Mavandadi S., Coskun A.F., Yaglidere O., Ozcan A. 'Optofluidic fluorescent imaging cytometry on a cell phone'. *Analytical Chemistry*. 2011;**83**(17):6641–7.

[38] Chan W.O., Crabb M., Sia D., Taranath D. 'Creating a digital retinoscope by combining a mobile smartphone camera and a retinoscope'. *Journal of AAPOS: The Official Publication of the American Association for Pediatric Ophthalmology and Strabismus/American Association for Pediatric Ophthalmology and Strabismus*. 2014;**18**(4):387–8.

[39] Darma S., Zantvoord F., Verbraak F.D. 'The quality and usability of smartphone and hand-held fundus photography, compared to standard fundus photography'. *Acta Ophthalmologica*. 2014;**93**(4): e310–1.

[40] Haddock L.J., Kim D.Y., Mukai S. 'Simple, inexpensive technique for high-quality smartphone fundus photography in human and animal eyes'. *Journal of Ophthalmology*. 2013;**2013**:518479.

[41] Maamari R.N., Keenan J.D., Fletcher D.A., Margolis T.P. 'A mobile phone-based retinal camera for portable wide field imaging'. *The British Journal of Ophthalmology*. 2014;**98**(4):438–41.

[42] Grant B.D., Schwarz R.A., Quang T., Schmeler K.M., Richards-Kortum R. 'High-resolution microendoscope for the detection of cervical neoplasia'. *Methods in Molecular Biology*. 2015;**1256**:421–34.

[43] Louie J.S., Richards-Kortum R., Anandasabapathy S. 'Applications and advancements in the use of high-resolution microendoscopy for detection of

gastrointestinal neoplasia'. *Clinical Gastroenterology and Hepatology: The Official Clinical Practice Journal of the American Gastroenterological Association*. 2014;**12**(11):1789–92.

[44] Vila P.M., Park C.W., Pierce M.C., *et al.* 'Discrimination of benign and neoplastic mucosa with a high-resolution microendoscope (HRME) in head and neck cancer'. *Annals of Surgical Oncology*. 2012;**19**(11):3534–9.

[45] Shin D., Protano M.A., Polydorides A.D., *et al.* 'Quantitative analysis of high-resolution microendoscopic images for diagnosis of esophageal squamous cell carcinoma'. *Clinical Gastroenterology and Hepatology: The Official Clinical Practice Journal of the American Gastroenterological Association*. 2015;**13**(2):272–9.

[46] Bradley J., Jiang N., Levy L., Richards-Kortum R., Sikora A., Smouha E. 'High-resolution microendoscope images of middle ear cholesteatoma and surrounding tissue: evaluation of interobserver concordance'. *Otolaryngology—Head and Neck Surgery: Official Journal of American Academy of Otolaryngology-Head and Neck Surgery*. 2014;**150**(4):654–8.

[47] Yu B., Nagarajan V.K., Ferris D.G. 'Mobile fiber-optic sensor for detection of oral and cervical cancer in the developing world'. *Methods in Molecular Biology*. 2015;**1256**:155–70.

[48] Zhu H., Ozcan A. 'Wide-field fluorescent microscopy and fluorescent imaging flow cytometry on a cell-phone'. *Journal of Visualized Experiments*. 2013;**74**:50451.

[49] Balsam J., Bruck H.A., Rasooly A. 'Webcam-based flow cytometer using wide-field imaging for low cell number detection at high throughput'. *Analyst*. 2014;**139**(17):4322–9.

[50] Balsam J., Bruck H.A., Rasooly A. 'Mobile flow cytometer for mHealth'. *Methods in Molecular Biology*. 2015;**1256**:139–53.

[51] Balsam J., Bruck H.A., Rasooly A. 'Cell streak imaging cytometry for rare cell detection'. *Biosensors and Bioelectronics*. 2015;**64**:154–60.

[52] Wei Q., Qi H., Luo W., *et al.* 'Fluorescent imaging of single nanoparticles and viruses on a smart phone'. *ACS Nano*. 2013;**7**(10):9147–55.

[53] Coskun A.F., Nagi R., Sadeghi K., Phillips S., Ozcan A. 'Albumin testing in urine using a smart-phone'. *Lab on a Chip*. 2013;**13**(21):4231–8.

[54] Navruz I., Coskun A.F., Wong J., *et al.* 'Smart-phone based computational microscopy using multi-frame contact imaging on a fiber-optic array'. *Lab on a Chip*. 2013;**13**(20):4015–23.

[55] Zhu H., Sencan I., Wong J., *et al.* 'Cost-effective and rapid blood analysis on a cell-phone'. *Lab on a Chip*. 2013;**13**(7):1282–8.

[56] Zhu H., Yaglidere O., Su T.W., Tseng D., Ozcan A. 'Cost-effective and compact wide-field fluorescent imaging on a cell-phone'. *Lab on a Chip*. 2011;**11**(2):315–22.

[57] Quinley K.E., Gormley R.H., Ratcliffe S.J., *et al.* 'Use of mobile telemedicine for cervical cancer screening'. *Journal of Telemedicine and Telecare*. 2011;**17**(4):203–9.

[58] Zouridakis G., Wadhawan T., Situ N., *et al.* 'Melanoma and other skin lesion detection using smart handheld devices'. *Methods in Molecular Biology*. 2015;**1256**:459–96.

[59] Maier T., Kulichova D., Schotten K., *et al.* 'Accuracy of a smartphone application using fractal image analysis of pigmented moles compared to clinical diagnosis and histological result'. *Journal of the European Academy of Dermatology and Venereology*. 2015;**29**(4):663–7.

[60] Ferrero N.A., Morrell D.S., Burkhart C.N. 'Skin scan: a demonstration of the need for FDA regulation of medical apps on iPhone'. *Journal of the American Academy of Dermatology*. 2013;**68**(3):515–16.

[61] Stoecker W.V., Rader R.K., Halpern A. 'Diagnostic inaccuracy of smartphone applications for melanoma detection: representative lesion sets and the role for adjunctive technologies'. *JAMA Dermatology*. 2013;**149**(7):884.

[62] Long K.D., Yu H., Cunningham B.T. 'Smartphone instrument for portable enzyme-linked immunosorbent assays'. *Biomedical Optics Express*. 2014;**5**(11):3792–806.

[63] Abbas T., Dutta A. 'p21 in cancer: intricate networks and multiple activities'. *Nature Reviews Cancer*. 2009;**9**(6):400–14.

[64] Greenbaum A., Zhang Y., Feizi A., *et al.* 'Wide-field computational imaging of pathology slides using lens-free on-chip microscopy'. *Science Translational Medicine*. 2014;**6**(267):267ra175.

[65] Zhang G., Li C., Lu Y., *et al.* 'Validation of a mobile phone-assisted microarray decoding platform for signal-enhanced mutation detection'. *Biosensors and Bioelectronics*. 2011;**26**(12):4708–14.

[66] Lee D., Chou W.P., Yeh S.H., Chen P.J., Chen P.H. 'DNA detection using commercial mobile phones'. *Biosensors and Bioelectronics*. 2011;**26**(11):4349–54.

[67] Choi J., Yoo J., Lee M., *et al.* 'A rapid antimicrobial susceptibility test based on single-cell morphological analysis'. *Science Translational Medicine*. 2014;**6**(267):267ra174.

[68] Watkins N.N., Hassan U., Damhorst, G., *et al.* 'Microfluidic CD4+ and CD8+ T lymphocyte counters for point-of-care HIV diagnostics using whole blood'. *Science Translational Medicine*. 2013;**5**(214):214ra170.

[69] Pollock N.R., Rolland J.P., Kumar S., *et al.* 'A paper-based multiplexed transaminase test for low-cost, point-of-care liver function testing'. *Science Translational Medicine*. 2012;**4**(152):152ra29.

[70] Balsam J., Rasooly R., Bruck H.A., Rasooly A. 'Thousand-fold fluorescent signal amplification for mHealth diagnostics'. *Biosensors and Bioelectronics*. 2014;**51**:1–7.

[71] Balsam J., Bruck H.A., Rasooly A. 'Orthographic projection capillary array fluorescent sensor for mHealth'. *Methods*. 2013;**63**(3):276–81.

[72] Balsam J., Bruck H.A., Rasooly A. 'Capillary array waveguide amplified fluorescence detector for mHealth'. *Sensors and actuators B: Chemical*. 2013;**186**:711–17.

[73] Balsam J., Bruck H.A., Kostov Y., Rasooly A. 'Image stacking approach to increase sensitivity of fluorescence detection using a low cost complementary metal-oxide-semiconductor (CMOS) webcam'. *Sensors and Actuators B: Chemical*. 2012;171–172:141–7.

[74] Sapsford K.E., Francis J., Sun S., Kostov Y., Rasooly A. 'Miniaturized 96-well ELISA chips for staphylococcal enterotoxin B detection using portable colorimetric detector'. *Analytical and Bioanalytical Chemistry*. 2009;**394**(2):499–505.

[75] Yang M., Sun S., Kostov Y., Rasooly A. 'A simple 96 well microfluidic chip combined with visual and densitometry detection for resource-poor point of care testing'. *Sensors and Actuators B: Chemical*. 2011;**153**(1):176–81.

[76] Rasooly A., Bruck H.A., Kostov Y. 'An ELISA lab-on-a-chip (ELISA-LOC)'. *Methods in Molecular Biology*. 2013;**949**:451–71.

[77] Balsam J., Bruck H.A., Rasooly A. 'Two-layer lab-on-a-chip (LOC) with passive capillary valves for mHealth medical diagnostics'. *Methods in Molecular Biology*. 2015;**1256**:247–58.

[78] Balsam J., Ossandon M., Bruck H.A., Lubensky I., Rasooly A. 'Low-cost technologies for medical diagnostics in low-resource settings'. *Expert Opinion on Medical Diagnostics*. 2013;**7**(3):243–55.

[79] Engvall E., Perlmann P. 'Enzyme-linked immunosorbent assay (ELISA). Quantitative assay of immunoglobulin G'. *Immunochemistry*. 1971;**8**(9):871–4.

[80] Van Weemen B.K., Schuurs A.H. 'Immunoassay using antigen-enzyme conjugates'. *FEBS Letters*. 1971;**15**(3):232–6.

[81] Yang M., Kostov Y., Bruck H.A., Rasooly A. 'Carbon nanotubes with enhanced chemiluminescence immunoassay for CCD-based detection of staphylococcal enterotoxin B in food'. *Analytical Chemistry*. 2008;**80**(22):8532–7.

[82] Yang M., Kostov Y., Rasooly A. 'Carbon nanotubes based optical immunodetection of staphylococcal enterotoxin B (SEB) in food'. *International Journal of Food Microbiology* 2008;**127**(1–2):78–83.

[83] Yang M., Kostov Y., Bruck H.A., Rasooly A. 'Gold nanoparticle-based enhanced chemiluminescence immunosensor for detection of staphylococcal enterotoxin B (SEB) in food'. *International Journal of Food Microbiology*. 2009;**133**(3):265–71.

[84] Laksanasopin T., Guo T.W., Nayak S., *et al.* 'A smartphone dongle for diagnosis of infectious diseases at the point of care'. *Science Translational Medicine*. 2015;**7**(273):273re1.

[85] Yang M., Sun S., Kostov Y., Rasooly A. 'lab-on-a-chip for carbon nanotubes based immunoassay detection of staphylococcal enterotoxin B (SEB)'. *Lab on a Chip*. 2010;**10**(8):1011–17.

[86] Yang M., Sun S., Kostov Y., Rasooly A. 'An automated point-of-care system for immunodetection of staphylococcal enterotoxin B'. *Analytical Biochemistry*. 2011;**416**(1):74–81.

[87] Rasooly R., Balsam J., Hernlem B.J., Rasooly A. 'Sensitive detection of active Shiga toxin using low cost CCD based optical detector'. *Biosensors and Bioelectronics*. 2015;**68**:705–11.

[88] Balsam J., Ossandon M., Kostov Y., Bruck H.A., Rasooly A. 'Lensless CCD-based fluorometer using a micromachined optical Soller collimator'. *Lab on a Chip*. 2011;**11**(5):941–9.

[89] Balsam J., Rasooly R., Bruck H.A., Rasooly A. 'Thousand-fold fluorescent signal amplification for mHealth diagnostics'. *Biosensors and Bioelectronics* 2014; Jan 15;**51**:1–7.

[90] Balsam J., Bruck H.A., Rasooly A. 'Smartphone-based fluorescence detector for mHealth'. *Methods in Molecular Biology*. 2015;**1256**:231–45.

Chapter 2
Paper-based diagnostic devices

Spencer A. Schultz[1], Isabelle C. Noxon[1], Tyler A. Sisley[1] and Andres W. Martinez[1]

2.1 Introduction

Paper, broadly defined as any porous hydrophilic membrane, has served as a platform for diagnostic assays for millennia. One of the earliest recorded examples of a paper-based assay comes from *c.* 77 A.D. when Pliny the Elder, a Roman naturalist, described a qualitative test for detecting the presence of adulterants in verdigris using papyrus impregnated with an extract of gallnuts [1,2]. Starting with simple qualitative colorimetric assays and moving all the way to the current field of paperfluidics, paper offers several unique advantages as a platform for conducting a wide variety of diagnostic assays as well as for collecting, storing and transporting samples. This chapter will provide an overview of existing diagnostic devices made primarily out of paper and then focus on paper-based microfluidic devices, the next generation of paper-based diagnostic devices that promises to extend the use of paper as a material for fabricating diagnostic devices well into the future.

Paper is defined formally as a flexible sheet made from pressed cellulosic fibers (Figure 2.1) [3]. The process of making paper was developed in China during the Han Dynasty (206 B.C. to 220 A.D.) [4]. Written records credit Tsai Lun, an official in the Han Court, with developing the modern form of paper as well as the papermaking process in 105 A.D. [5]. Although the equipment used for making paper has evolved over time, the essential steps of the papermaking process have remained unchanged: first, a dilute suspension of cellulosic fibers is prepared in water; then, the fibers are deposited onto a screen forming a mat; and, finally, the mat of fibers is pressed and dried to produce a sheet of paper [5]. The cellulosic fibers used to make paper can come from many sources, but most diagnostic devices are made with paper produced from cotton fibers made primarily of cellulose (Figure 2.2(a)) [4]. Many other types of porous hydrophilic membranes have been developed over time and have been incorporated into diagnostic devices [3]. Among these, porous membranes made from nitrocellulose are used most commonly because of their high protein-binding affinity

[1]Department of Chemistry and Biochemistry, California Polytechnic State University, San Luis Obispo, CA, USA

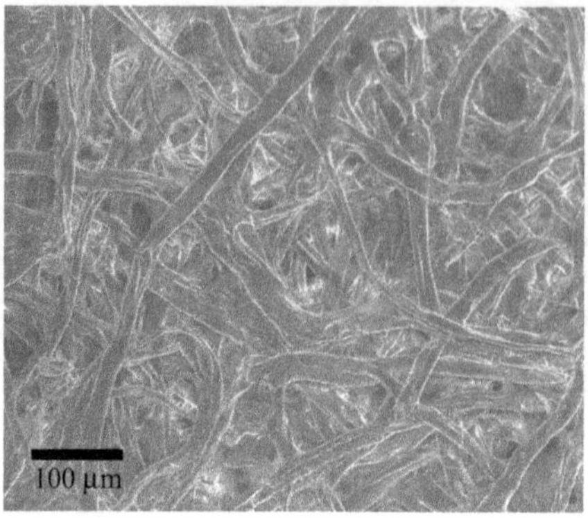

Figure 2.1 SEM micrograph (400×) of the surface of Whatman No. 1 chromatography paper

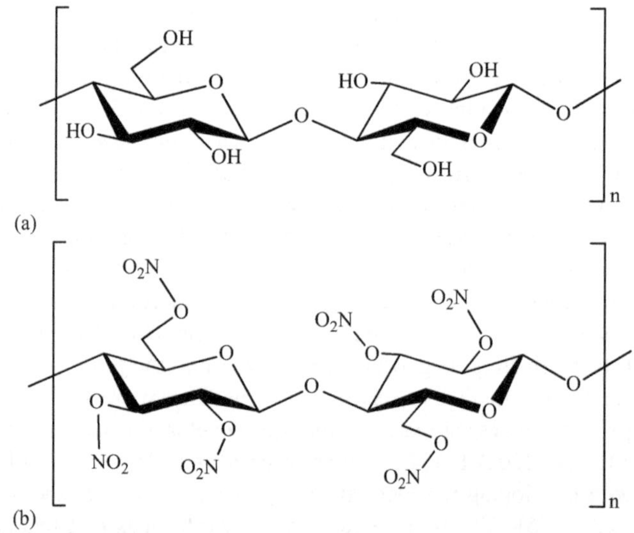

Figure 2.2 Chemical structures of cellulose (a) and nitrocellulose (b)

(Figure 2.2(b)) [3]. An enormous diversity of papers with varying chemical composition, thickness, and pore size are available, and each has unique characteristics that could be harnessed for specific applications in paper-based diagnostic devices.

Paper has many inherent characteristics that make it attractive as a platform for diagnostic assays. Paper is inexpensive and widely available, so scientists across the globe and throughout history have been able to work with paper and develop tests that are compatible with paper [2,4,6]. Paper is typically white and provides an excellent background for assays that produce a colored product or a change in color. Another attractive characteristic of paper is that it has a large surface-to-volume ratio, so a small volume of fluid (~1 μL) can typically produce a spot on paper that is large enough (~0.5 cm in diameter) to be seen with the naked eye. The large surface-to-volume ratio of paper also allows for reagents and samples to be dried and stored on paper-based devices. Paper can even be modified chemically in order to covalently bond reagents to the fibers [7]. Paper can also serve as the stationary phase for chromatographic separations of analytes [8]. Paper is disposable—one simple method for disposing paper-based tests is through incineration. Paper wicks fluid by capillary action, so fluids will move through a paper-based device without any external sources of power. And finally, as mentioned previously, paper can be produced from a wide variety of fibers and with a wide variety of characteristics (e.g., pore size, wicking speed, protein binding ability), which can be harnessed for specific applications and can even be combined in a single device to enable particular capabilities.

2.2 Current paper-based diagnostic devices

One of the best-known examples of a paper-based test is litmus paper, which is used to determine whether a solution is acidic or basic [7]. Scientific reports dating back to the nineteenth century describe the use of litmus paper for analyzing samples, and it is still sold commercially and used routinely by professional scientists and students [6,9]. In addition to litmus paper, a huge variety of other colorimetric tests have been developed to detect the presence of analytes on paper, but most of these tests have not been developed into commercial products [2,10–12]. Currently, there are two primary forms of commercially available paper-based diagnostic devices: dipstick assays and lateral-flow assays (LFAs) [3]. A third form of paper-based devices known as paper-based arrays are used primarily to collect, store, and transport samples but are also used in academic research to conduct assays.

2.2.1 Dipstick devices

Dipstick assays are one of the first examples of paper-based diagnostic devices for clinical use [4]. Dipstick devices comprise one or more small test pads of paper that have been impregnated with reagents for specific colorimetric assays and that are adhered to a plastic strip, which serves as a handle for the device (Figure 2.3). Dipstick assays are performed by quickly immersing the dipstick into the sample to be tested and then observing the colors that are produced on the test pads. The colors produced on the test pads can often be compared to a color-coded chart provided with the dipstick in order to make a semiquantitative determination of the concentration of the analyte in the sample [3].

The first pH dipstick test strips were patented and commercialized in the 1920s [4]. These devices can be purchased to determine the pH of a solution within a given range with varying amounts of sensitivity, both of which are determined by the number of test pads on the dipsticks as well as the specific indicators used on the test pads [13]. The first dipstick assays for clinical use were developed in the 1950s and were introduced commercially in the 1960s [3]. These initial dipstick assays were developed to quantify the concentration of glucose in urine by harnessing enzymatically catalyzed reactions involving glucose oxidase and horseradish peroxidase [14]. Urinalysis dipstick tests now include as many as ten

Figure 2.3 Dipstick device for analysis of water [18]

test pads to detect or analyze glucose, bilirubin, ketones, nitrite, urobilinogen, protein, blood, leukocytes, pH, and specific gravity [3,15,16]. These tests allow for the diagnosis of a variety of conditions and diseases including kidney disease, urinary tract infections, carbohydrate metabolism disorders, and liver disease [4,15,16]. For more sophisticated urinalysis tests, the dipsticks can be analyzed by a detector that analyzes the color of the test pads in order to improve the accuracy and precision of the tests [3,17]. Another common application for dipstick tests is for the analysis of water samples. Dipstick devices for testing drinking water and swimming pools can be used to quickly determine pH, total hardness, total chlorine, total bromine, free chlorine, total alkalinity, and cyanuric acid levels [18].

The main advantages of dipstick devices are that they are inexpensive, easy to use and provide immediate results. Dipstick tests are also simple to design and manufacture. The disadvantages of dipstick assays are that they are typically qualitative or semiquantitative, they often have limited sensitivity and selectivity, and they can only be developed for analytes that can react to produce a color change or be linked to a reaction that produces a color change [6]. Dipstick assays often also require a relatively large volume of sample in order to wet the test pads. For these reasons, dipstick devices are most practical for testing relatively simple samples, such as urine and water, that are abundant and contain relatively high concentrations of analytes, such as glucose or chlorine. In order to overcome some of the limitations with dipstick assays, LFAs were developed to expand the range of analytes that could be detected as well as the sensitivity and selectivity of the assays.

2.2.2 Lateral-flow devices

As the name suggests, lateral-flow devices rely on the capillary wicking of fluids across a porous membrane in order to detect an analyte [19]. Lateral-flow devices are made typically of several different types of paper that are treated with different reagents and held in contact with each other to create a single fluidic path that wicks a sample from a sample port to an absorbent pad, which serves as a waste-collection zone (Figure 2.4) [20]. In a typical LFA, a few drops of sample are introduced to the sample port, and the sample then wicks across a conjugate pad in

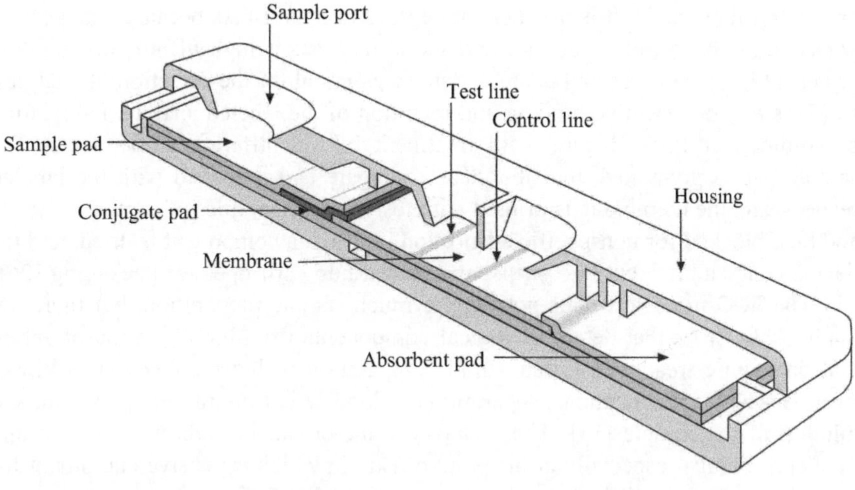

Figure 2.4 Schematic diagram of a lateral-flow device. Reproduced by permission of EMD Millipore Corporation. © 2002, 2008 EMD Millipore Corporation

which it interacts with a labeling reagent—typically an antibody bound to a colored particle—that binds specifically to the analyte. The labeled analyte then wicks across a membrane containing a capture reagent—typically a second antibody—that is immobilized on the device in a narrow test line and that captures selectively the labeled analyte. This process generates a colored line or band on the device when the analyte is present. All the other components of the sample as well as any excess labeling reagent continue to wick along the device into the waste zone. Most commercial devices also include a control line containing a second capture reagent on the membrane that binds only to the labeling reagent and serves to confirm that the capture reagents are working properly [3].

LFAs are widely used in laboratories, hospitals, and in the home where they can detect conditions such as pregnancy, disease, drug use, contamination, infection, and organ malfunction [3,21–23]. The first LFAs were developed in the late 1960s and have become a user-friendly, robust, affordable, point-of-care platform for diagnostics [19]. The first commercial LFAs were developed to detect human chorionic gonadotropin (hCG); a high level of this biomolecule is a positive indicator for pregnancy [24]. This test was fast (1–3 min), simple to use even for untrained users, required little user input, used small sample sizes, and was subsequently made commercial by 1988 [24]. The hCG test is now used globally to detect pregnancy and can be purchased for less than US$ 1 [25].

The three main facets of LFAs are the device or platform, sample preparation, and the immunochemical reaction. The LFA platform is composed of a sample pad, a fiberglass conjugate pad for storage of the labeled reagent, a membrane that is typically nitrocellulose, an absorbent waste pad, and often a small plastic holder that encases the assembled device (Figure 2.4) [19]. Although nitrocellulose is brittle and flammable, it is the membrane used in most LFAs because it is porous, wicks fluids by capillary action, and its surface has a high affinity for binding proteins [4,23]. This binding characteristic is essential for the detection of analytes, as LFAs rely on selective binding and retention of the labeled analyte. Fabrication of commercial LFAs begins with treatment of the different components. The sample pad is pretreated, the fiberglass conjugate pad is loaded with the labeled reagent, and the membrane is printed with test and control lines of capture reagents and then blocked for nonspecific adsorption [20]. Each component is dried, and the device is assembled, cut into strips, and placed into cartridges for packaging [26].

The hCG urine test does not require much sample preparation, but there are many other LFAs that detect biological components from blood plasma or serum that do require treatment. Often sample preparation includes collection of blood, treatment with anticoagulants, separation of blood cells from plasma or serum, and dilution of the sample [19]. These steps are important, but are cumbersome and time consuming—especially at the point of care in which expensive equipment for sample treatment is difficult to obtain or not available [27].

The immunochemical reaction is arguably the most important component of LFAs. There are two main types: immunoassays that rely on antibody–antigen interactions, and nucleic acid-based assays that rely on nucleic acid aptamers and target molecule binding or nucleic acid hybridization [19]. These immunochemical reactions rely heavily on the sensitive and specific interactions that consequently determine the sensitivity and selectivity of the assay itself [19]. The component of the immunochemical reaction that allows for visual detection is the colorimetric probe, which is conjugated to an antibody, protein, or DNA sequence to make the labeling reagent. The probes are most commonly latex beads, gold nanoparticles, fluorescent tags, quantum dots, and enzymes. The type of colorimetric probe is determined by several factors including the type of sample, the analyte, pH, and device storage conditions. Immunochemical reagents can be synthesized against a myriad of proteins and biomolecules. The heterogeneity of immunochemical combinations is part of what makes LFAs such a widely studied and versatile platform [19].

The main advantages of LFAs, as with most other paper-based devices, are that they are inexpensive and easy to use—the user typically only has to apply a few drops of sample to the device and then read the resulting colored lines, unless more sophisticated sample preparation is required [3]. What sets LFAs apart from other paper-based devices, like dipsticks, is that by harnessing the wicking capabilities of paper, LFAs perform what is effectively a chromatographic separation of the analyte from the other components of the sample as well as a concentration of the analyte along the test line. Both these actions lead to highly sensitive and specific assays. Some more sophisticated LFAs include a second step in which an amplification reagent is added to the device in order to enhance the color of the test line and further improve the sensitivity of the device [3]. Another advantage of LFAs is that as they rely on immunochemical reactions to detect analytes, the same basic device design can be used to detect a huge range of analytes by simply changing the labeling reagent and the capture reagent [19]. On their own, LFAs are typically qualitative and only provide yes/no results, but more recent LFAs are compatible with electronic readers that analyze the intensity of color that develops at the test line or rely on fluorescent labels in order to make quantitative determinations [3].

2.2.2.1 Vertical-flow devices

Vertical-flow devices rely on the same analyte-detection principles as lateral-flow devices and can be used to conduct the same types of assays, but, instead of lining up the different components of the device horizontally as in LFAs, the various components are stacked vertically. The advantage of stacking the layers vertically is that the path length for the sample to wick through the device is significantly reduced; however, these assays typically require multiple steps by the user in order to be performed correctly [3].

2.2.3 Paper-based arrays

Paper-based arrays are used primarily for the collection, storage, and transportation of samples. Paper-based arrays typically comprise a card marked with multiple circular zones that can be filled with liquid samples. The zones are often pretreated with reagents to improve the stability of the analytes or to disinfect the samples. Samples, such as blood, can be spotted on the array and dried, and then the array can be shipped to a central laboratory where the samples can be eluted from the array and analyzed [28]. This process allows for large numbers of samples to be collected and stored in a convenient format that does not require refrigeration [29]. The sample collection can also be performed in the field with minimal instrumentation [29].

One common application for paper-based arrays is in dried blood spot (DBS) testing [28]. This technique was developed in the early 1900s but gained widespread use in the 1960s and 1970s when Robert Guthrie used DBS to develop a technique for the systematic screening of newborns for metabolic diseases [30]. Now known as the Guthrie test, newborns around the world are tested for a series of metabolic disorders and diseases by collecting their blood from a heal prick on a DBS card [28]. In fact, DBS can be used to detect a wide range of analytes and diseases including HIV, dengue, hepatitis, and small molecule drugs [28,29,31].

An alternative application of paper-based arrays is for the synthesis of libraries of short peptides or small molecules via a technique known as spot synthesis [32–35]. Once prepared, these arrays can be used to conduct a wide range of assays including binding assays, enzyme assays, and studies with living organisms [36].

2.3 Paper-based microfluidic devices

In 2007, the Whitesides group at Harvard University published an article describing the first example of what would become known as paper-based microfluidic devices or microfluidic paper-based analytical devices (microPADs) (Figure 2.5) [37]. Having contributed a substantial body of work to the field of conventional microfluidic devices, including the development of soft lithography [38,39], George

Figure 2.5 A paper-based microfluidic device for detecting glucose (left zones) and protein (right zones)

Whitesides was looking to develop a new class of devices that married the advantages of lateral-flow and dipstick devices, such as capillary wicking and low cost, with the advantages of conventional microfluidic devices, such as low volumes of sample, multiplexing, and controlled sequential steps. The ultimate goal of the Whitesides group was to develop a new class of ultra-low-cost point-of-care diagnostic devices that could be deployed to resource-limit settings and improve access to healthcare to populations around the globe.

Paper was a natural choice as a material for fabricating this new class of devices for all the same reasons mentioned in the introduction. The main innovation described in the initial publication from the Whitesides group was the development of a method to pattern paper into a network of hydrophilic channels and test zones bounded by hydrophobic barriers [37,40]. The channels could wick samples and distribute them into the test zones where assays would take place. Although examples of patterning paper using waxes had been described previously [2,41], what set the work from the Whitesides group apart was the focus on the development of a new class of low cost, very simple to use portable diagnostic devices [37,42]. This work catalyzed a surge of interest in the development of paper-based assays and resulted in the establishment of the new field of paperfluidics [3,6,12].

2.3.1 Fabrication of paper-based microfluidic devices

Fabrication of paper-based microfluidic devices involves the patterning of a piece of paper using hydrophobic inks to define hydrophilic channels and test zones bounded by hydrophobic barriers [12]. Several methods for patterning paper have been developed since 2007, each with their own set of advantages and disadvantages [6,12]. Although the methods are very diverse in terms of equipment and inks, they can be classified broadly into two categories: indirect and direct patterning. In indirect patterning, the first step of the process involves coating the entire piece of paper with a hydrophobic ink or reagent such as a wax, a polymer, or a resin. In the second step, the hydrophobic reagent is etched away selectively to

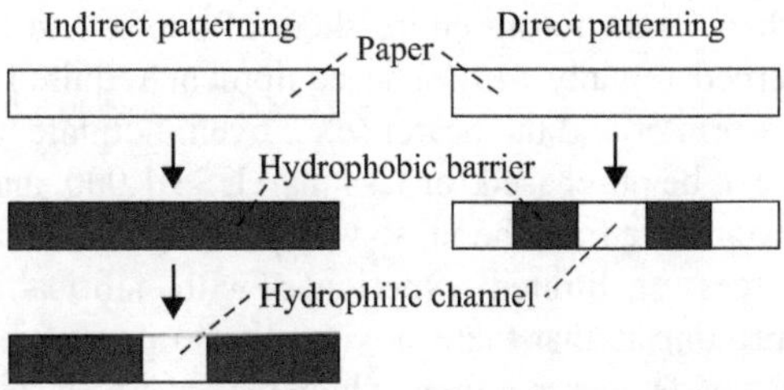

Figure 2.6 Cross-sectional schematic diagram of indirect and direct patterning of paper

produce hydrophilic channels. In direct patterning, the hydrophobic ink is printed directly on the paper in the desired pattern (Figure 2.6).

The first published method for fabricating paper-based microfluidic devices was an example of indirect patterning and was an adapted form of photolithography [37]. In this approach, the paper is first impregnated with SU-8 photoresist, a negative photoresist, and dried. The paper is then exposed to UV light through a photomask in order to cross-link the exposed portions of the photoresist. Finally, the unexposed portions of the photoresist are dissolved and removed from the paper using a solvent such as acetone to yield the desired pattern of channels and test zones in the paper [37,40,42]. Other photopolymers, in addition to SU-8, have also shown to be useful for patterning paper via indirect patterning [3].

Indirect patterning was also achieved using an inkjet printer [43]. The process involves first coating a piece of paper with polystyrene to make it hydrophobic. The polystyrene is then selectively removed from specific areas of the paper by printing toluene onto the paper using an inkjet printer in the desired pattern. The toluene washes the polystyrene out of the paper and returns the washed paper to its original hydrophilic state.

A third example of indirect patterning using plasma treatment involves impregnating a sheet of paper with alkyl ketene dimer, a reagent used commonly for sizing paper to make it more hydrophobic [44]. The reagent is then selectively etched away by placing the paper between two metal stencils and treating it in a vacuum plasma reactor. The areas of the paper that are exposed to the plasma, through the metal stencils, are made hydrophilic and the areas of paper that are protected by the stencil remain hydrophobic.

In direct patterning, hydrophobic agents are applied onto the paper in the desired pattern. Hydrophobic agents are applied typically through conventional printing technologies, such as inkjet printing, screen printing, flexographic printing, and stamping, but could even be applied by hand. Each technique has its own set of advantages and disadvantages in terms of cost, resolution, throughput, and technical difficulty. One of the most popular methods of direct patterning is wax printing [45–47]. In wax printing, a solid-ink printer is used to print wax directly on the surface of paper in any desired pattern. The printed paper is then heated to reflow the wax and create hydrophobic barriers within the paper. The entire process can be completed in less than 5 min for an entire sheet of paper, and features a small as 1 mm can be patterned reliably [46]. The equipment required for wax printing, namely the solid-ink printer and the heater (e.g., oven, hotplate, thermal laminator, heat gun, and iron) can be purchased for less than US$ 1,000, and the process does not require a clean room or a fume hood, so wax printing can be performed by users with limited resources or limited experience with fabrication [12]. Another advantage of wax printing is that there is virtually no production-cost penalty for making small batches of devices at a time, which makes it well suited for laboratory-scale prototyping.

In comparing indirect and direct patterning methods, indirect patterning methods are generally more time consuming as they involve at least two steps, but often several more. Indirect patterning methods also consume larger quantities of reagents and solvents as the entire piece of paper is treated and made hydrophobic. Direct patterning methods are generally faster, less expensive, and less technically challenging than indirect methods. One advantage of indirect patterning, however, is that it tends to allow for higher resolution patterns to be produced. Specifically in the case of photolithography, the resolution of the pattern is controlled by the resolution of the exposure to UV light, so smaller features can be produced using this method compared to wax printing [12]. In one instance, direct and indirect methods of patterning were combined to pattern paper with Teflon in order to generate microPADs that were compatible with organic solvents [48].

A third method of preparing microPADs is via shaping or cutting. In this approach the desired network of channels and test zones is cut out of a sheet of paper, either manually or using a laser cutter or cutting plotter [49,50]. This method is simple to perform and does not require any reagents. The devices fabricated via cutting usually have to be encased in plastic to protect the paper, provide structural support to the paper, and facilitate manipulation of the devices [12].

Once the patterning process is complete, devices can be further modified to add additional capabilities. One common post-patterning step is to encase devices in plastic using either tape, thermal lamination sheets or toner [49,51,52]. Another option is to add additional features to the device such as electrodes. Electrodes are typically printed onto the devices after they are patterned using conductive inks [53–57]. Devices can also be modified to control the wicking in the device either by incorporating valves [58], adding reagents or shunts that slow down wicking [59,60], or adding additional layers of paper or plastic to speed up wicking [52,57,61]. Finally, multiple layers of patterned paper can be stacked on top of each other to produce three-dimensional (3D) microPADs (Figure 2.7) [62–65]. The layers of paper in the 3D device can either be bonded to each other permanently using permanent adhesives or tapes, or they can be held together temporarily using a manifold or removable adhesives or tapes.

To complete the fabrication process, reagents for the assay must be applied to the device. Typically reagents are deposited onto microPADs from solution, and the solutions are then allowed to dry leaving the reagent behind on the device. Reagents can be deposited manually using a pipet or capillary tube, or can be

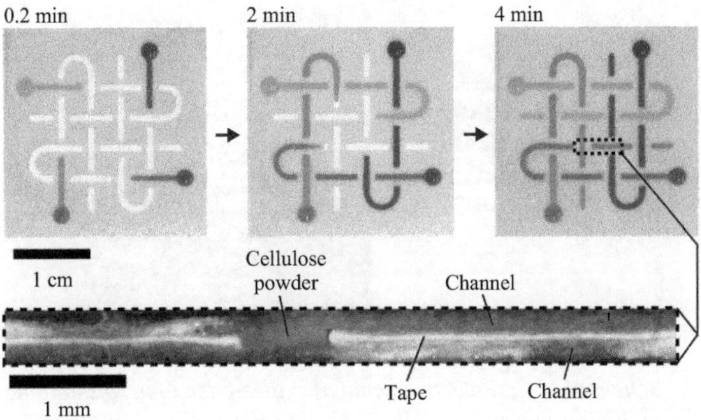

Figure 2.7 Top view and cross-sectional view of a three-dimensional microPAD assembled from two layers of patterned paper and a layer of double-sided tape. Reproduced by permission of The National Academy of Sciences from Reference 62. © 2008 National Academy of Sciences, USA

deposited using an inkjet printer or liquid-dispensing robot. More recently, a method for solid deposition of reagents using custom-made pencils was described, which led to improved shelf life of the reagents on the devices [66].

2.3.2 Applications of paper-based microfluidic devices

Initial research on paper-based microfluidic devices focused primarily on the development of methods of fabricating devices. The first devices consisted generally of a sample inlet that directed the sample into one more channels and then into multiple test zones in which reagents for colorimetric assays were deposited (Figure 2.5). These initial devices required small volumes of fluid (≤ 20 μL) and could perform multiple assays simultaneously from a single sample-addition step [37,42]. As work with paper-devices expanded, new applications and unique assays were developed, many of which were enabled by the unique capabilities of microPADs. This section will highlight four areas of applications for microPADs that show promise for future diagnostic devices.

Colorimetric assays, assays that result in a color change in response to the presence of analyte, remain one of the most common types of assays conducted on microPADs [3]. A wide range of colorimetric assays have developed for microPADs including indicator-based assays, enzymatic assays, sorbent-based assays, as well as combinations as these approaches such as enzyme-linked immunosorbent assays (ELISAs) [3]. These assays have been shown to be useful for detecting a wide range of analytes including ions, small molecules, proteins, and DNA. Concurrent with the development of colorimetric assays on microPADs, several methods for quantifying the resulting colorimetric signal on the devices were

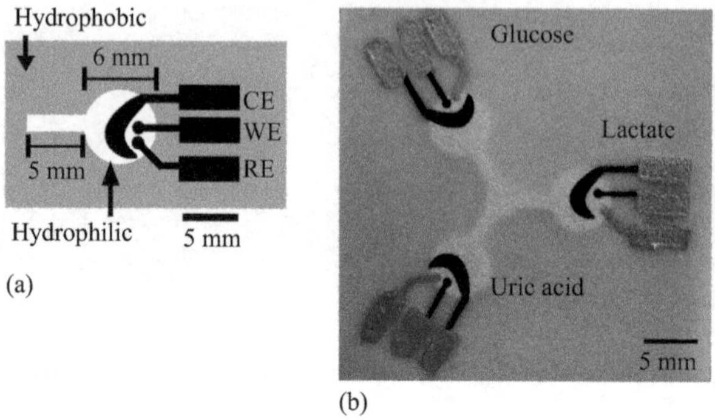

Figure 2.8 Schematic of an electrochemical paper-based microfluidic device including three electrodes (a), and image of an actual device for detecting glucose, lactate, and uric acid (b). Adapted with permission from Reference 53. © 2009 American Chemical Society

developed. One promising approach is the use of digital image colorimetry, in which a digital image of the results of a colorimetric assay is obtained, and then the pixel intensity of the color is measured [67,68]. The images could be obtained using a scanner, a digital camera or, most interestingly in the context of portable point-of-care diagnostic assays, a camera phone [42]. The results can be calibrated through either external calibration or standard addition and can typically produce results with relative errors and relative standard deviations within ten percent [68].

As an alternative to colorimetric assays, paper-based assays that rely on electrochemical detection offer many advantages, especially when it comes to performing quantitative tests [53,55,56]. Electrochemical paper-based assays have been demonstrated for a wide range of analytes with high accuracy and precision, and with the ability to detect analytes at femtomolar concentrations [6,69,70]. The electrodes for electrochemical assays are typically printed directly on the microPADs (Figure 2.8), and an external reader is used typically to perform the assay [6].

A third type of paper-based assays that eliminate the need for sophisticated external readers are chronometric assays, in which the signal from the assay is the time that it takes the sample to wick across a channel in a device (Figure 2.9) [64,71–74]. These assays rely on the unique properties of microPADs to wick fluids along channels in paper as well as the ability to pattern reagents in multiple layers of paper and then assemble these layers into a single 3D device. In these devices, a reagent that creates and impermeable or semi permeable barrier is added to a channel in the device. The presence of the analyte triggers a chemical reaction that results in the degradation of the barrier. So, when the analyte is present, the sample wicks across the channel more quickly than when the analyte is absent. Chronometric assays stand out for their simplicity, sensitivity, and low limits of detection, which rival traditional ELISA [74].

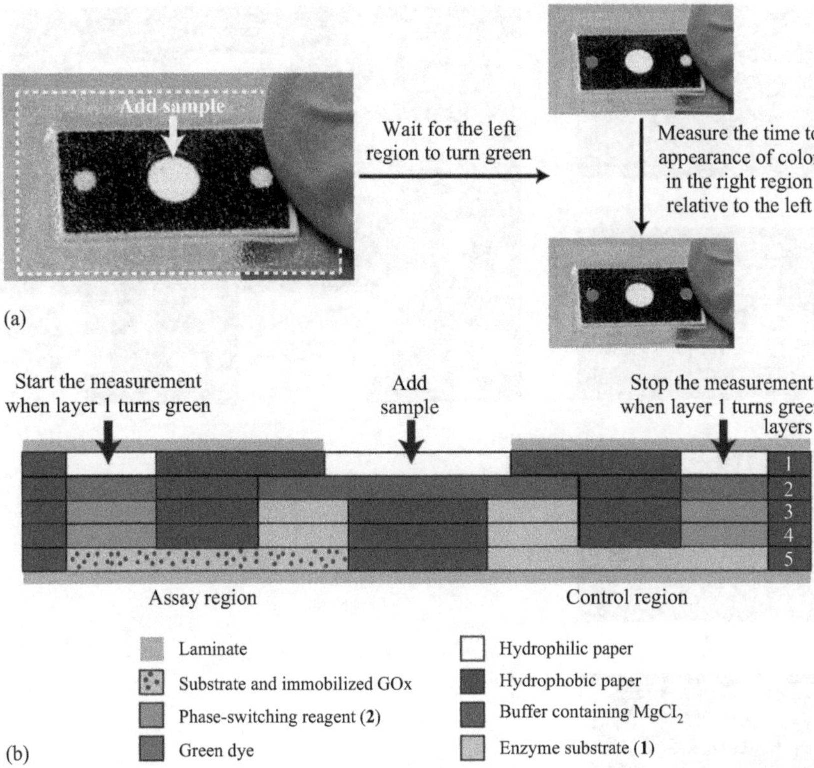

Figure 2.9 A chronometric paper-based assay (a) and a cross-sectional schematic of the layers and components making up the device (b). Adapted with permission from Reference 74. © 2013 American Chemical Society

In addition to developing specific types of quantitative assays, the paperfluidics community has devoted considerable effort toward developing devices capable of performing multiple sequential steps in order to automatically perform multistep assays, such as ELISAs or enzyme-inhibition assays. For example, a device that performs a typical lateral-flow immunoassay and then automatically adds an amplification reagent to enhance the intensity of the test line on the device and improve the limit of detection of the assay was developed by designing a device with multiple fluid inlets leading to a single-test zone (Figure 2.10) [50,75,76]. Alternatively, a similar type of assay was achieved by carefully designing a series of channels that would wick fluid from a single sample inlet and deliver it to a test zone at different times [77]. The development of fluidic diodes provided a third option for achieving automated multistep assays on microPADs [78,79]. Finally, paper-based enzyme-inhibition tests that require incubation of an enzyme with a sample followed by introduction of the enzyme substrate were automated by using erodible polymeric bridges, which allowed for the introduction of single-use timed shut-off valves into microPADs [58].

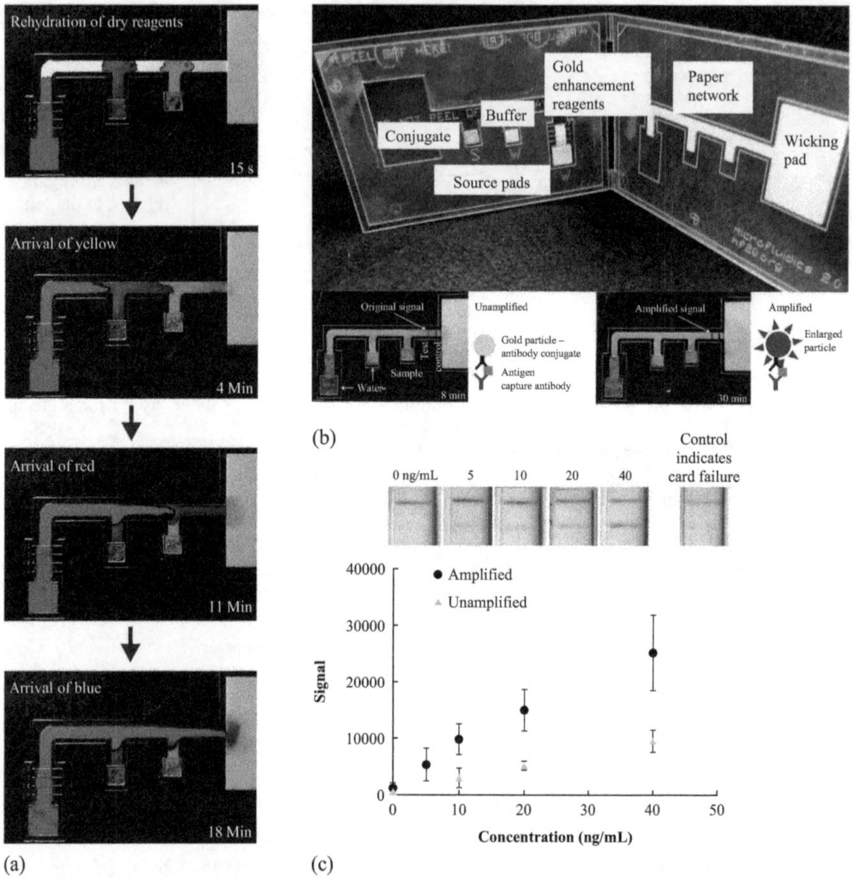

Figure 2.10 Multistep paper-based assay. (a) Images of the device performing multiple sequential delivery of reagents using dyes as sample reagents. (b) Image of actual device for detecting Malaria. (c) Results of the test showing the improved signal obtained using the multistep device (amplified). Adapted with permission from Reference 76. © 2012 American Chemical Society

2.4 Conclusions

Paper offers several advantages as a platform for conducting simple, point-of-care diagnostic assays. Current, commercially available paper-based diagnostic devices, such as dipstick assays and LFAs, will continue to be used widely due to their low cost and simplicity. A new generation of paper-based devices is currently being developed and promises to extend those applications of paper for point-of-care diagnostics.

The concept of making diagnostic devices out of paper is being adopted by a growing community of researchers. MicroPADs were first developed as diagnostic devices for use in developing countries but are now being developed to detect a wide range of analytes and could be used to monitor air, soil, and water quality; they could be used as diagnostic devices for animals and plants; they could be used in home healthcare to diagnose disease or monitor drug levels; and they could be used by the military and first-responders to assess a person's health status or detect toxins, biohazards, or explosives. MicroPADs also have many potential applications in basic research.

One of the great aspects of paper-based microfluidic devices is that there is a very low barrier to entry, both in terms of the cost of equipment and the technical expertise that is required to fabricate devices. A pair of scissors and a paper towel is all that is really needed to make simple paper-based devices. So, much like open-source software, we look forward to seeing contributions to this field from all kinds of scientists from all over the world and seeing where we can go with this simple yet-powerful technology.

References

[1] Pliny, B.J., Riley H.T. *The Natural History of Pliny*. 1857. Vol VI. pp. 140–1; 235–8; 362–4.

[2] Yagoda H. 'Applications of confined spot tests in analytical chemistry: Preliminary paper'. *Industrial Engineering Chemistry and Analytical Edition*. 1937;**9**(2):79–82.

[3] Yetisen A.K., Akram M.S., Lowe C.R. 'Paper-based microfluidic point-of-care diagnostic devices'. *Lab on a Chip*. 2013;**13**(12):2210–51.

[4] Hu J., Wang S., Wang L., *et al.* 'Advances in paper-based point-of-care diagnostics'. *Biosensors and Bioelectronics*. 2014;**54**:85–97.

[5] Hunter D. *Papermaking: The History and Technique of an Ancient Craft*. New York, NY: Dover Publications Inc.; 1978.

[6] Cate D.M., Adkins J.A., Mettakoonpitak J., Henry C.S. 'Recent developments in paper-based micro fluidic devices'. *Analytical Chemistry*. 2015;**87**:19–41.

[7] Pelton R. 'Bioactive paper provides a low-cost platform for diagnostics'. *TrAC – Trends in Analytical Chemistry*. 2009;**28**(8):925–42.

[8] Block R.J., Durrum E.L., Zweig G. *A Manual of Paper Chromatography and Paper Electrophoresis*. New York: Academic Press; 1955.

[9] Davy J. 'On gaseous compound of organic oxide and chlorine'. *Philosophical Transactions of the Royal Society of London*. 1812;**102**:144–51.

[10] Feigl F., Anger V., Koslow R.H. 'Spot tests in inorganic analysis'. *Journal of The Electrochemical Society*. 1973;**120**:261C.

[11] Hoffman J.H., Weaver J.W. 'Qualitative spot tests', in *Analytical Methods for a Textile Laboratory*. Research Triangle Park, NC: American Association of Textile Chemists and Colorists; 1984. pp. 156–206.

[12] Martinez A.W., Phillips S.T., Whitesides G.M., Carrilho E. 'Diagnostics for the developing world: Microfluidic paper-based analytical devices'. *Analytical Chemistry*. 2010;**82**(1):3–10.

[13] 'EMD Millipore pH Indicator Strips' [Online]. Available: http://www.emdmillipore.com/Web-US-Site/en_CA/-/USD/ViewParametricSearch-SimpleOfferSearch?SearchTerm=*&SingleResultDisplay=SFProductSearch&SearchContextCategoryUUIDs=UJWb.qB.jtAAAAE_9A93.Lxj. [Accessed: 30-Jun-2016].

[14] Free A.H., Adams E.C., Kercher M.L., Free H.M., Cook M.H. 'Simple specific test for urine glucose'. *Clinical Chemistry*. 1957;**3**(3):163–8.

[15] Boyd J., Barratt J. 'Interpretation and management of abnormal dipstick urinalysis'. *Medicine (Baltimore)*. 2011;**39**:(6):312–16.

[16] Roberts J.R. 'Urine dipstick testing: Everything you need to know urinalysis: A comprehensive review'. *Emergency Medicine News*. 2007;**29**:24–7.

[17] Peele J.D., Gadsden R.H., Crews R. 'Semi-automated vs. visual reading of urinalysis dipsticks'. *Clinical Chemistry*. 1977;**23**(12):2242–6.

[18] 'AquaChek Pool & Spa Test Strips' [Online]. Available: http://www.aquachek.com/pool-spa-owners/aquachek-select-7-in-1/. [Accessed: 15-Jun-2016].

[19] Bahadir E.B., Sezgintürk M.K. 'Lateral flow assays: Principles, designs and labels'. *TrAC Trends in Analytical Chemistry*. 2016;**82**:286–306.

[20] EMD Millipore. *Rapid Lateral Flow Test Strips*. Billerica, MA: Millipore Corporation; 2008.

[21] Chan C.P.Y., Sum K.W., Cheung, K.Y., *et al*. 'Development of a quantitative lateral-flow assay for rapid detection of fatty acid-binding protein'. *Journal of Immunological Methods*. 2003;**79**:91–100.

[22] Posthuma-Trumpie G.A., Korf J., Van Amerongen A. 'Lateral flow (immuno) assay: Its strengths, weaknesses, opportunities and threats. A literature survey'. *Analytical and Bioanalytical Chemistry*. 2009;**393**(2):569–82.

[23] Sajid M., Kawde A.N., Daud M. 'Designs, formats and applications of lateral flow assay: A literature review'. *Journal of Saudi Chemical Society*. 2015;**19**(6):689–705.

[24] Wild D. *The Immunoassay Handbook*, 4th ed. Amsterdam: Elsevier Science; 2013.

[25] 'Clinical Guard One Step HCG Pregnancy Test Strip' [Online]. Available: https://www.clinicalguard.com/one-step-hcg-pregnancy-test-strip-p-704.html. [Accessed: 28-Jun-2016].

[26] Sharma S., Zapatero-Rodriguez J., Estrela P., O'Kennedy R. 'Point-of-care diagnostics in low resource settings: Present status and future role of microfluidics'. *Biosensors*. 2015;**5**(3):577–601.

[27] Chin C.D., Linder V., Sia S.K. 'Lab-on-a-chip devices for global health: Past studies and future opportunities'. *Lab on a Chip*. 2007;**7**(1):41–57.

[28] Lehmann S., Delaby C., Vialaret J., Ducos J., Hirtz C. 'Current and future use of "dried blood spot" analyses in clinical chemistry'. *Clinical Chemistry and Laboratory Medicine*. 2013;**51**(10):1897–909.

[29] Meesters R.J., Hooff G.P. 'State-of-the-art dried blood spot analysis: An overview of recent advances and future trends'. *Bioanalysis*. 2013;**5** (17):2187–208.

[30] Guthrie R., Susi A. 'A simple phenylalanine method for detecting phenylketonuria in large populations of newborn infants'. *Pediatrics*. 1963;**32**: 338–43.

[31] Stevens R., Pass K., Fuller S., *et al*. 'Blood spot screening and confirmatory tests for syphilis antibody'. *Journal of Clinical Microbiology*. 1992;**30** (9):2353–8.

[32] Frank R. 'Spot-synthesis: An easy technique for the positionally addressable, parallel chemical synthesis on a membrane support'. *Tetrahedron*. 1992;**48** (42):9217–32.

[33] Frank R. 'The SPOT-synthesis technique: Synthetic peptide arrays on membrane supports – Principles and applications'. *Journal of Immunological Methods*. 2002;**267**(1):13–26.

[34] Blackwell H.E. 'Hitting the SPOT: Small-molecule macroarrays advance combinatorial synthesis'. *Current Opinion in Chemical Biology*. 2006;**10** (3):203–12.

[35] Kramer A., Reineke U., Dong L., *et al*. 'Spot synthesis: Observations and optimizations'. *Journal of Peptide Research*. 1999;**54**(4):319–27.

[36] Hilpert K., Winkler D.F.H., Hancock R.E.W. 'Peptide arrays on cellulose support: SPOT synthesis, a time and cost efficient method for synthesis of large numbers of peptides in a parallel and addressable fashion'. *Nature Protocols*. 2007;**2**(6):1333–49.

[37] Martinez A.W., Phillips S.T., Butte M.J., Whitesides G.M. 'Patterned paper as a platform for inexpensive, low-volume, portable bioassays'. *Angewandte Chemie International Edition*. 2007;**46**:1318–20.

[38] Xia Y., Whitesides G.M. 'Soft lithography'. *Annual Review of Materials Science*. 1998;**28**(1):153–84.

[39] Xia Y., Whitesides G.M. 'Soft lithography'. *Angewandte Chemie International Edition*. 1998;**37**(5):550–75.

[40] Martinez A.W., Phillips S.T., Wiley B.J., Gupta M., Whitesides G.M. 'FLASH: A rapid method for prototyping paper-based microfluidic devices.' *Lab on a Chip*. 2008;**8**(12):2146–50.

[41] Muller R.H., Clegg D.L. 'Automatic paper chromatography'. *Analytical Chemistry*. 1949;**21**:1123–5.

[42] Martinez A.W., Phillips S.T., Carrilho E., Thomas S.W., Sindi H., Whitesides G.M. 'Simple telemedicine for developing regions: Camera phones and paper-based microfluidic devices for real-time, off-site diagnosis'. *Analytical Chemistry*. 2008;**80**(10):3699–707.

[43] Abe K., Suzuki K., Citterio D. 'Inkjet-printed microfluidic multianalyte chemical sensing paper'. *Analytical Chemistry*. 2008;**80**(18):6928–34.

[44] Li X., Tian J., Nguyen T., Shen W. 'Paper-based microfluidic devices by plasma treatment'. *Analytical Chemistry*. 2008;**80**(23):9131–4.

[45] Lu R., Shi W., Jiang L., Qin J., Lin B. 'Rapid prototyping of paper-based microfluidics with wax for low-cost, portable bioassay'. *Electrophoresis*. 2009;**30**(9):1497–500.

[46] Carrilho E., Martinez A.W., Whitesides G.M. 'Understanding wax printing: A simple micropatterning process for paper-based microfluidics'. *Analytical Chemistry*. 2009;**81**(16):7091–5.

[47] Lu Y., Shi W., Qin J., Lin B. 'Fabrication and characterization of paper-based microfluidics prepared in nitrocellulose membrane by Wax printing'. *Analytical Chemistry*. 2010;**82**(1):329–35.

[48] Deiss F., Matochko W.L., Govindasamy N., Lin E.Y., Derda R. 'Flow-through synthesis on teflon-patterned paper to produce peptide arrays for cell-based assays'. *Angewandte Chemie International Edition*. 2014;**53**(25):6374–7.

[49] Fenton E.M., Mascarenas M.R., Lopez G.P., Sibbett S.S. 'Multiplex lateral-flow test strips fabricated by two-dimensional shaping'. *ACS Applied Materials and Interfaces*. 2009;**1**(1):124–9.

[50] Fu E., Ramsey S.A., Kauffman P., Lutz B., Yager P. 'Transport in two-dimensional paper networks'. *Microfluidics and Nanofluidics*. 2011;**10**(1):29–35.

[51] Schilling K.M., Lepore A.L., Kurian J.A., Martinez A.W. 'Fully enclosed microfluidic paper-based analytical devices'. *Analytical Chemistry*. 2012;**84**(3):1579–85.

[52] Camplisson C.K., Schilling K.M., Pedrotti W.L., Stone H.A., Martinez A.W. 'Two-ply channels for faster wicking in paper-based microfluidic devices'. *Lab on a Chip*. 2015;**15**(23):4461–6.

[53] Dungchai W., Chailapakul O., Henry C.S. 'Electrochemical detection for paper-based microfluidics'. *Analytical Chemistry*. 2009;**81**(14):5821–6.

[54] Nie Z., Nijhuis C.A., Gong J., *et al.* 'Electrochemical sensing in paper-based microfluidic devices'. *Lab on a Chip*. 2010;**10**(4):477–83.

[55] Nie Z., Deiss F., Liu X., Akbulut O., Whitesides G.M. 'Integration of paper-based microfluidic devices with commercial electrochemical readers'. *Lab on a Chip*. 2010;**10**(22):3163–9.

[56] Adkins J., Boehle K., Henry C. 'Electrochemical paper-based microfluidic devices'. *Electrophoresis*. 2015;**36**(16):1811–24.

[57] Renault C., Anderson M.J., Crooks R.M. 'Electrochemistry in hollow-channel paper analytical devices'. *Journal of the American Chemical Society*. 2014;**136**(12):4616–23.

[58] Jahanshahi-Anbuhi S., Henry A., Leung V., *et al.* 'Paper-based microfluidics with an erodible polymeric bridge giving controlled release and timed flow shutoff'. *Lab on a Chip*, 2013;**14**(1):229–36.

[59] Lutz B., Liang T., Fu E., Ramachandran S., Kauffman P., Yager P. 'Dissolvable fluidic time delays for programming multi-step assays in instrument-free paper diagnostics'. *Lab on a Chip*. 2013;**13**(14):2840–7.

[60] Toley B.J., McKenzie B., Liang T., Buser J.R., Yager P., Fu E. 'Tunable-delay shunts for paper micro fluidic devices'. *Analytical Chemistry*. 2013;**85**(23):11545–52.

[61] Jahanshahi-Anbuhi S., Chavan P., Sicard C., *et al.* 'Creating fast flow channels in paper fluidic devices to control timing of sequential reactions'. *Lab on a Chip*. 2012;**12**:5079–85.

[62] Martinez A.W., Phillips S.T., Whitesides G.M. 'Three-dimensional microfluidic devices fabricated in layered paper and tape'. *Proceedings of the National Academy of Sciences of the United States of America*. 2008;**105** (50):19606–11.

[63] Liu H., Crooks R.M. 'Three-dimensional paper microfluidic devices assembled using the principles of origami'. *Journal of the American Chemical Society*. 2011;**133**(44):17564–6.

[64] Lewis G.G., DiTucci M.J., Baker M.S., Phillips S.T. 'High throughput method for prototyping three-dimensional, paper-based microfluidic devices'. *Lab on a Chip*. 2012;**12**(15):2630.

[65] Schilling K.M., Jauregui D., Martinez A.W. 'Paper and toner three-dimensional fluidic devices: Programming fluid flow to improve point-of-care diagnostics'. *Lab on a Chip*. 2013;**13**:628–31.

[66] Mitchell H.T., Noxon I.C., Chaplan C.A., *et al.* 'Reagent pencils: A new technique for solvent-free deposition of reagents onto paper-based microfluidic devices'. *Lab on a Chip*. 2015;**15**:2213–20.

[67] Byrne L., Barker J., Pennarun-Thomas G., Diamond D., Edwards S. 'Digital imaging as a detector for generic analytical measurements'. *TrAC – Trends in Analytical Chemistry*. 2000;**19**:517–22.

[68] Chaplan C.A., Mitchell H.T., Martinez A.W. 'Paper-based standard addition assays'. *Analytical Methods*, 2014;**6**(5):1296.

[69] Cunningham J.C., Kogan M.R., Tsai Y.J., Luo L., Richards I., Crooks R.M. 'Paper-based sensor for electrochemical detection of silver nanoparticle labels by galvanic exchange'. *ACS Sensors*. 2015;**1**(1):acssensors.5b00051.

[70] Scida K., Cunningham J.C., Renault C., Richards I., Crooks R.M. 'Simple, sensitive, and quantitative electrochemical detection method for paper analytical devices'. *Analytical Chemistry*. 2014;**86**(13):6501–7.

[71] Lewis G.G., Ditucci M.J., Phillips S.T. 'Quantifying analytes in paper-based microfluidic devices without using external electronic readers'. *Angewandte Chemie International Edition*. 2012;**51**(51):12707–10.

[72] Lewis G.G., Phillips S.T. 'Quantitative point-of-care (POC) assays using measurements of time as the readout: A new type of readout for mHealth'. in *Mobile Health Technologies: Methods and Protocols*. New York, NY: Humana Press; 2015. pp. 213–229.

[73] Lewis G.G., Robbins J.S., Phillips S.T. 'A prototype point-of-use assay for measuring heavy metal contamination in water using time as a quantitative readout'. *Chemical Communications*. 2014;**50**(40):5352–4.

[74] Lewis G.G., Robbins J.S., Phillips S.T. 'Point-of-care assay platform for quantifying active enzymes to femtomolar levels using measurements of time as the readout'. *Analytical Chemistry*. 2013;**85**(21):10432–9.

[75] Fu E., Liang T., Ramachandran S., Lutz B., Yager P. 'Two-dimensional paper network format for amplified lateral flow assays'. In *Proceedings of*

the 15th International Conference on Miniaturized Systems for Chemistry and Life Sciences.. 2011;1891–3.

[76] Fu E., Liang T., Spicar-Mihalic P., Houghtaling J., Ramachandran S., Yager P. 'Two-dimensional paper network format that enables simple multistep assays for use in low-resource settings in the context of malaria antigen detection'. *Analytical Chemistry*. 2012;**84**(10):4574–9.

[77] Apilux A., Ukita Y., Chikae M., Chailapakul O., Takamura Y. 'Development of automated paper-based devices for sequential multistep sandwich enzyme-linked immunosorbent assays using inkjet printing'. *Lab on a Chip*. 2013;**13**(1):126–35.

[78] Chen H., Cogswell J., Anagnostopoulos C., Faghri M. 'A fluidic diode, valves, and a sequential-loading circuit fabricated on layered paper'. *Lab on a Chip*. 2012;**12**(16):2909.

[79] Gerbers R., Foellscher W., Chen H., Anagnostopoulos C., Faghri M. 'A new paper-based platform technology for point-of-care diagnostics'. *Lab on a Chip*. 2014;**14**:4042–9.

Chapter 3

Advanced lateral flow technology for point-of-care and field-based applications

Brendan O'Farrell[1]

3.1 Introduction

There is a continuum in diagnostic technologies that ranges from methods that require infrastructure and a centralized approach to testing, to technologies that can be used in a decentralized testing strategy and that require little to no supporting infrastructure. The latter categories are often termed "point of care," "point of contact," or "point of need." Historically, those technologies employed at the point of contact have been considered to have less diagnostic accuracy than those used in centralized testing environments, so the use of such tests was assumed to imply a compromise of performance. One major goal of advancing the design of point-of-contact technologies over the course of the past decade has been to extend the reach of more accurate technologies to decentralized testing scenarios, in which they can often be of most use.

Lateral flow has generally been considered to be an ideal technology for application in many point-of-contact testing environments, as it demonstrates many key attributes including the ability to deliver accurate, sensitive and in many cases quantitative results, with minimum complexity for the end user and without the need for expensive or complex infrastructure.

3.1.1 Advantages of lateral-flow assay systems

- Known and mature technology
- Relative ease of manufacture—equipment and processes already developed and available
- Easily scalable to high volume production
- Stable—shelf lives of 12–24 months often without refrigeration
- Ease of use: minimal operator-dependent steps and interpretation
- Can handle virtually any appropriately treated liquid or solid sample
- Can be integrated with reader systems and information systems
- Can have high sensitivity, specificity, and good stability
- Relatively low cost and short timeline for development and approval Market presence and acceptance—minimal education required for users and regulators

[1]DCN Carlsbad, CA, USA

Many of today's testing challenges require performance in point-of-contact tests that rivals that of much more costly central laboratory methods. The performance challenges that have limited the application of lateral flow in many highly demanding applications historically have been sensitivity and reproducibility and coupled with that, limitations in the ability to multiplex larger numbers of analytes.

However, in many cases the arguments that are used to dismiss the use of lateral flow in more demanding applications are based on historical design, development, and production methods and do not reflect the reality of today's knowledge at the cutting edge of lateral-flow technology.

3.1.2 The S-curve and lateral flow

As technologies evolve, they go through a familiar evolutionary cycle—an *S*-curve. At the start of the cycle, the basis of the technology is laid out, and there are few users. At a certain inflection point, the number of users of the technology goes exponential, and the number of applications for which it is applied grows at a rapid rate. All growth phases end, and that exponential phase levels off, plateaus, and may decline as the next technology takes off to replace it. *S*-curves apply to all technologies—phones, cars, computers, and point-of-contact diagnostic tests.

Today in point-of-contact diagnostic technology markets, we are in the middle of an *S*-curve. Many performance issues have been worked out of existing technologies, and we are reaching the point at which the application of the technology and what it can do for the user has become the central issue.

The primary technology used in every key point of need, rapid testing environment is the lateral-flow immunoassay. Lateral-flow tests are inherently simple devices, applicable in almost any environment from clinic to field, and applied in every conceivable application. They are used in human diagnostics, veterinary testing, agricultural testing, bio-terrorism detection, chemical and radiological threat detection, food safety, and consumer health testing. At an increasing rate, they are being applied in highly demanding applications such as testing for cancer, traumatic brain injury, cardiac infarction, kidney disease, and even genetic abnormalities.

Lateral-flow test technology is reaching the center of the *S*-curve—the point at which the curve inverts. At this "flip" in a technology life cycle, user-centered design takes center stage. Lateral-flow immunoassay technology is being incorporated into truly user-friendly designs, incorporating sample collection, and treatment devices that are minimally invasive and painless and reader systems that do all the work for the user. Data are being generated, interpreted, securely transmitted, archived, and mined for critical information, then translated to action, all invisibly to the user. The technology itself is becoming invisible. In the coming generation of rapid tests, the central technology will not be at issue; we know the immunoassay technology works. Instead, markets will demand devices that people will want to use and can use daily, easily and reliably.

There has been a tremendous amount of evolution in lateral flow technologies in the past decade, to the point at which the lateral-flow test strip sits at the heart of much more complex systems that are capable of extremely high performance. This trend has been driven by a convergence of market and technology forces, including the drive toward a digital, self-administered, and consumer-based model, coupled with the ability to readily capture, transmit, interpret, archive, and utilize data in both professional and consumer environments.

This technological advancement has been coupled with the growth of awareness of the importance of design for use. Users, whether in professional or consumer markets, have grown accustomed to an emphasis on user-centric design in the products that they handle on a daily basis. As a result, even in relatively mature testing markets, there has been a renewed emphasis that in order to succeed, developers and manufacturers must understand the basic premise that the fundamental testing technology is not the product.

This principle has been adequately demonstrated in every industry from automotive to communications over the past century; however, in diagnostics, the emphasis until recently has purely been on technical specification rather than user-centric design. That is changing. It is now understood that these testing devices need to fit into the users' workflow, be intuitive, easy to use, and in the best technological sense, be "sticky" to the end user. This requires a shift in mindset for the designers, developers, and manufacturers of lateral-flow test systems. To create products that are truly useful in point of contact, or low resource field environments that retain high performance, it is not possible to approach the development process in the same way as it has historically been done for lateral flow. For more highly specified applications involving ease of use, quantification, high sensitivity, and multiplexing, innovation is required in each of the key components of the lateral-flow system. In this chapter, the key components of a lateral-flow system will be discussed and key advances described that facilitate higher performance in field-based applications in any market space, from veterinary infectious disease testing to hormone and biomarker testing in animals and equally to human diagnostic, environmental, and biodefense applications of the technology. Some key principles of designing these devices for ease of use in field environments using user-centric design principles will also be discussed.

3.2 Lateral-flow assays from first principles: key elements of a high performance lateral-flow assay system

The key elements of a highly performing lateral-flow test system are in Figure 3.1.

Alongside the rational design and development of these system components, the design and implementation of carefully controlled manufacturing processes is critical to the ability to produce highly reproducible, quantitative, or multiplexed systems.

Many of these elements need to be addressed from first principles for best performance when designing and developing any given test system.

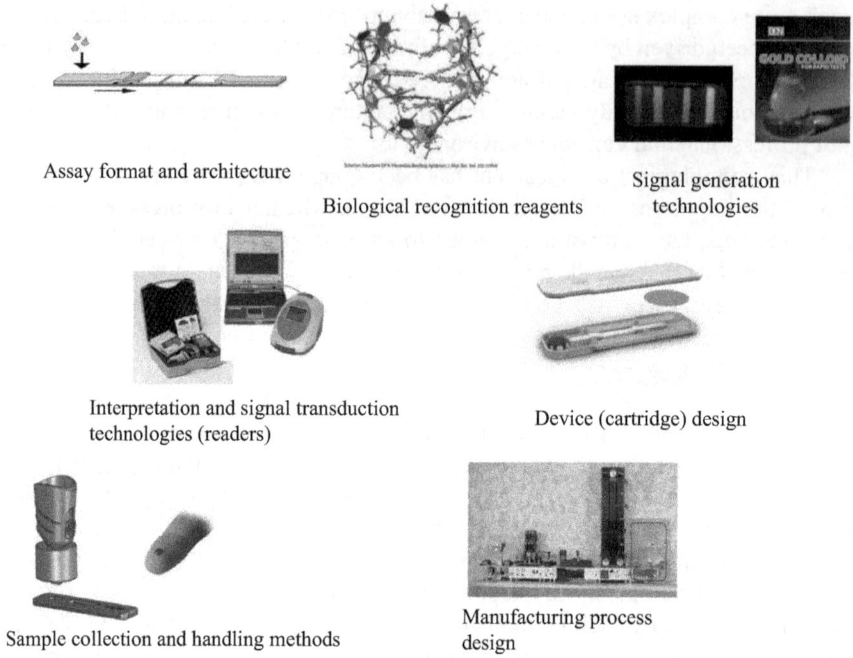

Figure 3.1 Key elements of high performance lateral-flow test systems

3.2.1 Lateral-flow assay architecture and formats—a brief introduction

Figure 3.2 shows a typical configuration for a standard lateral-flow assay. Traditionally designed assays are composed of a variety of materials, each serving one or more purposes, overlapping onto one another, mounted on a backing card using a pressure-sensitive adhesive.

The test device consists of several zones, typically constituted by individual segments of different materials, each of which will be briefly explained here.

When a test is run, a sample is added to the proximal end of the strip, onto a *Sample Application Pad*. Here, the sample is treated by means of added predetermined reagents to make it compatible with the rest of the test. Liquid-phase elements of the treated sample (which may be dissolved, suspended, emulsified, or any other liquidized formats) migrate to the next segment of the test device, the *Conjugate Pad*. Here, a detector reagent has been immobilized, typically consisting of a protein linked passively or covalently to a signal molecule or particle, typically a colloidal gold, cellulose nanobead, or a colored, fluorescent or paramagnetic monodisperse latex particle. The signal reagent can also be another reagent, including nonparticulates (e.g., soluble, directly labeled fluorophores). This label has been conjugated to one of the specific biological components of the assay, either an antigen or an antibody, depending on the assay format of the specific test

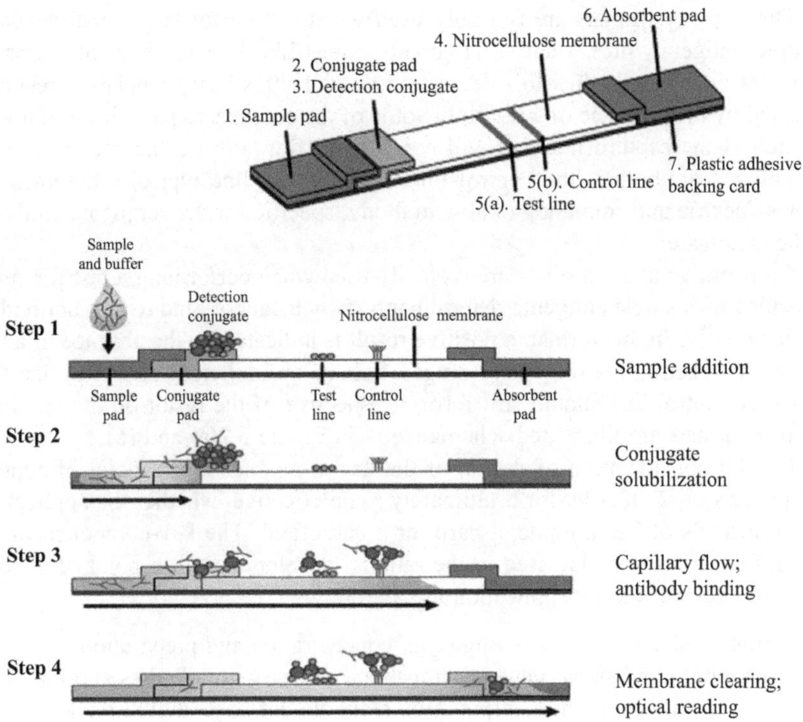

Figure 3.2 A typical lateral-flow test strip architecture and procedure

device. The liquid-phase sample remobilizes the dried conjugate material causing it to incorporate into the liquid-phase sample material, and analyte in the sample interacts with the conjugate as both migrate into the next section of the test strip, the *Reaction Matrix*, or "Membrane." The reaction matrix is typically a porous membrane with a hydrophilic, open structure for the purposes of transporting liquids to the reagent, and control areas, onto which the other specific biological capture reagents have been immobilized. These are typically proteins, either antibody or antigen, which have been laid down in bands or stripes in specific areas of the membrane in which they serve to capture the components of the liquid-phase sample, the analyte and conjugate, as they migrate past, through or over the capture lines. Excess liquid-phase materials (sample and reagents) continue to migrate across the strip, past the capture lines, and are entrapped in a *Wick* or absorbent pad. Test results are developed on the reaction matrix and are represented as the presence of absence of indicia (typically continuous lines) of captured conjugate which are read either by eye or using a reader device.

Assay formats can be either sandwich (direct) or competitive (competitive inhibition) in nature and can accommodate qualitative, semiquantitative, or in certain specific cases, fully quantitative assays.

Direct assay formats are typically used when testing for larger analytes with multiple antigenic sites, such as larger hormones like human chorionic gonadotropin (hCG) or antibodies to infectious agents. In this case, a positive result is indicated by the presence of a test line. Some of the conjugated particles will not be captured at the capture line and will continue to flow toward the second line of immobilized antibodies, the control line. This control line typically comprises a species-specific anti-immunoglobulin antibody, specific for the conjugate antibody on the conjugate.

Competitive assay formats are typically used when performing a test for small molecules with single antigenic determinants, which cannot bind to two antibodies simultaneously. In this format, a positive result is indicated by the absence of a test line on the reaction matrix. Many drugs of abuse or adulterant assays utilize this format. A control line should still form, irrespective of the result on the test line. The two formats are illustrated schematically in Figure 3.2(a) and (b).

Careful consideration of design of the key components of a point-of-contact system is required in order for it ultimately to be effective, whether the application is in a doctor's office, a home, a barn, or a battlefield. The key-component technologies that must be addressed in the rational development of a point-of-contact diagnostic device for any application are as follows:

- Sample collection and handling (e.g., concentration and preparation)
- Recognition and signal generation technologies (e.g., antibodies and labels)
- Signal transduction and interpretation technologies (e.g., digital readers)
- Assay architecture

3.2.2 Device design

3.2.2.1 Sample collection and handling

Traditional lateral-flow formats are capable of providing sufficient sensitivity for many applications. However, there is a growing demand for high sensitivity in many applications that make standard approaches to labeling and detection inadequate. Sample preparation is one major key to improved sensitivity and overall performance in many instances (Figure 3.3).

It should be remembered that a large element of the appeal of lateral flow and other point-of-contact testing systems is that they should provide, where possible, a complete "sample-to-answer" solution in a single step. It is therefore critical to consider the system as a whole, including the sample, the sampling method, the sample pretreatment methodology, and the concentration of analyte in the system. Analyte concentration can be a confounding factor, both when it is too high or too low for detection. Sample treatment can and must be used to overcome these related issues. Sampling and pretreatment methods, primarily concentration and the removal of potential cross reactive agents and reduction of background, are critical to determining the availability of many analytes for detection in an assay.

Analytical methods are traditionally divided into several steps: sampling, preliminary operations, measurement, calculation, and evaluation of results. The first step, sampling, refers to the generation of a representative sample of an

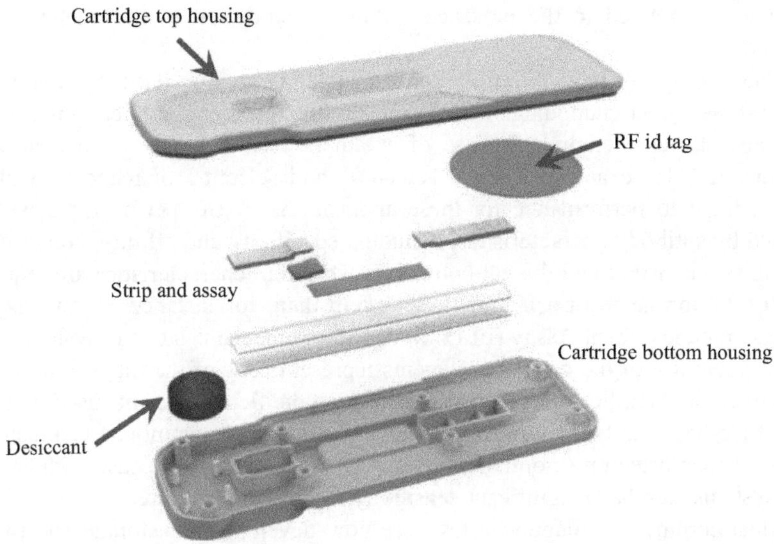

Figure 3.3 A typical lateral flow device, consisting of strip, cartridge, and other key components

inhomogeneous object. This lack of homogeneity presents a challenge to the success of the analytical method. As applied to highly sensitive rapid diagnostics, it is not the absolute sensitivity of the system that is the most critical factor, rather it is the ability to acquire as representative a sample as possible, and that, ultimately, it is the concentration of the analyte that one can detect in the primary sample that is critical [1].

In addition, in certain circumstances, high concentrations of analyte can be a confounding factor in an immunoassay. For a test to give accurate results, there must be an excess of antibodies, both capture and label, relative to the analyte being detected. If the concentration of analyte exceeds the amount of antibody, the dose response curve will plateau and, with further increase, may become negatively sloped, possibly resulting in a false negative ("hooking"). As a result, sample pretreatment may in some cases be necessary to reduce analyte concentration prior to the assay, to prevent hook effects.

In any assay system to be deployed in a low infrastructure setting, the sample collection, treatment, and delivery method must be simple, robust, foolproof, and ideally an integrated component of the test device. Minimal user-dependent steps should be required. As a result, careful definition of end-user requirements is a prerequisite to the development of successful products.

3.2.2.2 Recognition and signal generation

Along with sampling, the recognition and signal generation elements form the essential components of the analytical device in meeting high performance demands. These system components are responsible for producing a signal that is in some way related to the presence and/or concentration of the analyte to be detected.

The biological recognition systems of the assay are typically antibody–antigen-based in immunodiagnostic formats, although some applications call for other approaches including the use of aptamers, avidin–biotin, or nucleic acid hybridization, for example. Careful screening and selection of reagents is absolutely critical to performance for these applications. Assay performance will be affected by antibody characteristics, including specificity and affinity, and also the binding rate constant and the antibody class. Different characteristics are required for high performance in a lateral flow system than, for instance, in an Enzyme Linked Immunosorbent Assay (ELISA), due to the mechanics of the system.

The stability of the detection reagents represents one of the major issues with the delivery and application of diagnostic tests to the field. Reagents used in many current lab-based tests are typically liquid and require refrigeration. Even reagents that have been dried onto solid substrates, which is generally the case with lateral-flow systems, can have significant sensitivity to high temperatures.

Most commercial diagnostic tests are now developed for storage and use at 25–30 °C; however, many field applications can involve lower or higher ambient temperatures, which can lead to test degradation especially when taking into account the possible need for delivery to relatively remote locations without refrigeration. Thus, temperature stability and prolonged shelf-life of diagnostic kits are critical factors for improving field-based performance. Efforts to utilize alternative binding reagents such as aptamers, which have higher thermal stability that standard antibodies have met with limited commercial success in lateral-flow systems; however, this represents an interesting approach that bears further development. Antibodies from species such as camels and sharks, which can have improved thermal stability relative to murine antibodies, have also been evaluated in certain applications with some success.

The *Signal Generation* elements involve both the label and the reader or interpretation method used. Traditional point-of-contact immunoassays were designed to be qualitative, threshold assays, interpreted by the user without the aid of a reader. This method of interpretation brings with it issues of subjectivity and makes it virtually impossible to develop quantitative systems. In clinical and field environments, there is also the possibility of data loss, mistranscription of patient information, and user error. Developing assays with integrated reader systems is a critical feature of many next-generation Point of Care (POC) assay systems and represents a technical challenge, particularly for field-based diagnostic platforms.

For point-of-care immunoassays, the majority of detection is still done visually either by eye or using optical readers; however, there is growing use of fluorescent, chemiluminescent, and magnetic measurement systems. A variety of signal transducers are used, including fluorophores, either particle bound or direct labeled and colorimetric—colloidal gold, colored cellulose nanobeads, or colored latex particles (Table 3.1).

Colloidal gold remains the most commonly used label in lateral flow, due to its low cost and ease of use. Typically, gold particles used in lateral flow are in the

Table 3.1 Commonly used visual labels

Visual (colorimetric) labels
• Common visual labels
• Colloidal gold
• Cellulose nanobeads
• Latex
• Colloidal carbon

Figure 3.4 nanoAct^{TM} Cellulose nanobeads (image courtesy of Asahi Kasei Corporation)

range of 40–80 nm in diameter. Multiple suppliers are available, including DCN Diagnostics, and Innova. The quality of the colloid used is critical to the performance and stability of the ultimate product, and cost and quality can vary significantly among manufacturers so careful selection is critical to success (Figures 3.4 and 3.5; Table 3.2).

Colored latex particles are also common, although often they will generate less sensitivity than a gold particle, their benefit being in the multiple colors available and in the ability to conjugate to them via covalent methods as against the passive conjugations used for colloidal gold in most cases. Multiple suppliers are available including Bangs Labs, Merck Estapor, and Thermo Fisher. Typically, particles used are polystyrene, with diameters of around 100–300 nm (Table 3.3).

A new option in the lateral-flow arena that shows extreme promise in terms of generating high performance in sensitive and quantitative visual systems is

Figure 3.5 *Ultragold*^{TM} *Colloidal gold for lateral flow assays (image courtesy of DCN Diagnostics)*

Table 3.2 *Colloidal gold attributes for lateral flow*

Strengths	**Weaknesses**
• Well characterized handling methods • Well characterized conjugation methods • Relatively high sensitivity • Several good commercial sources of high quality gold	• Lack of color choices • Relatively unstable to pH and salt • Can be difficult to re-suspend when dried • Difficult to scale up and reproduce production in-house • Generally passive conjugation

Table 3.3 *Colored latex attributes for lateral flow*

Strengths	**Weaknesses**
• Conjugations can be done using a variety of techniques and can be well characterized	• More expensive than in-house produced gold
• Particles are highly uniform and reproducible	• More difficult to optimize and stabilize conjugates
• Available from a number of high-quality commercial sources	• Can require higher amounts of antibody than gold
• Several good commercial sources of high-quality gold	• Less sensitive in visual assays than gold due to lower color intensity and higher particle size
• A variety of detection modalities are available	

Table 3.4 Cellulose nanobead attributes for lateral flow

Strengths	Weaknesses
• Hydrophobic—release well and clear very well from commonly used pad materials	• Can be more expensive on a mass basis than gold or latex (balanced by a requirement for less mass per test)
• Very high color intensity	• Passive particles currently (activated particles for covalent conjugations in development)
• Multiple colors available—useful for multiplexing in visual assays	
• Very reproducible and homogeneous particle size	
• Require significantly less antibody in most applications than gold or latex	
• Require significantly less mass of particles to achieve same or better sensitivity than other visual labels	

NanoAct™, a new, high performance cellulose nanobead label for use in lateral-flow immunoassays, developed by Asahi Kasei Corporation (Japan).

Cellulose nanobeads, or CNBs for short, are highly stable, deeply colored particles that have demonstrated improved performance over standard labels in lateral-flow applications. Sensitivity has been shown to be greater than colloidal gold, and particle stability is excellent. Assay coefficient of variation (CVs) have also been demonstrated to be reduced relative to the use of colloidal gold, allowing for utility in quantitative assays. These particles also have the advantage of being available in multiple colors, allowing for easy multiplexing. These particles additionally allow for significant reductions in the amount of binding reagents (up to 90 per cent in certain assays) relative to other labels, and for the reduction in the use of surfactants in the system, all attributes that lead to higher sensitivity and better cosmetic results in an assay (Table 3.4).

3.2.2.3 Fluorescence

There are two general approaches to implementing fluorescent labels in lateral flow—particle-based and direct labeled fluorescence, where fluorophores are directly linked to the binding reagent (Table 3.5).

When working with particle-based fluorescence, most commonly polystyrene beads are used, containing Europium (Eu) or dyes that emit in ranges in which there is minimal background fluorescence either in the plastics or the biological samples in the system. Dyes such as TideFluor 5 are commonly used. Eu has the advantage of a huge Stokes shift resulting in an excellent signal/noise ratio, but the dark red dyes can be very effective as well, emitting in a range at which there is no background fluorescence in biological samples or most materials used in the assays. Using these labels developers consistently achieve sensitivities 1–2 log improved over colloidal gold or colored latex. Care should be taken when selecting dyes with narrow Stokes shift that the cost of reader optics can be high relative to the cost of optics for dyes, such as Eu, with broad Stokes shift.

Table 3.5 Common fluorescent labels used in lateral flow

Common fluorescent labels
• Organic dyes • Metal–ligand complexes • Fluorescent proteins • Semiconductor quantum dots • Lanthani decomplexes • Dye-doped polymer nanoparticles • Fluorescent silica nanoparticles

Traditional approaches to lateral flow involve the dispersion of conjugated particles onto a fibrous conjugate pad such as glass fiber or polyester. These glass fibers are hydrophobic materials that are treated with hydrophilizing reagents in order to make them hydrophilic enough to accept and release the conjugate. The conjugate is dried in place on the conjugate pads using high temperatures or lyophilization, typically stabilized in high concentrations of sugars or polymers.

This process requires multiple steps, each of which inherently contains sources of variability, which in turn result in variability in the test results. The efficiency of release from conjugate pads is typically not high and is always highly variable for particle conjugates. This is a major contributor to the variability seen in many lateral flow assays, when signal strength is measured. This in turn impedes our ability to be quantitative. In order to overcome these issues direct labeled, or nonparticulate fluorescent labels can be used—in other words antibodies are directly labeled with fluorophores. Because there is no particle involved, there is less binding of conjugates to the materials in the assay, and release is vastly improved. So as well as improved sensitivity over visual labels, these systems generate vastly improved CVs—allowing for better quantification—and may also have some IP advantages over the use of particulate labels in lateral flow systems (Table 3.6).

In choosing which label to use in a lateral flow system, it is important to consider the application; particularly the requirements for sensitivity, reproducibility, stability, complexity (the need for readers), and cost in making the decision. For low-cost qualitative applications, a high-quality colloidal gold may still be the best option. For high sensitivity visual systems or for multiplexed assays, CNBs will be optimal. For quantification and where the application will support a reader, fluorescence will likely be the best choice for most applications.

3.2.2.4 Reader systems in lateral flow

The signal transduction and analytical processing elements of the assay system involve the detection of the signal generated by the reagents, translation into a numerical or alphanumerical output, and delivery of that output to the user.

Traditional point-of-care immunoassays were designed to be qualitative, threshold assays, interpreted by the user without the aid of a reader. This method of

Table 3.6 Attributes of fluorescent labels for lateral flow

Strengths	Weaknesses
• Conjugations can be performed using a variety of techniques and can be well characterized	• Requires the use of a reader
• Particles (if used) can be highly reproducible and uniform	• Can be more difficult to optimize and stabilize conjugates
• More sensitive in most cases than visual labels	• Sometimes cost
• Very reproducible and homogeneous particle size	
• Require significantly less antibody in most applications than gold or latex	
• Require significantly less mass of particles to achieve same or better sensitivity than other visual labels	

interpretation brings issues of subjectivity into play and makes it virtually impossible to develop quantitative systems. In clinical and field environments, there is also the possibility of data loss, mistranscription of patient information, and user error. Developing assays with integrated reader systems is a critical feature of many next generation POC assay systems. The attributes of the reader system need to be carefully considered in the context of both regulatory requirements and the ultimate needs of the end user.

In developing instrumentation for application in the field, the goal is to design tests and devices that have the appropriate level of complexity required by the biology of the analysis process and that meet the needs of the user for a wide variety of other relevant specifications including cost, robustness ease of use, and connectivity.

Numerous clinical reader technologies exist for use in lateral-flow applications. These units are ideal for application in many environments, being robust, relatively low cost, calibrated, high performance units with associated customer service, and support available. Examples of producers of validatable, portable benchtop units include Qiagen Lake Constance, Axxin and LRE Medical. Producers of custom solutions exist, such as Planet Innovation. Producers of phone-based systems include Cellmic and FIO (Figures 3.6–3.8).

This latter approach is gaining in momentum in certain applications. The use of smart phone camera capabilities to transform analog test results into digital format and transmit the data to the cloud for analysis and further use is growing. The approach enables many potential benefits, including archiving of data, integration with medical records, or in certain applications linking more social media-driven uses. These generic approaches to the reading of tests in the field face some challenges, however, including the following:

- Lack of a coordinated approach with test developers: This can lead to issues of interpretation. For example, many qualitative tests exhibit high signal strength variability. This can lead to interpretation issues with digital readers.

Figure 3.6 DCN fluorescent assay visualizer

Figure 3.7 Qiagen Lake Constance LR3 reader

Coordination with test strip manufacturers to reduce this level of variation in qualitative test results will be beneficial.

- Image quality: Many phone-based reader applications allow for no control over image quality. Issues related to lighting and focal length may occur. Manufacturers may also regularly upgrade or change camera quality or position on their devices leading to issues of image quality and reproducibility.
- Automatic upgrades: The ability of phone manufacturers to upgrade their systems remotely is a detriment to the ability of test developers to validate this type of reader and ensure that validation remains relevant.
- Potential platform obsolescence: The technology is based on third party platforms in an industry where rapid obsolescence is common.
- Regulation: The regulatory pathway for cell phone-driven readers in diagnostic applications is not completely clear, and their use in medical applications may carry some risk as a result.

Figure 3.8 LRE Medical cPOC reader

To conclude in relation to inclusion of reader systems in lateral-flow assays, the decision whether or not to use one will depend on a variety of technical and market issues, including

- Cost
- Need for quantification
- Training and experience level of users
- Sensitivity requirements and the type of label that is chosen as a result

There are numerous options available, in terms of technology and format, however very careful consideration should be given to the choice of supplier based not only on the performance of the reader, but on their ability to manufacture and deliver a high quality product, on time and on budget, and assist with the documentation and validation of the reader system based on the ultimate regulatory requirements for the whole assay system.

3.2.3 User-centered design of devices for field-based applications

Lateral-flow assays are a relatively old technology, but not enough effort has historically been put into the design and development of assay systems that are based on well-defined user needs. This concept of user-centric design is common in many tech industries; however, the diagnostic field has historically focused more on technical performance than on making devices that are truly "sticky" to users, in the best sense, meaning they fit users' workflow needs, lifestyle needs, are easy to use, mistakes are easy to recover from, and where people focus more on the experience of using the device than on its technical performance. Technical performance is simply assumed to be adequate for the application.

For a user-centered design and development approach to work, it requires a deep understanding of the requirements of the end user of the assay system. Often the specifications that developers receive from clients tend to focus on technical performance. This is completely understandable; as technologists this is where we always tend to focus first. However, we have to remember that the test strip is not the product, and that there are other keys to commercialization. That is why the development process should allow for more focus on the requirements for the actual product, so that commercial needs, as well as technical needs, are met. The areas of product design in which a user-centric focus can play a critically important role in decentralized tests include sample preparation and cassette design for use.

3.2.3.1 Sample preparation devices

Lateral-flow devices often consist of three elements; the strip, the cassette and the sample collection, and handling device. Appropriate sample collection and preparation is one key to improved sensitivity, reproducibility, user friendliness, and overall performance in many instances. It should be remembered that a large element of the appeal of lateral flow and other point of need assay systems is that they should provide where possible a complete "sample-to-answer" solution in a single step. It is therefore critical to consider the system as a whole, including the sample, the sampling method, the sample pretreatment methodology, and the concentration of analyte present in the sample matrix.

Sample treatment can come in many forms depending on the sample type; for example, plasma separation from a fingerstick or venous whole blood, filtration, and breakup of mucins in saliva or respiratory samples, removal of fats from whole milk, or changes in pH of urine.

In any assay system to be deployed in a decentralized testing environment, the sample collection, treatment, and delivery method must be simple, robust, foolproof, and ideally an integrated component of the test device. Minimal user-dependent steps should be required. The creation of solutions that achieve several or all these steps in an intuitive, user-friendly way is a constant challenge, and the needs of every assay system are different.

3.2.3.2 Cassette design

In addition, the design of the overall device, which can include both the cassette and sample handling device and the means for the two components to integrate with the test, can be a great opportunity to produce a whole range of benefits for your device, customers, and users. The cassette is a very functional component of the assay, providing control of sample application, flow rates, flow patterns, and has a big impact on assay reproducibility (Figure 3.9).

Getting the design of the cassette and other components right can allow us to influence the user experience in a positive way, making the product attractive and "sticky" to users. It allows us to do things like improve hand feel, have it fit better into client workflow, add functionality without complexity, and in general can create a higher value proposition for the product. With the right design, we can add other valuable functionality like on-board communication of information through bar codes or radio frequency identification (RFID) tags and allow us to better integrate the test to reader systems, facilitating better data collection, interpretation, and storage.

The key in designing these devices is to remember that the strip, the cassette, the sample handling device, and a reader if included, all have to work together as a system. Cassettes, for instance, will affect the results in an assay through the control of flow in the system and will have a huge impact on accuracy and

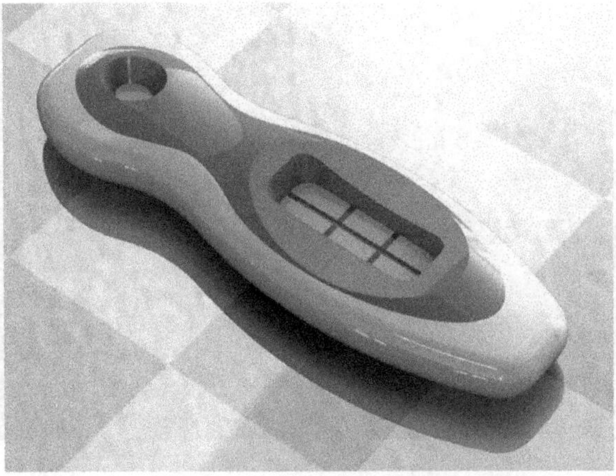

Figure 3.9 Example of a duplex assay cassette, designed for use, and appearance (courtesy DCN Diagnostics)

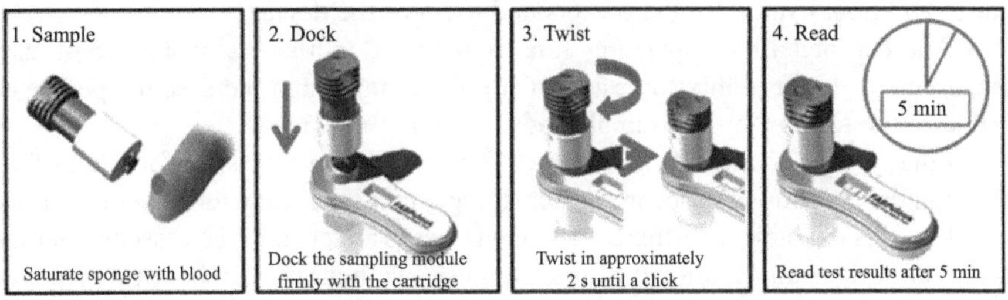

Figure 3.10 Illustration of operation of the FABPulous testing device for cardiac FABP (FABPulous, BV, The Netherlands)

reproducibility. Everything in this system is an engineered part, including the strip itself, and must work as part of a fully integrated device.

Figure 3.10 is an illustration of some of these principles, in an assay that was developed for a field-based human diagnostic application, although all of the same principles apply to a field-based point-of-contact diagnostic assay for any application. The assay is for cardiac fatty acid binding protein (FABP) for determination of myocardial infarction. The device was designed for use in an emergency environment, quantitatively collects whole blood from a fingerstick, then mates with a cassette, delivers running buffer, filters out red blood cells, and delivers a quantitatively diluted sample of plasma to the test, allowing for fast, reproducible easy determination of cardiac FAPB levels at the point of need. This device development program illustrates the points that, for best usability and performance, the entire device had to be designed in parallel, based on a deep understanding of user needs. The final device meets all of the input requirements of speed, accuracy,

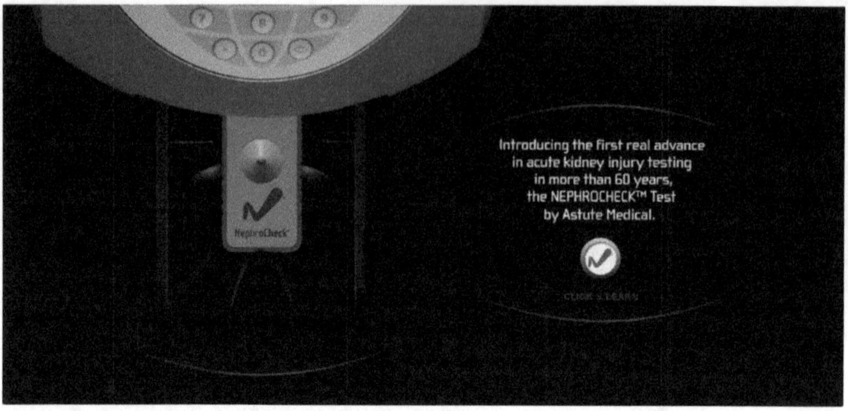

Figure 3.11 Image of the Nephrocheck™ system for acute kidney injury testing from Astute Medical (San Diego)

and ease of use in a high pressure, decentralized testing environment, and created a lot of additional value for the owners and users of the device.

Another benefit of spending time and effort in the establishment of user requirements is the ability to design in strong branding and increase the perceived value of the test, as in the example shown in Figure 3.11.

Finally, it is possible to add a lot of functionality to a cassette for lateral flow through the addition of on-board buffer storage, making it easy for a user to operate a test. In this example, a buffer storage container is attached to the cassette after the addition of the sample, allowing a user to push two buttons to deliver two separate buffers for dilution of the sample and delivery of a conjugate in sequence, without the need for additional pipetting steps or separate buffer bottles (Figure 3.12).

Appropriate user-centric design ultimately means adding value to the product while removing complexity for the user. It does not necessarily mean adding cost or complexity to the product.

3.2.3.3 Arraying in lateral flow formats

In the point-of-contact rapid diagnostic market, the concept of multiplexing is becoming increasingly important. Multiplexing is a key element of many diagnostic applications now, be it for autoimmune disease, allergy, infectious disease, or biomarker quantification for any application from biodefense to cancer diagnostics, to traumatic brain injury.

The first question asked relative to multiplexing should generally be "Why ... ?" before we try to address "How?". Traditional approaches to multiplexing can add a lot of complexity of a development program, and it is important to be clear on not only the technical requirements of the system from the outset, but also on the practicalities of reimbursement and utility in the field in the actual testing environment. Questions of cost, access to intellectual property, manufacturability, and the ability of proposed manufacturers to actually produce products of high

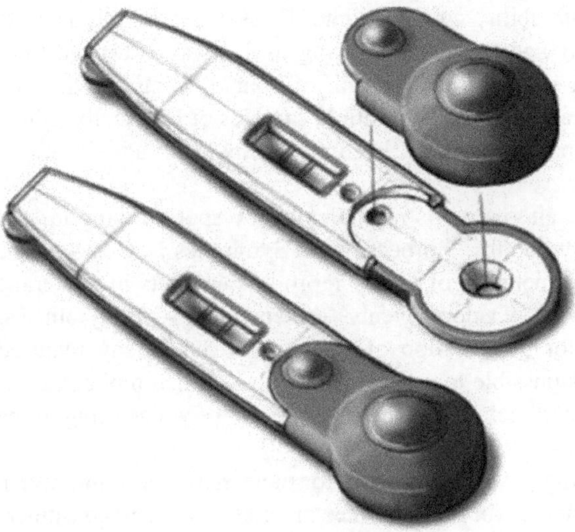

Figure 3.12 On-board buffer storage and delivery (image courtesy of DCN Diagnostics)

Option 1: *Multiple strips, one sample addition*

Option 2: *Multiple strips, multiple sample additions*

Option 3: *Array on a single strip, one sample addition*

Figure 3.13 Different approaches to design LFAs for multiplex analyte detection

quality and high complexity should all be addressed transparently before development begins.

When the determination is made that multiplexing is the right approach for the application, a limited number of viable technical options exist for successful development of highly multiplexed lateral flow assays, either qualitative or quantitative. Multiplexing in lateral flow can be a technically challenging task, particularly when any degree of quantification is required.

Lateral-flow sandwich assays can be multiplexed in three ways (Figure 3.13): (1) multiple strips, one for each analyte within a Cassette, run from a single sample application; (2) multiple strips, one for each analyte within a Cassette run from individual sample additions; and (3) a single strip with multiple detection zones in a Cassette, one for each analyte.

Standard approaches to multiplexing lines in a normal lateral-flow architecture are limited by the number of lines that can physically fit onto a lateral-flow membrane of reasonable length as well as the amount of conjugate required in the system. Kinetics change as flow slows down along the flow path, and any number of sources of variability can creep into the system, making quantification challenging. Practically speaking, multiplexing in this way is limited to only a relatively few lines on a single strip. We can, of course, design devices that will contain multiple strips and either take multiple samples or physically split a single sample; however, these devices can become costly and add to the complexity of manufacture.

One viable alternative is multiplexing by spot arraying in lateral-flow fields. The patented Symbolics™ process from Symbolics LLC (www.symbolicsdx.com) allows for the generation of highly reproducible spots in a lateral-flow field. The fact that each spot develops evenly—independently of any other spot in the flow field—allows for the creation of arrays of virtually any required density. This means that it is possible to multiplex analytes, utilize replicates for improved predictive power, and include internal controls, all of which bring tremendous value to the assay system.

This arraying power also allows for the generation of intuitive results—letters, words, and symbols—in place of lines or spots, for use in consumer applications or other appropriate assays.

The Symbolics process is based on the concept of pixilation of reagents on the analytical membrane in a fashion that allows for the even development of each pixel. This appears to be a rather simple concept, similar to dot matrix printing. The issue in lateral flow, however, is that the formation of one feature in a flow path causes flow perturbations in the system that prevents the even formation of features behind it. Or so most people have to date assumed. The key to the pixilation approach is in the controlled dispensing of the reagent spots for size and pitch (the center-to-center distance between spots in any axis), factors that must be balanced for each reagent based on the binding characteristics of the reagents. When well optimized and controlled, it is possible to position individual spots of the correct size and pitch in the flow path of a lateral-flow assay in such a way that each feature develops evenly and, most importantly, does not prevent the development of the features surrounding it.

Some examples are shown below:

Case 1: developing a true "plus/minus" symbol in a hCG assay

The traditional +/− symbol that is developed in several marketed lateral flow hCG assays utilizes a pre-printed ink line in one axis (the long axis of the strip), and a second reagent line printed perpendicular to it, such that if no reaction happens, the line in the long axis of the test looks like a "+−" and completion of the test line in the event of a positive creates a "+" symbol. This necessitates the use of a separate control line. What is demonstrated here is a true +/− using test and control reagents, which removes the necessity for a separate control area, reducing the complexity of the read for the user (Figure 3.14).

In this example, each feature in the flow path develops evenly without disturbing the development of the feature behind it. It is obvious that with this method, much more complex features can potentially be formed.

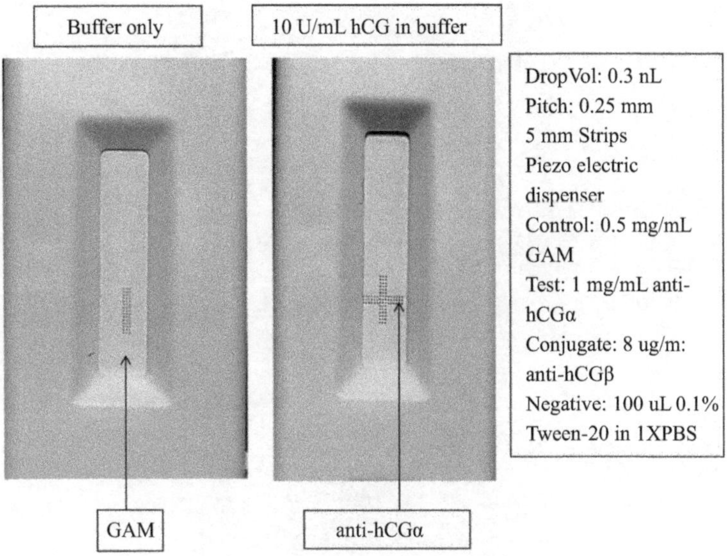

Figure 3.14 Intuitive result generation in lateral flow

Case 2: more complex features: letters and words

Two examples are shown, of the development of letters and words in lateral flow; one in a sandwich for HIV 1/2 antibody detection, and the second in multiplexed drugs of abuse assay. In each case, the pixels are used to form easily understood indicia that are much more intuitive to the end user than the standard line perpendicular to the direction of flow. Note that the binding characteristics of this reagent allowed for the use of much larger drop sizes at bigger pitches than were optimal for the hCG reagents in the example above. The flexibility to optimize around the binding characteristics of each reagent is a key characteristic of the Symbolics processing technology (Figure 3.15).

Case 3: multiplexed arraying

Multiplexing in lateral-flow formats is generally difficult for a variety of reasons. Traditional multiplexing involves laying multiple capture lines one after the other in the direction of flow. This results in the potential for a number of adverse conditions. First, the possibility of line bleed exists, in which conjugate from one line can bleed across into the next line or can move into the area between lines. In reader-based systems, these results in a condition in which background does not return to baseline between lines, resulting in issues of discrimination and sensitivity. It also increases the potential for nonspecific binding, as conjugates cross multiple test lines. In addition, real estate is an issue, with most membranes being 25–35 mm in length, only a finite number of lines can physically be striped onto the membrane with significant enough distance between them to overcome some of the previously listed issues. Finally, and possibly most importantly when it comes to

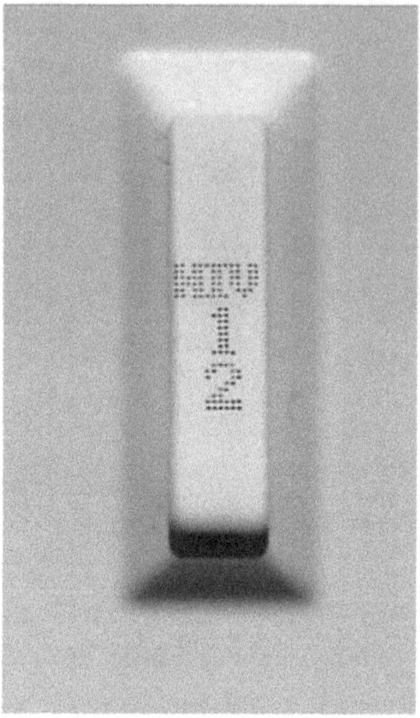

Figure 3.15 Use of letters and words in lateral flow

quantitative multiplexed systems, the flow rate of nitrocellulose decreases relative to distance from origin in a nonlinear manner. As a result, for a multiplexed assay with lines formed downstream of the origin of fluid flow, the time spent for the reaction to reach the last test line can be extended significantly over the time it takes the first line to form. As a result, the kinetics of the last assay in the flow field are very different from that of the first (Figure 3.16).

The Symbolics process allows for a number of methods designed to overcome these issues.

Multiplexing using Symbolics™

The pixilation approach allows for the formation of multiple independent spots that can form micro-scale patterns to create macro-scale features, as in the approach to dispensing words or symbols shown above. Alternatively, the pixilation pattern can produce discrete macro-scale spots to allow for two-dimensional (2-D) arraying in the flow field. An example of a 2-D array using multiplexed drugs of abuse reagents is shown below, with a 2×2 array plus control for each shown (Figure 3.17).

It is a simple step to envisage a lateral-flow assay with discrete control features on-board that will allow for the generation of more internal controls than is possible for the current lateral-flow format, including true positive or negative controls, or even the generation of gradient binding patterns on the strip, which will allow for different methods of quantification or even calibration for reader systems.

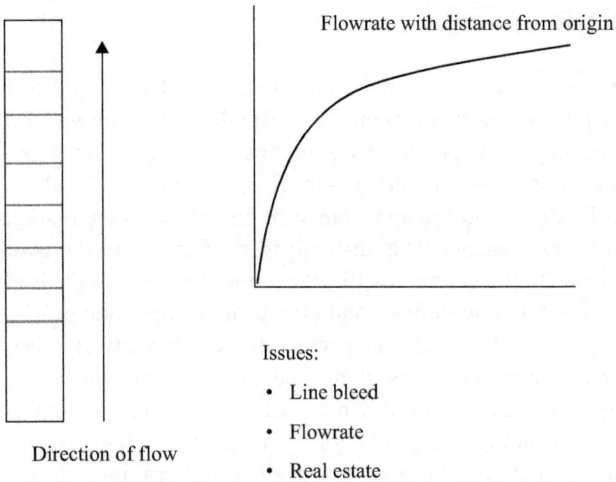

Figure 3.16 Illustration of the pattern of decreasing flow rate from point of origin seen in lateral flow, which can impact the ability to multiplex and be quantitative in standard multi-line architectures

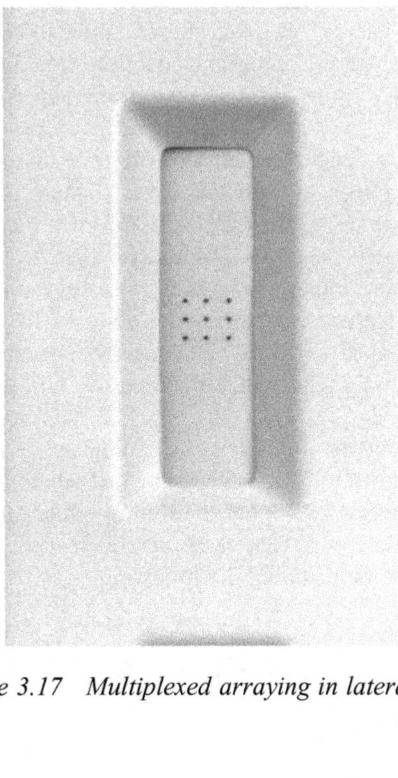

Figure 3.17 Multiplexed arraying in lateral flow

3.3 Conclusion

Extrapolating the future requirements and direction of the rapid testing markets based on past performance and trends is difficult. For example, there are shifts underway in the ways in which products are bought and paid for, in regulation, the emergence of new and powerful companies from within and outside the traditional diagnostic marketplace and the appearance of completely new markets and applications. For all these reasons, it is difficult to predict where the evolution of the market will take us in the coming 5–10 years, other than to say the trend is positive in terms of market size and number and type of new applications.

However, from a technology perspective, there is a relatively clear form to the way lateral flow is evolving. It is likely that the focus on the implementation of continuous improvements with the aims of better sensitivity, reproducibility, quantification, and multiplexing capability will continue. The evolution of applications into areas such as consumer diagnostics will continue to drive the move toward integrated design features that lead to intuitive use by minimally or untrained users. Additional applications will also continue to be found. In general terms, lateral-flow assays are capable of much more than the market has given them credit for historically, but for highest performance, consideration must be given to every major component of the test system, and improvements to key assay components such as membranes and label technologies are central to potential performance improvements. There is a lot of opportunity for generating real value-added tests in many market spaces: low cost, simple applications or high value more complex tests but in each case, appropriate attention should be given to the design and development of the entire system, both from a performance and a user-experience perspective.

User-centered design, long the touchstone of Steve Jobs and Apple products, is more than ever making its way into the consciousness of product developers, marketers, and technologists in diverse fields. In order for a product to be truly “sticky” to an end user, it needs to be intuitive, easy to use, difficult to make mistakes with, and—most importantly—the technology on which the device is predicated needs to disappear into the background of the application. In other words, the technology must perform well enough and consistently enough that users never think about whether it will work when we turn it on or how it is doing what it is doing for us.

When thinking about new products, technologists, product developers, and marketers alike can’t afford to ignore the context of what is going on in point-of-contact diagnostic markets. Developing the right product for the right application will require more than copying what came before. Instead, we need to focus on much more relevant factors including the following:

- Performance
- Manufacturability
- Quality
- User-centric design

- Market desire
- Regulation
- Intellectual property
- Channels to market
- Competitive environment

Lateral-flow represents the most mature, broadly applicable diagnostic immunosensor available on the market today but when a technology matures, the market expects more from the technology than just performance. Design, development, and production standards must change and even the best of the old generation will give way. Designing the right product for the target market demands that we have a much broader focus. With adequate focus on the design and development of the right sample handling methods, assay architecture and device design, coupled with implementation of the best labels, reader technologies and manufacturing process technologies, the implementation of lateral flow tests in numerous new and demanding applications, from infectious disease diagnosis to biomarker monitoring, is likely to continue unchecked.

Reference

[1] O'Farrell B. 'Introduction to lateral flow immunoassays', in Wong R., Tse H. (eds.). *Lateral Flow Immunoassay*. 1st ed. Humana Press; 2009.

Chapter 4

Point-of-care electrochemical sensors for antibody detection

Robert L. Rubin[1] *and Konstantin N. Konstantinov*[1]

4.1 Introduction

Most biosensors are designed to detect biologically and/or clinically important molecules, and an immunosensor generally refers to a biosensor in which antibody/antigen interactions are the basis for the detector specificity. By far, the most common type of immunosensor employs antibodies as a reagent to capture their cognate antigen, usually in conjunction with a secondary, labeled antibody to a different epitope on the analyte. Often these antibodies are raised in animals to produce high-affinity reagents, which can be generated in large quantities in a pure and monomolecular form using hybridoma technology, or produced by recombinant DNA technology. In contrast, immunosensors in which the antibody is the analyte of interest require a target antigen as the reagent for capturing the antibody and may employ a labeled secondary antibody specific to the Fc portion of the primary antibody to measure the immune complex.

Valuable clinical information can be obtained by measuring blood-borne antibodies because such reactivities can provide a highly sensitive and specific history of the individual's exposure to a foreign antigen or to the presence of an inappropriate immune response to potentially antigenic material such as nonpathogenic environmental antigens, medications, or self-antigens. For assessment of patients, antibody reactivity is generally measured in licensed clinical laboratories using approved methods in order to produce information that is clinically relevant. These assays require *appropriate* sensitivity and specificity, non-trivial issues because antibodies are induced or elicited in vivo through processes that are inherently random, resulting in a wide range of affinities, concentrations, and potentially misleading cross-reactions, as well as consisting of several immunoglobulin classes of unique biologic and clinical significance. For these reasons, the commonly accepted criteria for assessing the quality of immunoassays, such as limit of detection and dynamic range, may be largely irrelevant to measurement of antibodies in clinical medicine. Rather,

[1]Department of Molecular Genetics and Microbiology, University of New Mexico, Health Sciences Center Albuquerque, New Mexico, USA

the protective, diagnostic, or pathogenic significance of a particular antibody activity as assessed by clinical correlative studies is generally more useful, which can be quantified by positive and negative predictive value metrics.

Antibody assays are relevant to a wide range of clinical issues, although there remain large gaps in their availability in clinical laboratories and are rarely offered for patient or near-patient access. Specific antibody assays are useful for monitoring infectious, autoimmune, inflammatory, immune deficiency, and allergic diseases as well as for spontaneous responses to tumor antigens in cancer diagnosis, to alloantigens in organ transplantation, and to therapeutic macromolecules. Within each clinical setting, antibody testing can provide insight into general immune status, response to treatment, disease stage or presence of hypersensitivity immune reactions. Large clinical laboratories are the usual setting for antibody assays because of the general need for substantial instrumentation, infrastructure, and expertise. In these settings, samples may be processed individually in single or multiplex detection systems, or multiple samples can be efficiently processed simultaneously.

However, antibody assays performed in centralized clinical laboratories can be prolonged, labor intensive, and costly, which slows the diagnostic process and may restrict its use for a large segment of the population. An alternative would be the availability of testing capability while the patient remains in a clinic or even a remote or possibly at-home setting. The general advantages, characteristics, and challenges of point-of-care (POC) testing have been well delineated [1].

4.2 Challenges in antibody measurement by POC technology

POC testing for specific antibody presents some problems unique to this class of analytes for the following reasons:

1. The analyte (specific antibody) is usually a tiny fraction of the total immunoglobulin in the sample. Irrelevant antibody consists of physicochemically identical macromolecules except for a relatively small antigen-binding site. For the biosensor to show appropriate specificity, the large excess of irrelevant immunoglobulin must ignore or be removed from the sensing platform.
2. Related to list item (1), irrelevant antibodies are usually highly heterogeneous. Often these would include a subset of immunoglobulins, capable of binding any particular antigen under some conditions. Such a "nonspecific antibody" may have substantial affinity for the targeted antigen under certain assay conditions due to its cross- or multi-reactivity or as a consequence of its hydrophobic or electrostatic interaction with some component of the sensor. Minimizing such nonspecific interactions by appropriate washing, blocking, antibody dilutions and diluting solutions, and/or duration of the binding reaction can be difficult without compromising the sensitivity of the sensor for the antibody of interest. Since most tests for antibody detection produce continuous outputs (antibody "titers"), it is generally necessary to assign a positive/negative cut-off value to eliminate such background antibody binding in the reported results.
3. Human serum antibodies can exist in four major classes or isotypes due to their different Fc regions while having identical or nearly identical Fab

(the antigen binding site) regions. Frequently, only one immunoglobulin class has well-defined clinical significance, but sensors that lack isotype specificity for the clinically important class of specific antibodies may not be as useful.

4. By definition specific antibody detection requires recognition of a specific antigen. The molecular form of the antigen such as its nativeness and purity as well as the physicochemical basis of its immobilization on solid phase surfaces may affect its capacity to capture antibody. These variables can be unique to the antigen and may require considerable effort to optimize.

4.3 Detection of antibody utilizing electrochemical methodology

Harnessing electrochemical phenomena to measure antibody/antigen interactions has seen an explosion of applications because of its many advantages over alternatives for signal generation. While all electrochemical antibody assays involve charge transfer between an electrode and a redox active compound in solution or on a hydrophilic solid phase in an aqueous environment, the design of the electrochemical cell determines how the reactive compound is detected [2]. Amperometric sensors detect current flow as a diffusing electroactive analyte is transiently captured on the electrode. Coulometric sensors measure the charge between two electrodes as the electroactive analyte is oxidized or reduced at one of the electrodes. Potentiometric sensors measure voltage generated as the electroactive analyte accumulates in solution. Impedimetric sensors measure gain or loss of solution conductivity as the concentration of charge-carrying analyte changes. The majority of immunosensors for both antigen analytes [3] and antibody analytes (Table 4.1) employ amperometric detection principles and typically anchor the capturing agent on or in close proximity to the electrode.

In most sensors the analyte being directly measured is not the antibody; rather a molecular mediator is employed, whose concentration is proportional to the amount of antigen-bound antibody. Successful mediators have favorable electrochemical activity within the design of the electrochemical cell as well as stoichiometric, physical and/or chemical relationship with the antigen-bound antibody. Commonly, mediators are products of the transformation of substrates of an enzyme covalently linked to a detecting antibody, which recognizes the Fc region of the primary antibody when bound to its cognate antigen. This design allows for greatly enhanced sensitivity of the immunosensor for two reasons: (1) multiple molecules of the enzyme-conjugated secondary (detecting) antibody can bind to each primary antibody/antigen immune complex and (2) the amount and rate of substrate → electroactive product development generated by each enzyme molecule is limited only by the enzyme's catalytic properties, the duration of the reaction and the concentration of the substrate, although the reaction environment and enzyme inhibition by product accumulation can result in sub-optimal signal generation.

Measurement of specific antibodies using an enzyme-conjugated secondary antibody for production of an electroactive mediator has several advantageous features. In addition to its potential for high sensitivity as mentioned above, this design

Table 4.1 Electrochemical sensors for specific antibodies

Specificity	Design characteristics			Assay features		Method validation[a]			Reference
	Electroactive mediator	Antigen platform	Transducer	Duration (min)	Isotype specific	vs. standard	vs. multiple + controls[c]	vs. multiple – controls[c]	
S. japonicum	Enzyme substrate	Electrode	Amperometric	50	No	No	Rabbit	No	[4]
S. typhi	Enzyme substrate	Electrode	Amperometric	75	No	Agglutination	Human	No	[5]
P. falciparum	Enzyme substrate	Au-electrode	Amperometric	40	Yes	Microscopy	Human	No	[10]
DNA, HIV[b]	Methylene blue	DNA-electrode	Amperometric	45	No	ELISA[c]	No	No	[11,12]
E. granulosus	Enzyme substrate	Electrode	Amperometric	26	Yes	No	No	No	[13]
Gliadin	Enzyme substrate	Electrode	Amperometric	30	Yes	SPR[d], ELISA[c]	Human	No	[14]
Human growth hormone	Enzyme substrate	Electrode	Amperometric	<60	Yes	SPR[d]	No	No	[15,16]
Gliadin; transglutaminase	Enzyme substrate	Electrode	Amperometric	>140	Yes	ELISA[c]	Human	Human	[17,18]
Transglutaminase	Enzyme substrate	Magnetic beads	Amperometric	70	Yes	ELISA[c]	Human	Human	[19]
DNA	Enzyme substrate	Membrane	Amperometric	20	Yes	ELISA[c]	Human	Human	[20]
Transglutaminase	Enzyme substrate	Electrode	Amperometric	>120	Yes	ELISA[c]	Human	Human	[21]
Influenza A virus	K ferro/ferri-cyanide	Electrode	Impedeometric	NS[f]	No	ELISA[c]	Avian	Avian	[22]

Digoxin	Ferrocene	DNA-electrode	Amperometric	45	No	No	No	?	[23]
Transglutaminase	Cadmium	Electrode	Impedeometric	>120	Yes	No	No	No	[24,25]
Intracellular autoantigens	Enzyme substrate	Membrane	Amperometric	20	Yes	IF[e]	Human	Human	[26]
Peptide autoantigen	K ferro/ferri-cyanide	Electrode	Impedeometric	40	No	ELISA[c]	Human	Human	[27]
Bovine herpes virus	None	Au strip	Potentiometric	10	No	ELISA[c], SPR[d]	Bovine	No	[28]
p53 cancer antigen	K ferro/ferri-cyanide	Au nano-particles	Amperometric	60	No	No	No	Mouse	[29]

[a]Method validation by comparison to a standard or alternative assay and use of positive and negative control serum samples.
[b]HIV, human immunodeficiency virus.
[c]ELISA, enzyme-linked immunsorbent assay.
[d]SPR, surface plasmon resonance.
[e]IF, immunofluorescence microscopy.
[f]NS, not stated

permits detection of class-specific antibodies of particular clinical or experimental importance. The flexibility of this approach derives from several choices on the type and concentration of conjugated enzyme and substrate as well as on the duration of product accumulation, allowing quantitative results in just a few minutes or over a wide dynamic range. It is adaptable to various portable platforms and electrochemical cell constructs with potential for miniaturization and low cost. Combined with a built-in digital reader, this type of system has the potential to be engineered into a portable device useable by health care professionals at the POC.

On the other hand, the electrochemical approach for class-specific antibody measurement involving an enzyme-conjugated secondary antibody and its substrate has potential drawbacks. Multiple steps – primary antibody incubation, wash, secondary antibody incubation, wash, substrate addition – can make for an awkward process if not fully automated. There are several variables to consider that may affect the output readings such as choice of diluents for the reagents, concentration of reactants, duration of incubations at each step and volume of washing solution. There may also be electrochemically active interfering substances in the sample, making for troublesome electron-transfer pathways. Shelf-life and stable storage conditions of labile components could be problematic as could the re-use and cleaning of contaminated surfaces of the non-disposable components. Overall, however, because electrochemistry is intrinsically flexible and adaptable, is relatively easy to miniaturize and inexpensive to manufacture the electrode and signal transduction reader, electrochemical sensors for antibody detection have particular promise for portable, POC applications.

4.4 Electrochemical biosensors for specific antibody with potential as POC instruments

Table 4.1 lists examples of electrochemical biosensors for antibodies to targets of particular interest to inventors of the devices. While most of these reports were published in the past few years, some go back well over ten years [4,5]. About one-third of these electrochemical immunosensors detect antibodies to infectious agents, about one-half to auto(self)-antigens, and several measure antibodies to non-infectious, foreign agents. Half the methods exploit electroactive mediators that are products of the action of enzymes, usually alkaline phosphatase or horseradish peroxidase, linked to a secondary antibody with specificity for the Fc region of the primary antibody that defines the sensor specificity. In addition to enhancing sensitivity, use of a secondary antibody permits the sensor to monitor a selected antibody isotype of clinical importance or interest (Section 4.3). However, sensors without isotype specificity capability have the advantage of label-free, reagentless simplicity, allowing direct measurement of antibody binding to antigen. In most cases the target antigen is bound directly to the electrode, minimizing the time for diffusible mediators to undergo oxidation/reduction reactions with the charged electrode surface. However, this format may place constraints on the amount, form, and/or orientation of the antigen on the metallic surface.

Ideally, new assays should include an initial validation using species-relevant samples and in which the antibody is derived from its natural source, although often diluted in a solution of designed composition. Validation is most easily done by comparison to a standard or reference assay in which samples with a wide range of antibody activities, including multiple non- and low-reactive samples, are tested in both assays. Many of the biosensors shown in Table 4.1 have not been thoroughly validated using multiple positive and negative controls, especially using antibody sources in the actual environment for which the sensor is intended.

The implied or explicit objective of antibody sensors in Table 4.1 is their potential utility as POC devices. Of particular importance for POC devices is assay time, which is included in Table 4.1. Assay duration for quantitative antibody measurement tends to be considerably longer than for antigen-detecting immunosensors, reflecting the heterogeneity and possibly low affinity of naturally-arising antibodies and the multiple steps in carrying out the assay, including several washing procedures to remove irrelevant antibody from natural sources. The average assay duration was about 57 min, varying from 10 min to upwards of 2 h. While these are considerably shorter than many current tests for antibodies performed in clinical pathology laboratories, antibody assay times for practical POC devices while the patient remains in the clinic will need to be minimized.

4.5 Perspectives on future development of POC devices for antibody measurement

Conversion and deployment of potential POC devices developed in research laboratories to actual field-based settings is challenging and has many hurdles to overcome [1]. Motivation for the development of POC tests is generally driven by societal needs for rapid, inexpensive and convenient ways for detecting or monitoring diseases or medical conditions. While technological limitations have up until recently resulted in a dearth of tests for specific antibodies, even purely qualitative POC antibody tests have proven useful. The wide use of line or spot immunoassays from multiple manufactures for rapid detection of antibodies to human immunodeficiency virus [6] speaks to the importance of societal, patient, and health care needs as the principal motivating force for POC assay development. However, while quantitative information on antibodies is commonly requested by health care professionals to aid diagnostic decisions, there may be resistance and obstacles to performing these antibody tests at the POC. With the rapidly evolving technological advances in electroimmunochemistry, electronics miniaturization and microfluidics exemplified by the reports in Table 4.1, quantitative measurement of specific antibody could begin to penetrate the barrier to the wide adoption of testing at the POC.

Transforming laboratory based testing devices to POC instruments should incorporate multiple desirable features [7]. The optimum technology should produce quantitative results in less than 30 min, preferably much quicker. While initial validation of a device can be adequately performed by the laboratory that developed the instrument [8], real clinical- or field-based settings in pre-approved clinical trials may be necessary for tests with high-risk implications [9]. Reasonable cost for non-disposable components of the instrument and low cost for the individual assay-specific materials are paramount. The technological complexity for performing the test can determine the extent of personnel training and proficiency oversight required to do the test as determined by regulatory agencies in some countries [8]. Handling and disposal of biohazardous and chemical fluids and cleaning of non-disposable components is another possible complication. While administrative issues such as transfer of results into patient charts and reimbursement requirements need to be addressed for POC testing within health care settings, generation of quantitative results by electrochemical transducers has the promise to facilitate direct electronic or wireless transmission of this data.

References

[1] *Bringing the Lab to the Patient: Developing Point-of-Care Diagnostics for Resource Limited Settings*. Washington, DC: American Society of Microbiology; 2012. Available from http://academy.asm.org/images/stories/documents/pointofcarediagnostics.pdf

[2] Wang Y., Xu H., Zhang J., Li G. 'Electrochemical sensors for clinic analysis'. *Sensors* (Basel). 2008;**84**(4):2043–81.

[3] Bahadir E.B., Sezginturk M.K. 'Applications of electrochemical immunosensors for early clinical diagnostics'. *Talanta*. 2015;**132**:162–74.

[4] Zhou Y.-M., Wu Z.-Y., Shen G.-L., Yu R.-U. 'An amperometric immunosensor based on Nafion-modified electrode for the determination of Schistosoma japonicum antibody'. *Sensors and Actuators B: Chemical*. 2003;**89**:292–8.

[5] Rao K.V., Rai G.P., Agarwal G.S., Suresh S. 'Amperometric immunosensor for detection of antibodies of Salmonella typhi in patient serum'. *Analytica Chimica Acta*. 2005;**531**(2):173–7.

[6] Tan W.S., Chow E.P., Fairley C.K., Chen M.Y., Bradshaw C.S., Read T.R. 'Sensitivity of HIV rapid tests compared to fourth generation enzyme immunoassays or HIV RNA tests – a systematic review and meta-analysis'. *AIDS*. 2016;**30**(12):1951–60.

[7] St John A., Price C.P. 'Existing and emerging technologies for point-of-care testing'. *Clinical Biochemist Reviews*. 2014;**35**(3):155–67.

[8] Liotta L.A., Petricoin E.F., 3rd. 'Regulatory approval pathways for molecular diagnostic technology'. *Methods in Molecular Biology*. 2012;**823**:409–20.

[9] Anonymous. 'Home-brew tests need regulation'. *Nature* [editorial]. 2014; **512**(7512):5.

[10] Sharma M.K., Agarwal G.S., Rao V.K., *et al*. 'Amperometric immunosensor based on gold nanoparticles/alumina sol–gel modified screen-printed electrodes for antibodies to *Plasmodium falciparum* histidine rich protein-2'. *Analyst*. 2010;**135**(3):608–14.

[11] Ricci F., Adornetto G., Moscone D., Plaxco K.W., Palleschi G. 'Quantitative, reagentless, single-step electrochemical detection of anti-DNA antibodies directly in blood serum'. *Chemical Communications (Camb)*. 2010;**46**(10): 1742–4.

[12] White R.J., Kallewaard H.M., Hsieh W., *et al*. 'Wash-free, electrochemical platform for the quantitative, multiplexed detection of specific antibodies'. *Analytical Chemistry*. 2012;**84**(2):1098–103.

[13] Pereira S.V., Bertolino F.A., Messina G.A., Raba J. 'Microfluidic immunosensor with gold nanoparticle platform for the determination of immunoglobulin G anti-Echinococcus granulosus antibodies'. *Analytical Biochemistry*. 2011;**409**(1):98–104.

[14] Rosales-Rivera L.C., Acero-Sanchez J.L., Lozano-Sanchez P., Katakis I., O'Sullivan C.K. 'Electrochemical immunosensor detection of antigliadin antibodies from real human serum'. *Biosensors and Bioelectronics*. 2011; **26**(11):4471–6.

[15] Ramanaviciene A., German N., Kausaite-Minkstimiene A., Voronovic J., Kirlyte J., Ramanavicius A. 'Comparative study of surface plasmon resonance, electrochemical and electroassisted chemiluminescence methods based immunosensor for the determination of antibodies against human growth hormone'. *Biosensors and Bioelectronics*. 2012;**36**(1):48–55.

[16] German N., Kausaite-Minkstimiene A., Kirlyte J., *et al*. 'Determination of antibodies against human growth hormone using a direct immunoassay format and different electrochemical methods'. *Analyst*. 2013;**138**(5): 1427–33.

[17] Neves M.M., Gonzalez-Garcia M.B., Nouws H.P., Costa-Garcia A. 'Celiac disease detection using a transglutaminase electrochemical immunosensor fabricated on nanohybrid screen-printed carbon electrodes'. *Biosensors and Bioelectronics*. 2012;**31**(1):95–100.

[18] Neves M.M., Gonzalez-Garcia M.B., Nouws H.P., Costa-Garcia A. 'An electrochemical deamidated gliadin antibody immunosensor for celiac disease clinical diagnosis'. *Analyst*. 2013;**138**(7):1956–8.

[19] Kergaravat S.V., Beltramino L., Garnero N., *et al*. 'Electrochemical magneto immunosensor for the detection of anti-TG2 antibody in celiac disease'. *Biosensors and Bioelectronics*. 2013;**48**:203–9.

[20] Rubin R.L., Wall D., Konstantinov K.N. 'Electrochemical biosensor for quantitation of anti-DNA autoantibodies in human serum'. *Biosensors and Bioelectronics*. 2014;**51**:177–83.

[21] Giannetto M., Mattarozzi M., Umilta E., Manfredi A., Quaglia S., Careri M. 'An amperometric immunosensor for diagnosis of celiac disease based on covalent immobilization of open conformation tissue transglutaminase for determination of anti-tTG antibodies in human serum'. *Biosensors and Bioelectronics*. 2014;**62**:325–30.

[22] Jarocka U., Sawicka R., Gora-Sochacka A., *et al*. 'Electrochemical immunosensor for detection of antibodies against influenza A virus H5N1 in hen serum'. *Biosensors and Bioelectronics*. 2014;**55**:301–6.

[23] Wei W., Zhang L., Ni Q., Pu Y., Yin L., Liu S. 'Fabricating a reversible and regenerable electrochemical biosensor for quantitative detection of antibody by using "triplex-stem" DNA molecular switch'. *Analytica Chimica Acta*. 2014;**845**:38–44.

[24] Martin-Yerga D., Gonzalez-Garcia M.B., Costa-Garcia A. 'Electrochemical immunosensor for anti-tissue transglutaminase antibodies based on the in situ detection of quantum dots'. *Talanta*. 2014;**130**:598–602.

[25] Martin-Yerga D., Costa-Garcia A. 'Towards a blocking-free electrochemical immunosensing strategy for anti-transglutaminase antibodies using screen-printed electrodes'. *Bioelectrochemistry*. 2015;**105**:88–94.

[26] Rubin R.L., Konstantinov K.N. 'Biosensor for total antinuclear antibody determination at the point-of-care'. *Biosensors and Bioelectronics*. 2016;**83**:306–11.

[27] Araujo G.R., Fujimura P.T., Vaz E.R., *et al.* 'A novel reactive epitope-based antigen targeted by serum autoantibodies in oligoarticular and polyarticular juvenile idiopathic arthritis and development of an electrochemical biosensor'. *Immunobiology*. 2016;**221**(5):634–40.

[28] Tarasov A., Gray D.W., Tsai M.Y., *et al.* 'A potentiometric biosensor for rapid on-site disease diagnostics'. *Biosensors and Bioelectronics*. 2016;**79**:669–78.

[29] Elshafey R., Siaj M., Tavares A.C. 'Au nanoparticle decorated graphene nanosheets for electrochemical immunosensing of p53 antibodies for cancer prognosis'. *Analyst*. 2016;**141**(9):2733–40.

Chapter 5

Portable magnetoelastic biosensors

Howard C. Wikle, III[1] *and Bryan A. Chin*[2]

5.1 Introduction

Magnetoelastic (ME) biosensors are a type of acoustic wave (AW) gravimetric sensor that are constructed from a resonating platform coated with a biorecognition layer that is specific and selective towards a target species. The resonating platform (resonator) is typically composed of an amorphous magnetostrictive ferromagnetic alloy – one that displays a strong coupling between the elastic and magnetic fields (ME behaviour). Because of the ME coupling, the vibration of ME biosensors may be actuated and the resonance properties determined by noncontact methods. ME biosensors do not require on-board power or any wires physically connected to the sensor to operate. As with all other AW gravimetric sensors, the mass sensitivity increases as the sensor dimensions decrease. ME biosensors have been fabricated in which the longest dimension is less than 50 μm. Detection times for target species are on the order of minutes compared with hours for state-of-the-art biomolecular techniques such as polymerase chain reaction and days for standard microbiological culturing protocols. ME biosensors have unique advantages for in situ or in vivo detection of biological species due to the wireless and passive features, very high mass sensitivity and low mechanical damping of the resonator platform, and the rapid capture and binding of target species by the biorecognition layer. ME biosensors can operate in air or liquids which permits them to be used to detect pathogens and other adulterants that may have contaminated our foods, pharmaceuticals or the environment in which we work and live. These advantages make ME biosensors a strong candidate technology for point-of-care diagnostic and detection systems for healthcare, agriculture, environmental monitoring and the food service industry.

In this chapter, we give an overview of the ME biosensor and the current state research on this platform. We begin by reviewing the origin of magnetostriction and ME coupling. This is followed by a review of the fabrication processes for the ME resonator platform and the biorecognition element. Three common resonant

[1]Department of Research, Materials Engineering, Auburn University, Auburn, AL, USA
[2]Materials Research and Education Center Director, Auburn University Detection and Food Safety Center Auburn, AL, USA

frequency measurement techniques are discussed. We close this chapter with some significant findings that have resulted from this investigation on ME biosensors.

5.1.1 Magnetostriction and magnetoelastic coupling

Magnetostriction is the change in shape of a material due to the ME coupling of the strain field within a material with an external magnetic field. Magnetostriction occurs in all pure substances but is only significant for ferromagnetic metals and their alloys. A ferromagnetic material can be considered to be divided into many magnetic domains which are regions with uniform magnetization, as shown in Figure 5.1. Within a magnetic domain, the magnetic moments are aligned along a particular direction; however, at the domain walls the magnetic moments vary, caused by the reorientation of the magnetic moments across the interface. In the demagnetized state, total magnetization is cancelled because of the random orientation of all of the magnetic domains.

Upon application of a low-strength external magnetic field, domains that are favourably aligned with the external field grow, due to domain wall migration, at the expense of domains that oppose the field. As the field strength increases, domain rotation becomes significant in which the magnetic moments in unfavourably aligned domains rotate from their original direction of magnetization to a preferred direction. For crystalline materials, the preferred direction lies along a particular crystallographic direction, termed the easy axis and giving rise to magnetocrystalline anisotropy. Amorphous materials, without the long-range order of crystalline materials, exhibit no magnetocrystalline anisotropy; however, an easy axis still exists, typically along the longitudinal axis of the thin sheet (amorphous metal alloys are typically only available as thin sheets <100 μm thick). At high field strengths (saturation), the magnetic moments rotate to align with the external field direction, resulting in a single magnetic domain.

Magnetostriction typically refers to the strain produced by the longitudinal change in length of a specimen as a result of the external magnetic field and is defined as

$$\lambda = \frac{\Delta l}{l_0} \tag{5.1}$$

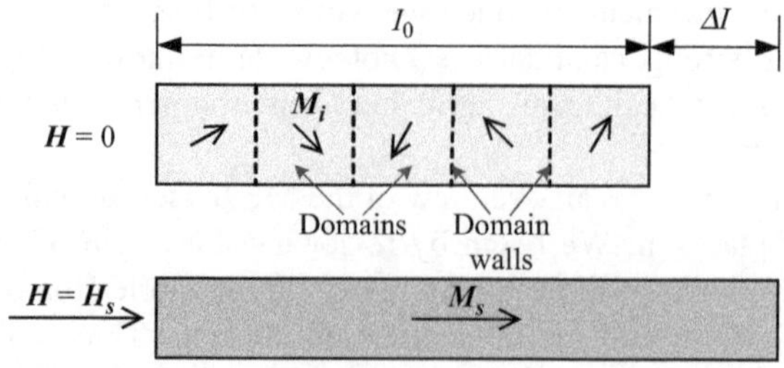

Figure 5.1 Depiction of the reorientation of magnetic moments and domain wall migration for a ferromagnetic material in the presence of an external magnetic field

where l_0 is the length of the demagnetized specimen, and $\Delta l = l - l_0$ is the change in length caused by the magnetic field. Most of the change in length is due to the rotation of the magnetic domains. Because there are no grain boundaries and no long-range crystal order or anisotropy in the metallic glasses, domain walls can migrate with relative ease.

Amorphous ME alloys are characterized by soft magnetic properties including low coercive field strength H_c, low hysteresis loss and high permeability. Figure 5.2 shows the normalized magnetization (*M–H*) loops of an as-received Metglas 2826MB ribbon (28 μm thick) and an electrochemically deposited $Fe_{80}B_{20}$ alloy (12 μm thick) that were measured with a vibrating sample magnetometer [1]. The *H*-field was applied parallel to the longitudinal direction of the ribbon/films. The Metglas ribbon shows no discernible hysteresis in its magnetization curve whereas the FeB alloy shows slight hysteresis. The coercive field strength H_c and squareness (M_r/M_s) were, respectively, 0.05 Oe and 0.0011 for the ribbon and 9.37 Oe and 0.15 for FeB alloy.

Although the magnetization curves invert as the applied field is reversed, so that the material becomes magnetized in the opposite direction, the magnetostrictive strain begins to increase again. Figure 5.3 shows the magnetostriction as a function of the applied field for a hysteretic material. The variation of λ with H traces out a double loop, called a butterfly loop. A biasing DC magnetic field (H_{DC}) is typically used to shift the operation of an AW magnetostrictive sensor to a quasilinear region of the λ–H curve. In applying the H_{DC} bias field, the sensor will vibrate at the same frequency as the excitation field and with greater strain amplitude. An optimum H_{DC} may be determined in which the amplitude to the magnetostrictive strain is greatest.

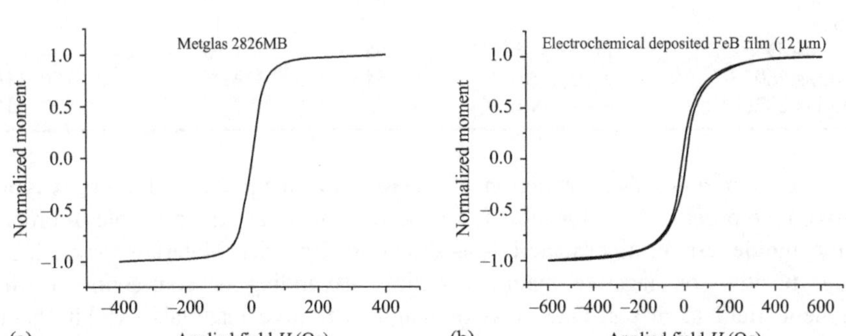

Figure 5.2 Magnetization loops of amorphous alloys. (a) METGLAS® 2826MB alloy ribbon (28 μm in thickness), obtained from Honeywell International; (b) $Fe_{80}B_{20}$ film (12 μm in thickness) fabricated by electrochemical deposition. The magnetization loops were measured with the H-field applied parallel to the surface of the ribbon/film. M–H hysteresis loops are normalized in order to compare the magnetization curves of these two materials

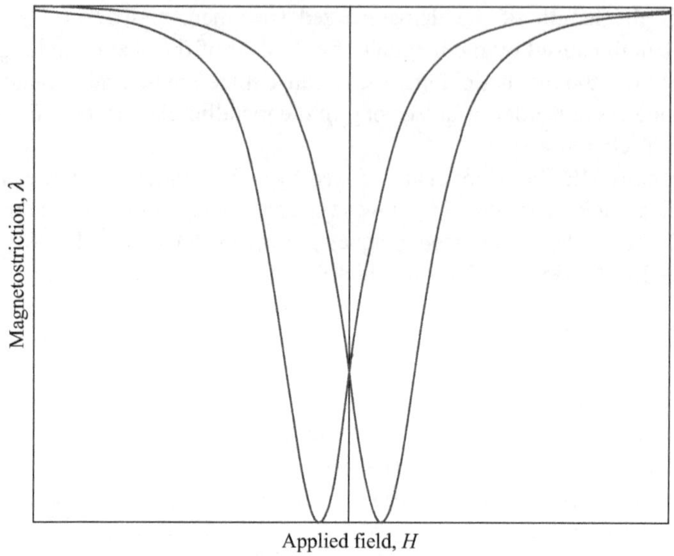

Figure 5.3 Magnetostriction hysteresis loop as a function of an externally applied magnetic field

Table 5.1 Properties of various magnetostrictive metals and alloys [2–4]

Material	Composition	T_c (K)	λs (10^{-6})	E (GPa)
Nickel		631	−36	210
Cobalt		1,400	−23	209
Terfenol-D	$Tb_{0.3}Dy_{0.7}Fe_2$	653	1,100	25–35
Metglas 2605SC	$Fe_{81}B_{13.5}Si_{3.5}C_2$	643	27	100–110
Metglas 2826MB	$Fe_{40}Ni_{38}Mo_4B_{18}$	626	12	100–110

The saturation magnetostriction λ_s, measured at magnetic saturation, is typically on the order of 10^{-5} for most ferromagnetic materials and a couple of orders of magnitude less for other materials as shown in Table 5.1. Materials may exhibit either positive or negative magnetostriction, expanding or contracting as the magnetic field increases. One class of magnetostrictive materials exhibit 'giant magnetostriction', over 100 times greater (e.g. Terfenol-D).

5.1.2 Magnetostrictive ribbons

When a magnetostrictive material is subjected to an alternating (excitation) magnetic field, a corresponding oscillating shape change occurs in the material. During vibration, the material stores and transfers energy between potential and kinetic energy modes. Resonance occurs at particular frequencies in which the material can easily exchange potential and kinetic energy resulting in greater oscillation

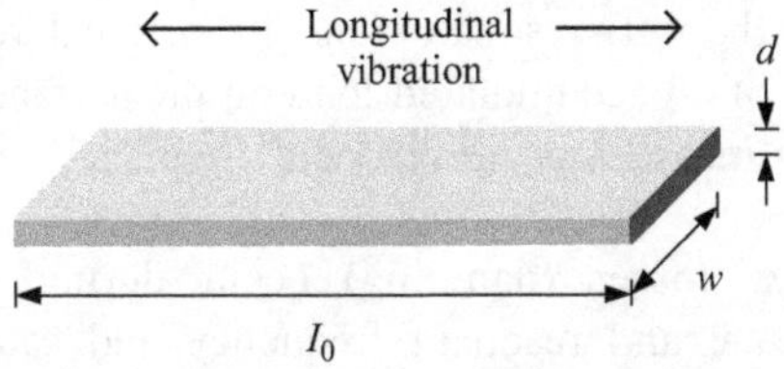

Figure 5.4 Depiction of free-standing magnetostrictive resonator

amplitude. The natural resonant frequency f_n of a freestanding thin ribbon, as shown in Figure 5.4 with $d \ll w$ undergoing longitudinal vibration in its basal plane is [5]

$$f_n = \frac{1}{2l_0}\sqrt{\frac{E}{\rho(1-\upsilon^2)}} \tag{5.2}$$

where E is Young's modulus of elasticity, ρ and υ are the density and the Poisson's ratio of the resonator material, respectively, and L is the length of the resonator. The term $2l_0$ is the wavelength of the compression wave propagating through the material and the square root term is the wave speed.

Vibrating structures have been routinely exploited as gravimetric AW sensors to measure the absorption or deposition of material on the surface of the resonator. If we let the initial measured frequency be the natural frequency ($f_0 = f_n$), then for a small, volumeless, uniformly distributed mass load Δm such that $\Delta m \ll m_0$, the resonant frequency for the thin ribbon can be expressed as

$$f = \sqrt{\frac{1}{[1 + (\Delta m/m_0)]}}f_0 \tag{5.3}$$

The resonant frequency will decrease with the addition of a small mass load so that the resonant frequency shift will be negative in sign, $\Delta f = f - f_0 < 0$.

The analytical sensitivity, which is defined as the shift in the resonant frequency due to the addition of a unit mass load, can be approximated as

$$\frac{\Delta f}{\Delta m} \approx -\frac{1}{2}\frac{f_0}{m_0} \tag{5.4a}$$

by taking the first, linear term of the Maclaurin series expansion of the radical in (5.3). From (5.4a), we can see that the sensitivity is proportional to the resonant frequency of the ribbon and inversely proportional to its mass. One of the most effective ways to increase the sensitivity is to reduce the length of the resonator. This decreases the resonator mass and increases the resonant frequency.

Rearranging (5.4a) to obtain the accumulated mass, we obtain

$$\Delta m \approx -2m_0\frac{\Delta f}{f_0} \tag{5.4b}$$

Thus, if the mass of the ME resonator and its original resonant frequency are known, then the mass of an accumulated load on the surface can be quantitatively determined by simply measuring the resonant frequency and calculating the frequency shift.

Dissipative energy losses (damping) occur during vibration that reduce the oscillation amplitude and resonant frequency and causing a broadening of the resonant frequency peak in the amplitude–frequency spectrum. Damping can arise from various sources, for example viscous, structural or magnetic forces. The damped resonant frequency f_d is given by

$$f_d = \sqrt{1 - \zeta^2} f_n \tag{5.5}$$

where ζ is the damping ratio. The Q-value, defined as the ratio of the energy stored in the resonant structure to the total energy loss per oscillation cycle, is related to the damping ratio by $Q = 1/2\zeta$. The Q-value is calculated by dividing the resonant frequency by its 3 dB bandwidth obtained from the amplitude–frequency spectrum. The Q-value reflects the sharpness of the resonant peak and thus a large Q-value permits greater resolution in determining the resonant frequency and the minimum detectable change in frequency. It was found that the ME biosensors exhibit a much higher Q-value (>1,000 in air and ~100 in water) than the best microcantilevers (MCS) [6]. The high Q-value of ME biosensors stems from their longitudinal resonance mode in which only the tiny cross-sectional areas of the two ends need to push against the surrounding medium. This advantage is more notable when the sensor is operated in liquid. Therefore, the ME resonator platform is a good candidate from which high performance biosensors can be developed since most of the biological analysis are conducted in liquids.

As different types of gravimetric AW sensors operate at different frequencies, it is necessary to normalize their analytical sensitivities in order to compare the mass sensitivities. The mass sensitivity S_m of a gravimetric AW sensor is defined in terms of the resonant frequency or the phase velocity [7]. In terms of resonant frequency, S_m is defined as

$$S_m = \frac{1}{f_0} \lim_{\Delta m'' \to 0} \frac{\Delta f}{\Delta m''} \tag{5.6}$$

where $\Delta m''$ is the areal density of a uniformly distributed mass. For a ribbon resonator that accumulates mass on one surface, it can be shown that S_m is

$$S_m = -\frac{1}{2\rho d} \tag{5.7}$$

5.1.3 Magnetostrictive microcantilevers

Another resonator configuration used with magnetostrictive alloys is the magnetostrictive microcantilevers (MSMC). Unimorph MSMCs have been investigated where an inactive layer (copper) is plated on one side of a ribbon Metglas 2826MB

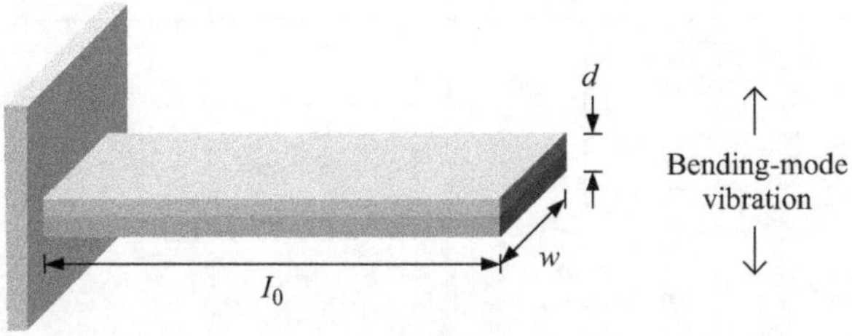

Figure 5.5 Depiction of magnetostrictive microcantilever resonator

magnetostrictive material as shown in Figure 5.5 [6,8,9]. The magnetostrictive layer lengthens when a magnetic field is applied to the MSMC, causing the cantilever to bend; in an AC excitation magnetic field, the MSMC will vibrate. The resonant frequency of the MSMC is

$$f_n = \frac{\lambda_n^2}{2\pi\sqrt{12}} \frac{d}{l_0^2} \sqrt{\frac{\tilde{E}}{\tilde{\rho}}} \tag{5.8}$$

where λ_n are the eigenvalues, $\tilde{E}$ is the effective elastic modulus, and $\tilde{\rho}$ is the effective density of the composite beam [10].

The mass sensitivity of the MSMC is

$$S_m = -\frac{1}{2\rho d} \tag{5.9}$$

Figure 5.6 shows the accumulated mass that was measured using acoustic resonance by several ME biosensors and a MSMC biosensor along with the theoretical minimum detectable mass (assuming a minimum measurable relative frequency change $\Delta f/f_0 = 2 \times 10^{-7}$ [11]). A ME biosensor with a length of 384 μm has the theoretical capability of detecting the mass of a single bacteria cell (1 pg).

5.1.4 Comparison to other AW devices

For comparison, the mass sensitivities of several AW devices are shown in Table 5.2. The mass sensitivity of ME resonators is expected to exceed that of the top performing MCS since the density of ME materials $(7,900\,\text{kg m}^{-3})$ is more than double that of silicon nitride $(3,400\,\text{kg m}^{-3})$ which is typically used for MCS. As mass accumulates on the largest surface (the active area) of the ME ribbon sensor, the sensitivity can be further increased by reducing the thickness of the resonator without changing the active surface area.

BAW, bulk AW; DL-MCS, distributed load microcantilever; EL-MCS, end load microcantilever; FPW (lamb), flexural plate wave; ME, magnetoelastic; SAW, surface AW; SH-APM, plate mode; TSM (QCM), transverse shear mode (quartz crystal microbalance).

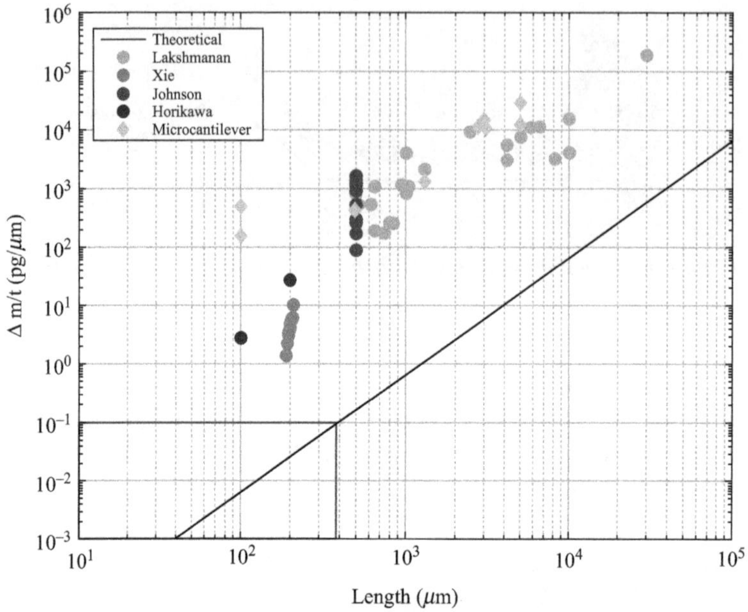

Figure 5.6 Accumulated mass measured by several investigators using ME sensors. Because of the different thicknesses of the ME resonators, the masses are normalized by the sensor thicknesses. The theoretical curve is based on a resonator with a l:w:d ratio of 1,000:200:28 and a relative frequency resolution of 2×10^{-7}

Table 5.2 Mass sensitivities of acoustic wave devices [11]

Device	f_0 (MHz)	S_m
SAW	112	$\propto -1/\rho\lambda$
TSM (QCM)	6	$\propto -1/\rho d$
SH-APM	104	$\propto -1/\rho d$
FPW (lamb)	2.6	$\propto -1/2\rho d$
BAW	6	$\propto -2/\rho\lambda$
DL-MCS	5–0.02	$\propto -1/\rho d$
EL-MCS	5–0.02	$\propto -1/2\rho d$
ME (1 surface)	2–38	$\propto -1/2\rho d$
(2 surfaces)		$\propto -1/\rho d$

5.2 ME biosensors

5.2.1 Commercially available magnetostrictive ribbons

Traditionally, ME resonators have been fabricated by mechanically dicing commercially available amorphous magnetostrictive alloy ribbons such as Metglas 2826MB. These magnetostrictive ribbons are cast using a melt spinning process where the molten alloy is sprayed onto a chilled rotating wheel in which cooling rates reach 10^6 degree s^{-1}. The rapid quenching freezes the microstructure of the alloy in its disordered state, suppressing the nucleation of crystalline phases and forming the amorphous alloy ribbons [12]. By employing this technique, amorphous alloy ribbons, typically in width of 10–100 mm and thickness of 20–30 μm, are produced. The ME resonators are fabricated from these ribbons.

The basic process of mechanical fabrication of ME resonators from the ME magnetostrictive ribbons [13–15] is described. The ME ribbons are first polished using fine-grit, metallographic polishing paper (1,000 and 2,000 μm) to reduce the thickness and to obtain smooth surfaces. The rectangular resonators are then cut from the polished ribbon using a computer-controlled automatic micro-dicing saw. Any debris or grease remaining from the dicing process is then removed by cleaning the diced sensors ultrasonically in acetone for 30 min. The cleaned resonators are then subjected to a thermal anneal in a vacuum oven at 200 °C for 2 h and oven cooled under a vacuum ($>10^{-3}$ Torr). Research has shown that the annealing process removes residual stresses and improves the morphology and performance of the sensor platform [16,17]. In order to construct the biosensor, a chromium layer and then a gold layer are sputtered onto all sides of the resonators. The layer of chromium is sputtered to improve the adhesion of the gold film to the substrate. The gold layer improves the sensors' resistance to corrosion and provides a ready surface for immobilization of the biorecognition element, such as antibody or phage.

5.2.2 Microfabrication

Photolithography and physical vapour deposition have been used to fabricate ME resonators in a micro-scale sizes with uniform, highly repeatable dimensions and excellent surface quality [17,18]. The microfabricated resonators are made from a binary alloy of iron and boron with a composition near 80/20 atomic per cent. The microfabrication process is shown schematically in Figure 5.7 [18]. A silicon

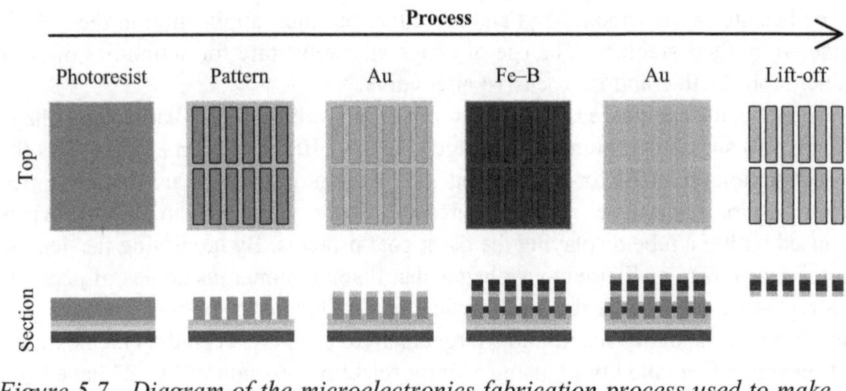

Figure 5.7 Diagram of the microelectronics fabrication process used to make micron-scale magnetoelastic resonators

wafer is used to support the patterning of rectangular shapes of the desired size using photolithography. The fabrication process begins with a plain silicon wafer sputter coated with a layer of Cr followed by a layer of Au. Then the wafer is patterned with the rectangular resonator shapes using a suitable photoresist. A thin gold film is then deposited onto the substrate. The gold serves as the bottom protective layer of the resonators. Fe and B sputter targets [19], placed on separate cathodes, are used to simultaneously deposit the iron–boron alloy onto the wafer. After the deposition of the Fe–B film, a final gold thin film is sputtered onto the substrate wafer as the top protective layer for the resonators. The gold layer also serves as a ready surface for immobilization of the biomolecular recognition element. A lift-off process using a solvent wash is then used to remove the resonators from the wafer. After removal, the resonators are cleaned with acetone.

5.2.3 Biomolecular recognition element

In order to form a biosensor to detect pathogens, a biomolecular recognition element is immobilized on the ME resonator surface to capture the target species upon contact. Pathogen capture results in an increase in mass of the biosensor and a corresponding decrease in its resonant frequency. Therefore, the presence of a target pathogen can be detected by monitoring the shift in the resonant frequency of ME biosensors. For an immunoassay biosensor, the immobilized biorecognition element captures the target pathogen, determining the specificity of the biosensor; on the other hand, the sensitivity and the detection limit are determined by the sensor platform.

To form a functional ME biosensor, a biomolecular recognition element must be immobilized onto the ME resonator surface to bind the specific target species. Antibodies have been traditionally used for this purpose [20]. The strengths and weaknesses of antibody binding are well known. An antibody is a relatively fragile species and subject to denaturation with consequential loss of sensitivity and other binding characteristics when exposed to unfavourable environments. Moreover, the quality of polyclonal antibodies can vary with different animals and production variables. To be used in biosensors, polyclonal antibodies require affinity purification and stabilization, which dramatically increase their cost. Monoclonal antibodies are more standardized and selective, but their application in the field is hindered by their stability. The use of phage as a substitute for antibodies offers a stable, reproducible and inexpensive alternative.

Genetically engineered 12600 lytic phage (Figure 5.8), E2 filamentous phage (Figure 5.9) and JRB7 filamentous phage (Figure 5.10) have been employed as the biorecognition elements for ME biosensors. Filamentous phages are thread-shaped bacterial viruses, whose genetic material (single stranded circular DNA) is enclosed within a tube displaying the outer coat proteins. By modifying the genetic material, a library of filamentous phages that display numerous copies of peptides in a repeating fashion on the viral surface can be created. The phage clones displaying peptides highly specific to a target analyte of interest can then be identified and separated from the library using affinity selection procedures, which have been described in detail by Petrenko *et al.* [21–27].

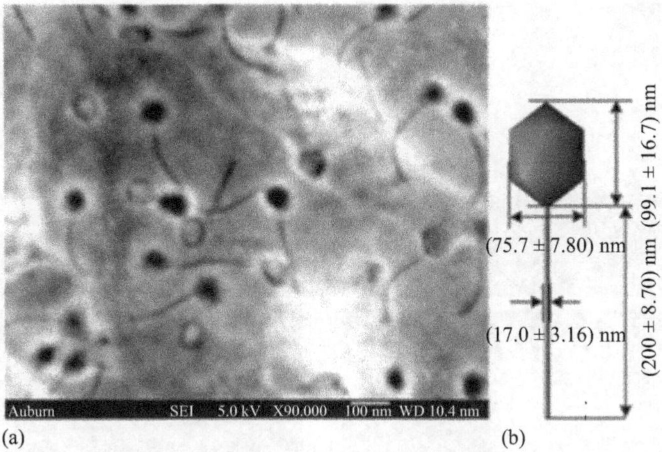

Figure 5.8 (a) Micrograph of lytic phages 12600 immobilized on the surface of a ME biosensor. (b) Morphology and dimension of the lytic phage [28]

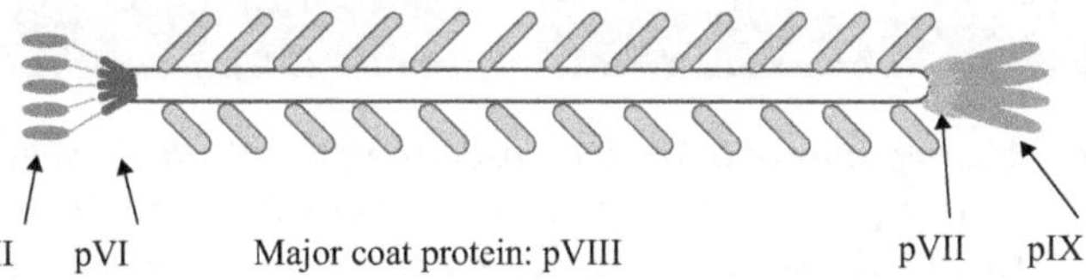

Figure 5.9 Composition of phage and arrangement of major surface proteins on E2 phage [29]

As a biorecognition element, phage has several key advantages over antibodies. Affinity-selected phages are obtained through a method similar to molecule synthesis, without requiring the immunization of an animal. Phages can be produced in large quantities at a relatively low cost. The phage serves as a three-dimensional support for bioreceptors against the target pathogen. The surface area density of phage is $300-400\,m^2\,g^{-1}$, out of which more than 50% of the surface and 90% of the mass is composed of peptides that form the active binding sites [23]. The three-dimensional recognition surface with multiple binding sites provides strong multivalent interactions with the target pathogens. More significantly, the structures of filamentous phage are very robust and resistant to degradation. Phages have been found to retain their infectivity after exposure to organic solvents, including 20% isopropanol and 30%–55% ethanol [30,31]. Thermostability studies of phage have shown that recombinant phage are resistant to heat up to 80 °C [27], whereas antibodies are known to lose their activity at 25 °C in less than 20 days [32–35].

The longevity of the E2 phage-based ME biosensors was studied at different temperatures. The results (Figure 5.11) showed that the biosensors retained 59%,

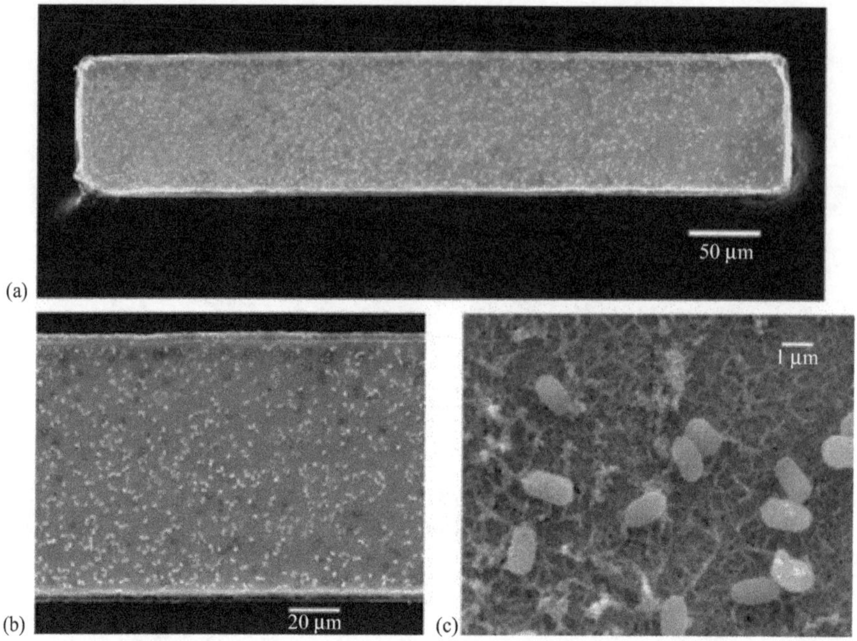

Figure 5.10 Micrographs of a ME biosensor coated with JRB7 bacteriophage following exposure to a Bacillus anthracis Sterne strain solution. Captured spores are visible on the surface of the biosensor, and phage filaments, are visible in (c) [18]

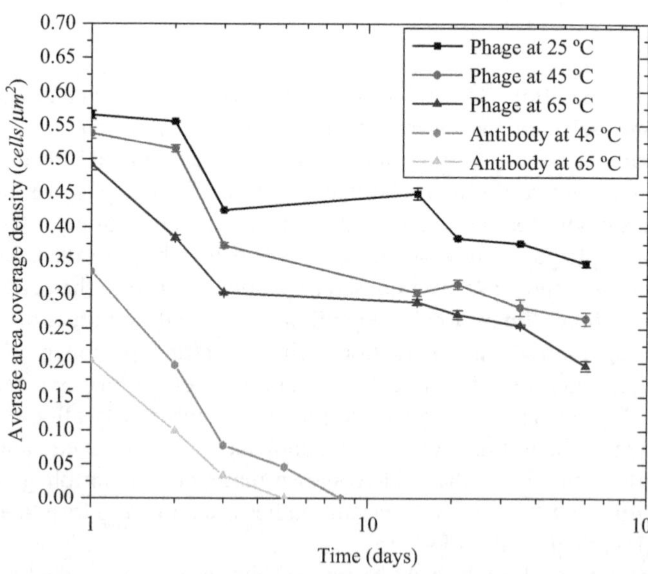

Figure 5.11 Longevity of phage (E2) as biomolecular recognition element. Antibody loses all binding affinity after five days storage at 65 °C, whereas E2 phage remains active after 60 days

45% and 33% of their binding affinity after storage at 25, 45 and 65 °C, respectively, after 62 days [36]. After only five days at 65 °C, the antibody-based sensors show no ability to bind while the binding affinity of the phage remains better than the initial binding affinity of the antibodies even after 40 days had elapsed [35]. Due to their excellent chemical resistance and thermal stability, phage can withstand harsh environments that are found in the field.

5.3 Measurement techniques

Figure 5.12 shows an embodiment of the operation of ME biosensor. An excitation coil generates an AC magnetic field that is imposed upon the ME biosensor, generating longitudinal elastic waves within the sensor. The applied AC magnetic field can be generated with a Helmholtz coil, a solenoid coil or a flat coil. At resonance, the mechanical impedance of the resonator is at a minimum so that its vibration amplitude is greatest. The mechanical vibration that results from the ME coupling generates a varying magnetic flux that is emitted from the sensor. The varying magnetic flux will generate an electrical current in a separate pickup coil that can be measured. Alternatively, the excitation coil may also serve as the pick-up coil. With a network analyser, the sensor response is measured in the S_{11} reflected mode. With an impedance analyser, the excitation coil and the sensor are mutually inductive coupled so that sensor response affects the electrical impedance of the excitation coil. In either case, a peak in the measured response signal occurs at the resonant frequency.

Three different techniques to measure the ME biosensor resonant frequency are described in more detail: (1) swept frequency, (2) transient response and (3) flat coil measurement techniques. These differ in the manner in which the AC magnetic field is applied and the sensor response is measured.

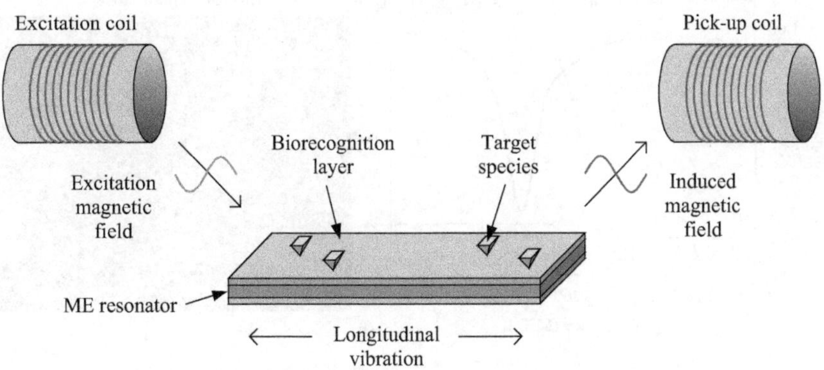

Figure 5.12 The principle of operation of ME biosensors. Because of the strong magnetoelastic coupling between the magnetostrictive resonator and the applied magnetic field, a longitudinal elastic wave is generated in the sensor. In turn, the elastic wave causes the magnetization of the sensor to vary

5.3.1 Swept frequency measurement technique

The swept frequency measurement technique is the most widely used method of measuring the resonant frequency of the sensor [13,16,37]. In this technique, an AC excitation signal is swept over a frequency range that includes the sensor's resonant frequency, continuously exciting the sensor vibrations. To obtain a relatively uniform magnetic field, the length of the coil is at least twice the length of the sensor. Before measurement, the analyser is calibrated to cancel any ambient electrical noise. A DC magnetic field is used to provide an initial magnetostrictive strain so that the sensor operates in the quasilinear region of the λ–H curve (Figure 5.3). A network analyser, impedance analyser or function generator/lock-in amplifier combination can be used to excite and measure the resonant frequency of the sensor. Figure 5.13(a) shows the measurement of the resonant peaks of an E2 phage-coated ME biosensor using the swept frequency technique using a network analyser before and after exposure to 5×10^8 CFU mL^{-1} *Salmonella typhimurium* on a tomato surface. A frequency change of 6.325 kHz was obtained. The micrograph, taken with a scanning electron microscope (SEM), shows the bound *Salmonella* cells on the surface of the biosensor.

Figure 5.14 shows the resonant frequency shift of a 500 μm long E2 phage coated ME biosensor (measurement sensor) and a control sensor (devoid of phage) during exposure to increasing concentrations of *S. typhimurium*. A measurable decrease in the resonant frequency was observed when the biosensor was exposed to a concentration of 50 CFU mL^{-1}. The resonant frequency continuously decreased with the introduction of each successive concentration (5×10^1 CFU mL^{-1} through 5×10^5 CFU mL^{-1}) of *Salmonella* for a change totalling ~100 kHz in the resonant frequency. The control sensor showed a very small frequency change due to non-specific binding.

The swept frequency technique enables an accurate, wireless and almost real-time detection. However, it requires an optimized DC bias magnetic field to obtain a

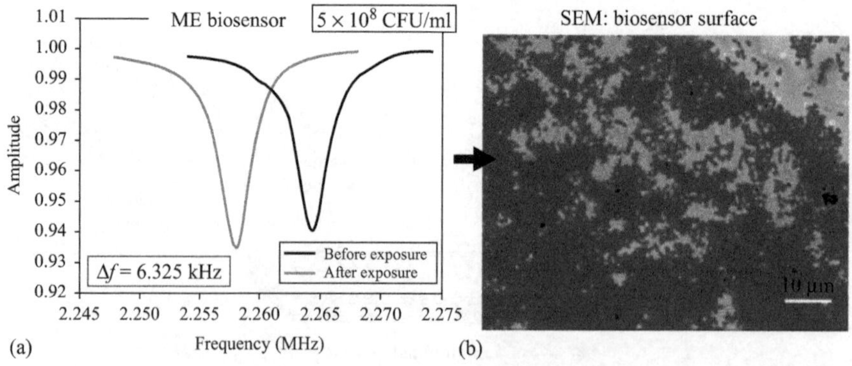

Figure 5.13 (a) Resonant frequency curves before and after exposure to 5×10^8 CFU mL^{-1} Salmonella typhimurium on the surface of a tomato. (b) SEM micrograph of the sensor surface showing a large number of bound Salmonella cells [38]

linear and maximized signal. The resonant frequency of the ME sensors is sensitive to the DC bias; thus, the bias field needs to be well controlled. Meanwhile, sweeping across a range of frequencies is relatively time consuming. In practice, the use of a solenoid coil limits the number of sensors that can be simultaneously monitored.

5.3.2 Transient response measurement technique

A transient response technique has also been used to measure the resonant frequencies of ME sensors. The transient response technique uses a magnetic pulse to excite the sensors into vibration and then the resonant frequency is determined from the ring-down decay of the vibration. This is similar to striking a bell and measuring the decay of the sound to determine the bell's resonant frequency.

Figure 5.15 shows the schematic diagram of the transient response detection system initially developed by Shen *et al.* [39–41] and improved by Xie *et al.* [42]. A function generator outputs a square wave at a specified repetition rate. The waveform is amplified and then fed to the primary side of a bridge transformer to excite the sensor. On the secondary side of the transformer, a pair of coils are wound in the opposite direction and connected in series to serve as the pick-up coil. The advantage of this design is that the two secondary coils cancel out the large transient current when the excitation pulse is applied, as well as ambient noise. When a ME sensor is placed inside one of the secondary coils, the bridge becomes unbalanced and an output signal is produced. The output signal is zero when there is no sensor in the core of the transformer. The ring-down signal is acquired from

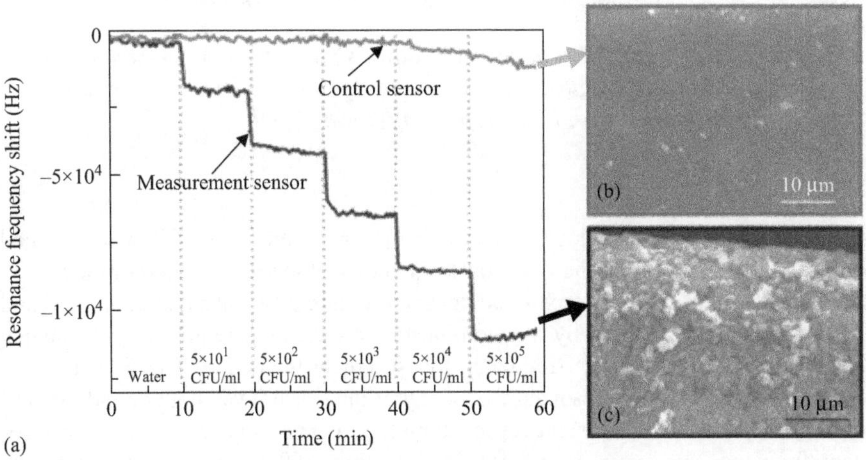

Figure 5.14 (a) Measurement and control sensors (devoid of phage) during exposure to increasingly greater concentrations of Salmonella typhimurium. (b) SEM image of the control sensor surface after the experiment, showing negligible binding of Salmonella cells. (c) SEM micrographs of the measurement biosensor at the end of the experiment, showing a very large number of bound Salmonella cells [1]

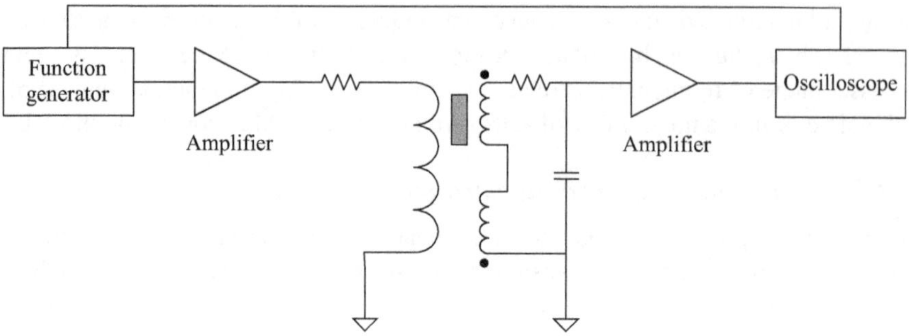

Figure 5.15 Block diagram of the transient response detection system

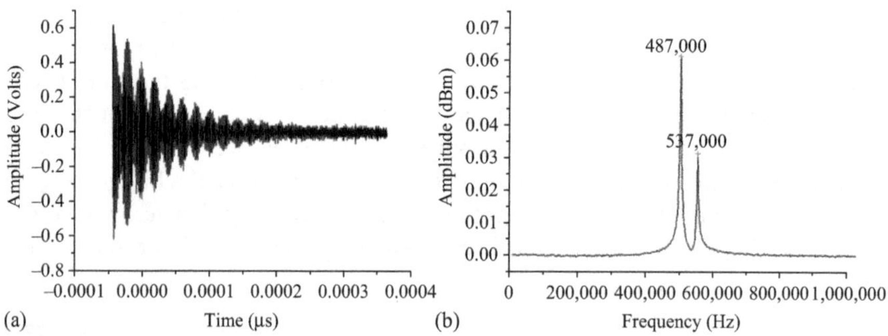

Figure 5.16 (a) Transient response ring-down signal of two ME resonators of different lengths. (b) The resonant frequencies were determined by performing an FFT on the ring-down signal

the transformer by an oscilloscope and a fast Fourier transform (FFT) is performed on the signal from which the resonant frequency of the sensor is determined.

The transient response technique does not require a DC bias magnetic field and provides quicker detection by overcoming the longer times required by frequency sweeping used in the swept frequency measurement technique. The FFT analysis can be used to obtain the resonant frequency of the sensor even under conditions of high noise levels. The transient response method is particularly suited to the resonant frequency measurement of sensors less than 500 μm in length due to the inherent noise cancellation. This technique is also suitable for monitoring multiple ME biosensors simultaneously. The ring-down response of two ME sensors, of slightly different lengths, that were placed together in the pick-up coil and excited simultaneously is shown in Figure 5.16. The measured signal is the superposition of the two ring-down responses. Following the FFT analysis, two resonance peaks appear in the frequency spectrum at the resonant frequencies of the sensors.

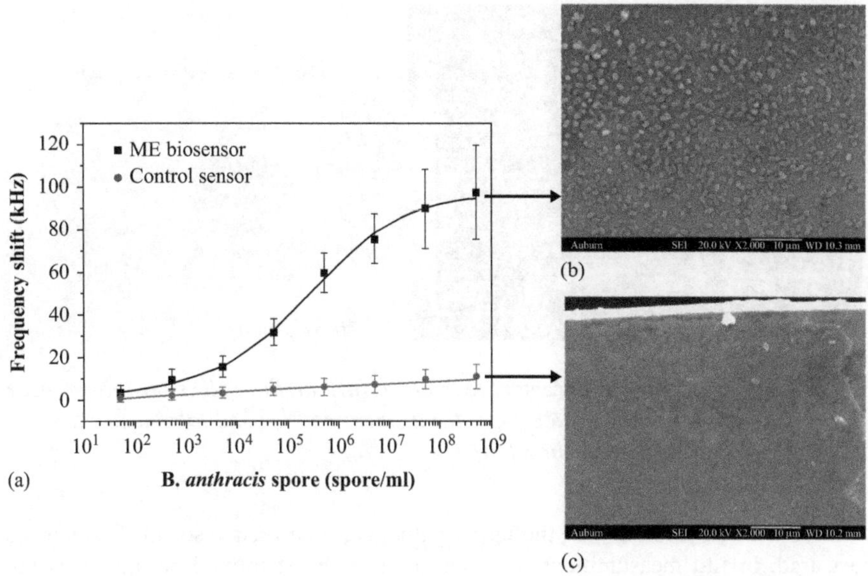

Figure 5.17 (a) Frequency shift vs. concentration of Bacillus anthracis Sterne spores. SEM images of measurement (b) and control (c) sensor surfaces [42]

The transient response technique was used to measure the capture of *Bacillus anthracis* Sterne spores on the surface of 200 μm long ME biosensors. Figure 5.17 shows the magnitude of the frequency shifts for 200 μm long ME biosensors coated with JRB7 phage (measurement sensors) and control sensors (devoid of phage). SEM analysis was used to confirm that the frequency shifts due to spore binding. The slight increase in the control sensors' frequency shift was due to nonspecific binding.

Another transient response technique was developed by Zeng *et al.* in the early 2000s [43,44]. This method applies a few cycles of a sinusoidal pulse with a known frequency to excite the sensor and then compares the transient response of the sensor to a threshold value. Each time the transient signal crosses the threshold level, a threshold-crossing pulse is generated, and the total number of pulses is counted for each excitation frequency. The frequency of the applied excitation signal is incremented over a range. The digital threshold-crossing pulse count from the sensor versus the applied pulse frequency is plotted, and the resonant frequency of the sensor is determined from the peak location on the plot. This method however still requires DC bias and frequency sweeping to determine the resonant frequency.

5.3.3 Flat coil measurement technique

Flat coils have recently been investigated as an alternative to solenoid coils for exciting and measuring the resonant frequencies of the ME biosensors [45–47]. With this configuration, any number of ME biosensors may be distributed upon a

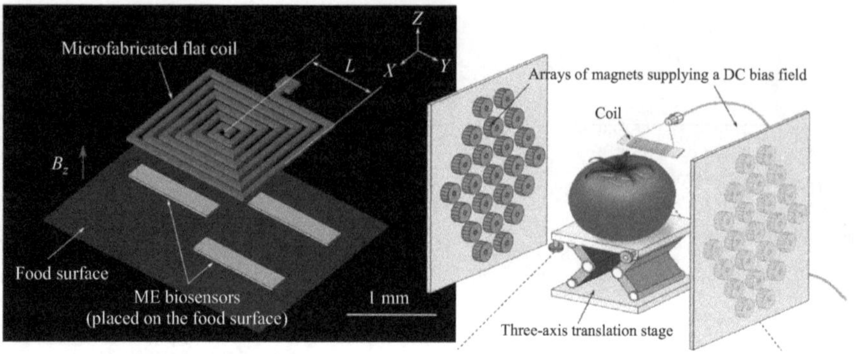

Figure 5.18 Flat coil measurement technique. (a) The flat coil is positioned near the tomato surface [45], (b) a spiral coil is used to measure the resonant frequency [46]

surface, allowed to bind with the target pathogen, then their resonant frequencies measured. In situ measurements may be more easily performed because it is not necessary to remove the sensors from the surface before measuring the resonant frequencies. The flat coils may be used to both excite the sensor vibration and measure the sensor response. Figure 5.18 shows a configuration for measuring ME biosensor placed upon the surface of a food item. A DC bias field is still used to bias the sensors. In this configuration, two parallel moveable plates with an array of strong (large BH_{max}) neodymium permanent magnets, oriented perpendicular to the plane of the flat coil, are used to generate the DC bias field. The flat coil is brought near the food item and swept across its surface to measure the resonant frequency of the ME biosensors.

Real-time, in situ detection of *S. typhimurium* on the surface of a watermelon was demonstrated by Chai *et al.* [46] using the flat-coil measurement technique as shown in Figure 5.19. Measurement and control sensors were prepared similarly except that the measurement sensors were coated with the E2 phage biorecognition element, and the control sensors were not. The watermelon surface was spiked with three different surface densities of *S. typhimurium* (1.5×10^2, 1.5×10^4 and 1.5×10^6 CFU mm^{-2}) and the measurement and control sensors placed in the spiked areas. For the measurement sensors, the absolute value of the resonant frequency shift increased as bacteria was captured and bound to the measurement sensors. The frequency shift saturated in approximately 10 min.

5.4 Additional results

5.4.1 Detection in the presence of masking bacteria

The E2 phage-based ME biosensors show excellent specificity and selectivity toward *S. typhimurium*. Lakshmanan *et al.* [14,36,37] studied the effect of masking bacteria on the detection of *S. typhimurium* by the E2 phage-based ME biosensor.

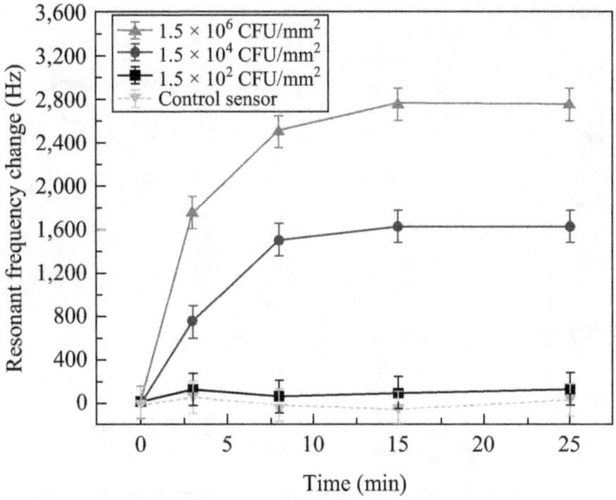

Figure 5.19 Real-time, in situ detection on a watermelon using a flat coil located above the surface [47]

The ME biosensors were exposed to *S. typhimurium* suspensions mixed with high concentrations of masking agents (microbial populations). Two different masking mixtures were used in the study: (1) a mixture of *S. typhimurium* with one masking bacteria (*Escherichia coli*) and (2) a mixture of *S. typhimurium* with two masking bacteria (*E. coli* and *Listeria monocytogenes*). The dilutions of *S. typhimurium* $\left(5 \times 10^1 - 5 \times 10^8\ \text{CFU mL}^{-1}\right)$ were prepared in such a way that all the suspensions contained a high concentration $\left(5 \times 10^7\ \text{CFU mL}^{-1}\right)$ of the masking bacteria. Figure 5.20 shows the comparison of response of the ME biosensors as they detect *S. typhimurium* in water, as well as in masking mixtures [48]. The sensitivity of the detection drops slightly at higher concentrations of *S. typhimurium* (10^6 and greater), but the detection limit remains unchanged at 5×10^2 CFU mL^{-1}. The biosensor was capable of detecting small amounts of *S. typhimurium*, even in the presence of high concentrations of masking bacteria.

5.4.2 Detection in liquid foods

It has been successfully demonstrated that phage-based ME biosensors were able to directly detect pathogens in liquid food product that can be a complex environment due to the presence of biomolecules (e.g. proteins, lipids) and other suspended solids. Figure 5.21 shows the results of E2 phage-based ME biosensor directly detecting *S. typhimurium* in liquid foods. The ME biosensors $(2 \times 0.4 \times 0.015\ \text{mm})$ coated with E2 phage were exposed to milk, apple juice and water spiked with increasing concentrations of *S. typhimurium* $\left(5 \times 10^1 - 5 \times 10^8\ \text{CFU mL}^{-1}\right)$ [36,37]. The dose response of the biosensors exposed to spiked apple juice and water samples was very similar, except at high concentrations.

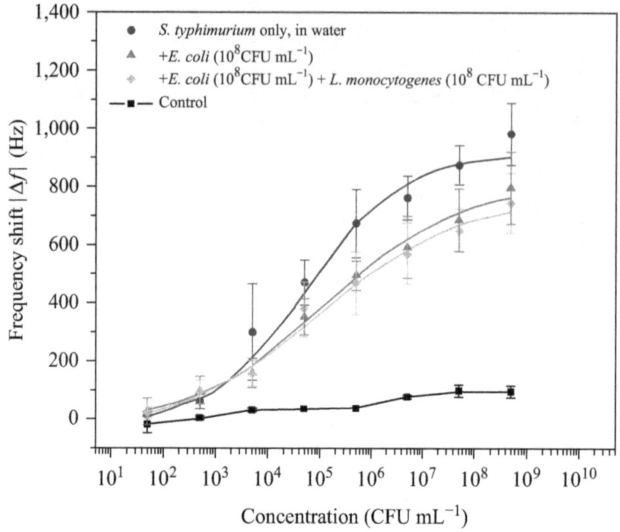

Figure 5.20 Specificity of E2 phage. The E2 phage is tested in solutions with masking agents of common Escherichia coli (10^7 CFU mL^{-1}) and Listeria monocytogenes (10^7 CFU mL^{-1}). The detection limit of 5×10^2 CFU mL^{-1} of the Salmonella ME biosensor is not affected by the masking agents. This shows the specificity of the E2 phage is better than 1 in 10^5 [48]

The resonant frequency shifts obtained for spiked milk samples were lower than spiked water and spiked apple juice samples. The biosensor's frequency shift was calculated by subtracting the final measured frequency from the initial frequency measured in the liquid food without spiked *S. typhimurium*. The dose response was linear over five aliquots of concentrations $\left(5 \times 10^3 - 5 \times 10^7 \text{ CFU mL}^{-1}\right)$ for the three different media. The sensitivity of the ME biosensor was calculated as the slope of the linear region of the dose response curve (Hz per decade of concentration change). Based on these results, the sensitivity of biosensors exposed to spiked water, apple juice and milk was 161, 155 and 118 Hz decade^{-1}, respectively. The control sensor had a negligible change in resonant frequency in response to even high concentrations of *S. typhimurium*. The control sensor showed a maximum resonant frequency shift of 50 Hz, whereas a maximum resonant frequency shift of 980 Hz was observed for the ME biosensor. This significant difference in the measured frequency shifts (control versus measurement sensor) indicates specific phage-bacteria binging versus negligible, non-specific binding of bacteria to the bare gold surface. SEM photomicrographs in Figure 5.21 provide visual verification of bacterial binding to the sensor surfaces.

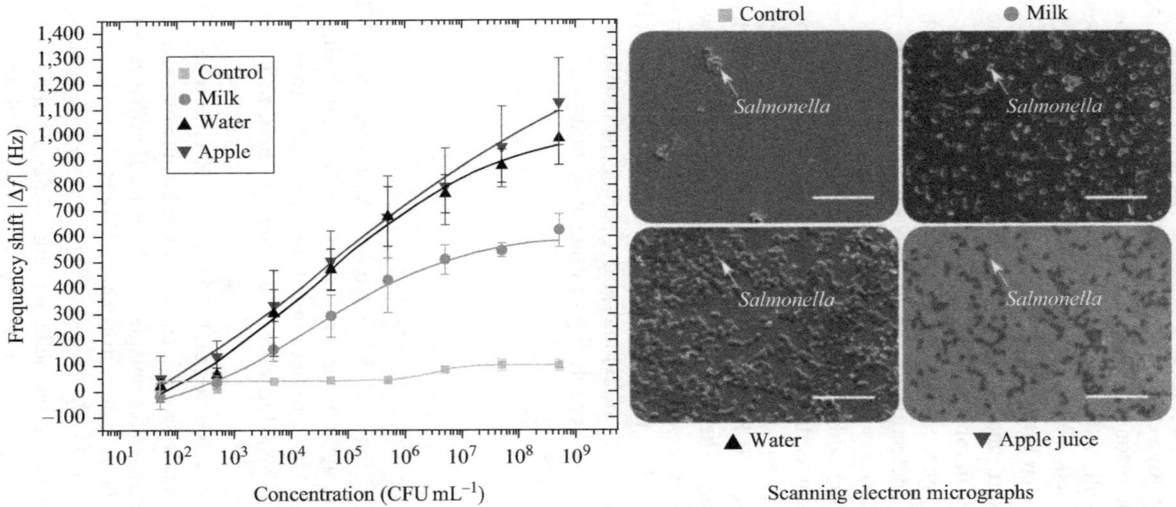

Figure 5.21 *Comparison of dose responses of phage-coated ferromagnetic particles (2 × 0.4 × 0.015 mm), when exposed to increasing concentrations ($5 \times 10^1 - 5 \times 10^8$ CFU mL $^{-1}$) of Salmonella typhimurium suspensions in water (▲), apple juice (▼) and milk (●). Control (■) represents the uncoated particle's response. The SEM micrographs (right) confirm that specific binding of Salmonella typhimurium on the phage-coated particles has occurred (scale bars: 10 μm)*

5.5 Outlook

Micron-scale ME biosensors show great promise as portable biosensors due to their magnetic interrogation and wireless operation. Although most research to date on ME biosensors has used relative expensive and complicated laboratory instruments to actuate and measure the resonant frequencies, the physical dimensions and frequencies of interest are well within the scale of the technologies that enable our consumer electronics and wireless communications. With continued advancement in this field, we envision that portable systems will allow a large number of ME biosensors to be deployed and monitored simultaneously and the resonant frequency changes of any number of sensors identified out of many. This will provide at least two unique capabilities for ME biosensor operation. First, since ME biosensors are capable of detecting extremely low concentrations of pathogens, deploying multiple ME biosensors (hundreds) significantly enhances the probability of the sensors binding with the target pathogen. The presence of even a few cells can be detected as long as the sensors bind or capture them. Second, multiple ME biosensors can simultaneously detect different pathogens in a single sample. Several groups of multiple ME biosensors can be employed for the detection of multiple pathogenic species. The different groups of ME biosensors would have different dimensions and different biological recognition elements immobilized upon their surfaces so that different groups could be distinguished by their resonant frequencies and their response to individual target pathogens monitored. The realization of these capabilities will enable high-throughput, multiplexed detection of pathogens with ME biosensors.

References

[1] Li S., Horikawa S., Park M.-K., Chai Y., Vodyanoy V.J., Chin B.A. 'Amorphous metallic glass biosensors'. *Intermetallics*. 2012;**30**:80–5.

[2] Anderson, III P.M. 'Magnetomechanical coupling, ΔE effect, and permeability in FeSiB and FeNiMoB alloys'. *Journal of Applied Physics*. 1982;**53**(11):8101.

[3] Le Bras Y., Lasheras A., Gutiérrez J., Mazaleyrat F. Greneche J.M. 'A new magneto-elastic resonance based technique to determine magneto-mechanical parameters of amorphous ferromagnetic ribbons'. *The Review of Scientific Instrumentscientific Instruments*. 2013;**84**(4):043904.

[4] Anon, n.d. *Metglas 2826MB Technical Data Sheet*. Available online at: http//metglas.com/products/magnetic_materials/2826mb.asp

[5] Landau L.D., Lifshitz E.M. *Theory of Elasticity*. 3rd ed. New York: Pergamon Press; 1986.

[6] Cheng Z.-Y., Li S., Zhang K., Fu L., Chin B.A. 'Novel magnetostrictive microcantilever and magnetostrictive nanobars for high performance biological detection'. *Advances in Science and Technology*. 2008;**54**:19–28.

[7] Wenzel S.W., White R.M. 'Analytic comparison of the sensitivities of bulk-wave, surface-wave, and flexural plate-wave ultrasonic gravimetric sensors'. *Applied Physics Letters*. 1989;**54**(20):1976–8.

[8] Li S., Orona L., Li Z., Cheng Z.-Y. 'Biosensor based on magnetostrictive microcantilever'. *Applied Physics Letters*. 2006;**88**(7):73507.

[9] Fu L., Li S., Zhang K., Chen I.-H., Petrenko V.A., Cheng Z.-Y. 'Magnetostrictive microcantilever as an advanced transducer for biosensors'. *Sensors*. 2007;**7**(11):2929–41.

[10] Yi J.W., Shih W.Y., Shih W.-H. 'Effect of length, width, and mode on the mass detection sensitivity of piezoelectric unimorph cantilevers'. *Journal of Applied Physics*. 2002;**91**(3):1680.

[11] Thundat T., Oden P.I., Warmack R.J. 'Microcantilever sensors'. *Microscale Thermophysical Engineering*. 1997;**1**(3):185–99.

[12] Du Trémolet de Lacheisserie E., Gignoux D., Schlenker M. *Magnetism: Materials & Applications*. New York: Springer-Verlag; 2005.

[13] Wan J., Shu H., Huang S., *et al.* 'Phage-based magnetoelastic wireless biosensors for detecting *Bacillus anthracis* spores'. *IEEE Sensors Journal*. 2007;**7**(3):470–7.

[14] Lakshmanan R.S., Guntupalli R., Hu J., *et al.* 'Phage immobilized magnetoelastic sensor for the detection of *Salmonella typhimurium*'. *Journal of Microbiological Methods*. 2007;**71**(1):55–60.

[15] Guntupalli R., Hu J., Lakshmanan R.S., Huang T.S., Barbaree J.M., Chin B.A. 'A magnetoelastic resonance biosensor immobilized with polyclonal antibody for the detection of *Salmonella typhimurium*'. *Biosensors & Bioelectronics*. 2007;**22**(7):1474–9.

[16] Huang S., Hu J., Wan J., Johnson M.L., Shu H., Chin B.A. 'The effect of annealing and gold deposition on the performance of magnetoelastic biosensors'. *Materials Science and Engineering: C*. 2008;**28**(3):380–6.

[17] Johnson M.L., Odum L., Yoon S.H., *et al.* 'Dual-cathode method for sputtering magnetoelastic iron-boron films'. *Vacuum*. 2009;**83**(6):958–64.

[18] Johnson M.L., Wan J., Huang S., *et al.* 'A wireless biosensor using microfabricated phage-interfaced magnetoelastic particles'. *Sensors and Actuators A: Physical*. 2008;**144**(1):38–47.

[19] Clark A.E., Restorff J.B., Wun-Fogle M., Lograsso T.A., Schlagel D.L. 'Magnetostrictive properties of body-centered cubic Fe–Ga and Fe–Ga–Al alloys'. *IEEE Transactions on Magnetics*. 2000;**36**(5):3238–40.

[20] Ivnitski D., Abdel-Hamid I., Atanasov P., Wilkins E. Biosensors for detection of pathogenic bacteria. *Biosensors & Bioelectronics*. 1999;**14**(7):599–624.

[21] Petrenko V.A., Smith G.P., Gong X., Quinn T. 'A library of organic landscapes on filamentous phage'. *Protein Engineering*. 1996;**9**(9):797–801.

[22] Smith G.P., Petrenko V.A. 'Phage display'. *Chemical Reviews*. 1997;**97**(2):391–410.

[23] Smith G.P., Petrenko V.A., Matthews L.J. 'Cross-linked filamentous phage as an affinity matrix'. *Journal of Immunological Methods*. 1998;**215**(1–2):151–61.

[24] Petrenko V.A., Vodyanoy V.J. 'Phage display for detection of biological threat agents'. *Journal of Microbiological Methods*. 2013;**53**(2):253–62.

[25] Petrenko V.A., Sorokulova I.B. 'Detection of biological threats: A challenge for directed molecular evolution'. *Journal of Microbiological Methods*. 2004;**58**(2):147–68.

[26] Sorokulova I.B., Olsen E.V., Chen I.-H., *et al.* 'Landscape phage probes for *Salmonella typhimurium*'. *Journal of Microbiological Methods*. 2007;**63**(1):55–72.

[27] Brigati J.R., Petrenko V.A. 'Thermostability of landscape phage probes'. *Analytical and Bioanalytical Chemistry*. 2005;**382**(6):1346–50.

[28] Hiremath N., Guntupalli R., Vodyanoy V.J., Chin B.A., Park M.-K. 'Detection of methicillin-resistant *Staphylococcus aureus* using novel lytic phage-based magnetoelastic biosensors'. *Sensors and Actuators B: Chemical*. 2015;**210**:129–36.

[29] Petrenko V.A. 'Evolution of phage display: From bioactive peptides to bioselective nanomaterials'. *Expert Opinion on Drug Delivery*. 2008;**5**(8):825–36.

[30] Olofsson L., Ankarloo J., Nicholls I.A. 'Phage viability in organic media: Insights into phage stability'. *Journal of Molecular Recognition*. 1998;**11**(1–6):91–93.

[31] Olofsson L., Ankarloo J., Andersson P.O., Nicholls I.A. 'Filamentous bacteriophage stability in non-aqueous media'. *Chemistry & Biology*. 2001;**8**(7):661–71.

[32] Van der Linden R.H.J., Frenken L.G.J., de Geus B., *et al.* 'Comparison of physical chemical properties of llama VHH antibody fragments and mouse monoclonal antibodies'. *Biochimica et Biophysica Acta, Protein Structure and Molecular Enzymology*. 1999;**1431**(1):37–46.

[33] Usami A., Ohtsu A., Takahama S., Fujii T. 'The effect of pH, hydrogen peroxide and temperature on the stability of human monoclonal antibody'. *Journal of Pharmaceutical and Biomedical Analysis*. 1996;**14**(8–10):1133–40.

[34] Dooley H., Grant S.D., Harris W.J., Porter A.J. 'Stabilization of antibody fragments in adverse environments'. *Biotechnology and Applied Biochemistry*. 1998;**28**:77–83.

[35] Guntupalli R., Lakshmanan R.S., Wan J., *et al.* 'Analytical performance and characterization of antibody immobilized magnetoelastic biosensors'. *Sensing and Instrumentation for Food Quality and Safety*. 2007;**2**(1):27–33.

[36] Lakshmanan R.S. *Phage-based Magnetoelastic Sensor for the Detection of Salmonella Typhimurium*. Ph.D. Thesis. Auburn: Auburn University; 2008.

[37] Lakshmanan R.S., Guntupalli R., Hu J., Petrenko V.A., Barbaree J.M., Chin B.A. 'Detection of *Salmonella typhimurium* in fat free milk using a phage immobilized magnetoelastic sensor'. *Sensors and Actuators B: Chemical*. 2007;**126**(2):544–50.

[38] Li S., Li Y., Chen H., *et al.* 'Direct detection of *Salmonella typhimurium* on fresh produce using phage-based magnetoelastic biosensors'. *Biosensors & Bioelectronics*. 2010;**26**(4):1313–19.

[39] Shen W., Mathison L.C., Petrenko V.A., Chin B.A. 'Design and characterization of a magnetoelastic sensor for the detection of biological agents'. *Journal of Physics D: Applied Physics*. 2010;**43**(1):015004.

[40] Shen W., Mathison L.C., Petrenko V.A., Chin B.A. 'A pulse system for spectrum analysis of magnetoelastic biosensors'. *Applied Physics Letters*. 2010;**96**(16):163502.

[41] Shen W., Zhang Z., Horikawa S., *et al.* 'Time domain characterization of magnetoelastic sensors: A pulse method for resonance frequency determination'. *Review of Scientific Instruments*. 2010;**81**(8):084702.

[42] Xie H., Chai Y., Horikawa S., Li S., Chin B.A., Wikle III, H.C. 'A pulsed wave excitation system to characterize micron-scale magnetoelastic biosensors'. *Sensors and Actuators A: Physical*. 2014;**205**:143–9.

[43] Zeng K., Ong K.G., Mungle C., Grimes C.A. 'Time domain characterization of oscillating sensors: Application of frequency counting to resonance frequency determination'. *Review of Scientific Instruments*. 2002;**73**(12):4375–80.

[44] Zeng K., Grimes C.A. 'Threshold-crossing counting technique for damping factor determination of resonator sensors'. *Review of Scientific Instruments*. 2004;**75**(12):5257–61.

[45] Chai Y., Horikawa S., Wikle III H.C., Wang Z., Chin B.A. 'Surface-scanning coil detectors for magnetoelastic biosensors: A comparison of planar-spiral and solenoid coils'. *Applied Physics Letters*. 2013;**103**(17):173510.

[46] Chai Y.,Wikle III H.C., Wang Z., *et al.* 'Design of a surface-scanning coil detector for direct bacteria detection on food surfaces using a magnetoelastic biosensor'. *Journal of Applied Physics*. 2013;**114**(10):104504–7.

[47] Chai Y., Horikawa S., Li S., Wikle III H.C., Chin B.A. 'A surface-scanning coil detector for real-time, in-situ detection of bacteria on fresh food surfaces'. *Biosensors & Bioelectronics*. 2013;**50**:311–17.

[48] Lakshmanan R.S., Guntupalli R., Hong J.W., *et al.* 'Selective detection of *Salmonella typhimurium* in the presence of high concentrations of masking bacteria'. *Sensing and Instrumentation for Food Quality and Safety*. 2008;**2**(4):234–9.

Chapter 6

Portable and handheld cell-based biosensors

Spyridon E. Kintzios[1]

6.1 Introduction

Cell-based biorecognition platforms represent a quantum leap in biosensor design due to the inherent advantages of high sensitivity and speed of assay, relative resistance to interference by non-target analytes and low cost. These features are particularly promising for developing high-throughput systems for screening pathogens and toxins. On the other hand, cell-based approaches are presenting scientists and inventors with challenges, mostly associated with the selectivity of the desired response as well as storability. In spite of these limitations, cellular biosensors have already established themselves in several applications and commercial systems. The present review is divided into two parts: First, novel approaches to improving the performance of cell-based sensors are presented. Second, an overview is provided of systems and assays already in use in toxicity testing, the most prominent application area for this particular group of biosensors.

6.2 Designer cells: better than nature?

A bold step towards the increase of the selectivity and specificity of cellular biorecognition elements could be the fabrication of entirely novel cell types, lying on the border between cells of natural origin or genetically engineered ones (albeit in a "traditional" sense) and more or less synthetic constructs, which bear functional traits aberrant from the original genetic and biochemical set-up. The development of these radically new cell types is primarily dependent on the availability of the so-called synthetic gene circuits. These can be used either as a means to achieve an improved way to regulate cell physiology and/or even to create novel biochemical pathways. The development of genetic switches has been boosted by bacteriophage-derived recombinases, which can bind and cut DNA sequences flanked with specific recognition sites and perform AND/OR-type logic functions [1–5]. In practical

[1]School of Food, Biotechnology and Development Director, Laboratory of Cell Technology, Department of Biotechnology, Agricultural University of Athens, Athens, Greece

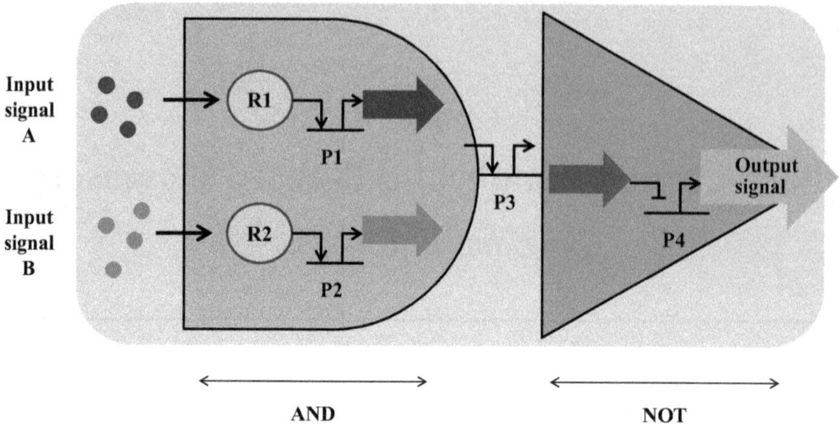

Figure 6.1 Schematic representation of a synthetic gene circuit operating as a pair of logical gates (AND, NOT) in tandem arrangement. R1, R2 = cellular receptors, P1–P4 = promoters (with the product of P3 acting as inhibitor to the expression of P4)

terms, such a genetic switch would operate depending on the level of one or more of the input signals, as shown in Figure 6.1.

The number of customized detection applications based on engineered transcription circuitries is steadily increasing. Weber *et al.* [6] were among the first to report a synthetic mammalian gene circuit that senses the binding of the repressor of the *Mycobacterium tuberculosis* ethionamide-producing (*ethA*) gene to its specific operator in human cells, resulting in quantitative reporter gene expression readout. Since ethionamide is a major factor in the expression of multidrug-resistance against tuberculosis, this approach could offer the possibility to assess the specificity, bioavailability and cytotoxicity of a candidate drug in a single, rapid assay. From a biosensor design perspective, the ultimate goal of synthetic cell technology is the ability to develop cells able to respond to external stimuli in a fully controllable way. Quite recently, Saxena *et al.* [7] reported the design of a synthetic transgenic and genomic network able to regulate the timely control of neurogenin 3 (*Ngn3*), pancreatic and duodenal homeobox 1 (*Pdx1*) and V-maf musculoaponeurotic fibrosarcoma oncogene homologue A (*MafA*) variants in order to program human induced pluripotent stem cells-derived pancreatic progenitor cells into glucose-sensitive insulin-secreting beta-like cells. Even more fascinating is the perspective of constructing miniaturized cell biosensors able to travel into the body, detecting chemical changes possibly associated with target diseased areas. Fundamental work in this direction has been already reported: for example, Kotula *et al.* [8] reported the construction of engineered bacteria designed to sense the presence of the antibiotic tetracycline in the gut. Their approach was based on a bistable switch for memorizing exposure to anhydrotetracycline (aTc). When aTc is present, the repression of a promoter (Ptet) by a repressor (TetR) is abolished, leading to the production of the cro repressor, which in turn allows for the quantitative expression of an output lacZ reporter. Another elegant paradigm is the

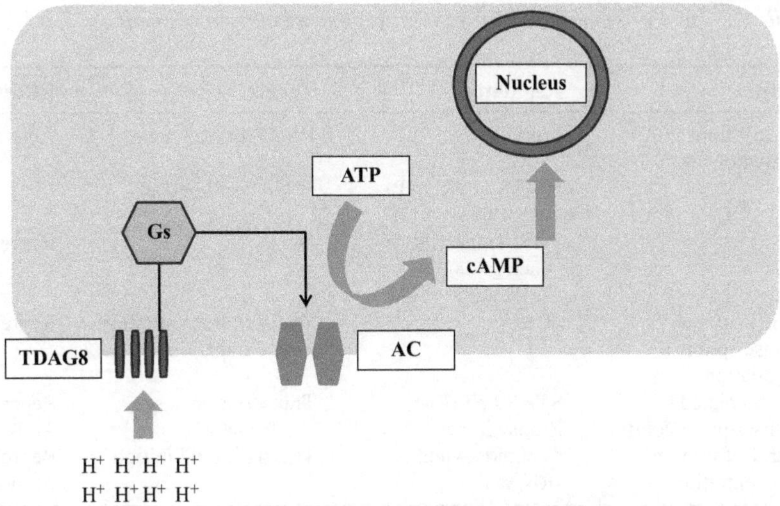

Figure 6.2 *Working principle of the pH biosensor developed by Ausländer et al. [3], such as changes of the extracellular proton concentration lead to the activation of a membrane-bound adenylyl cyclase (AC) coupled to the human proton-activated cell-surface receptor TDAG8 through a specific G protein (Gs) and constitutively expressed in different mammalian cell lines. This in turn leads to the conversion of ATP into the second messenger cyclic AMP (cAMP) and the activation of a cascade of cAMP-dependent responses*

creation of a mammalian cell-based pH sensor [3]. In this case, the human proton-activated cell-surface receptor TDAG8 was coupled to a specific G protein and constitutively expressed in different mammalian cell lines, such as HEK-293, HeLa and HT-1080. In response to a reduction of the extracellular pH value, the receptor construct activated a membrane-bound adenylyl cyclase that converted adenosine triphosphate (ATP) into the second messenger cyclic AMP (cAMP). This resulted in a cascade of cAMP-dependent responses (Figure 6.2).

The successful construction of complex synthetic gene circuits can be limited by the availability of part families characterized by specificity and orthogonality (i.e. interactions along the chain of logical process circuitry components but with minimal cross-talk with irrelevant elements). The fulfilment of these requirements may prove to be quite challenging in scale-up sensing specifications, due to the large number of design processes involved. Naturally, the more complexly regulator is a promoter, the greater the variety of control inputs in a single assay will be. The stability of a synthetic gene circuitry in the host cell is another issue, especially if its maintenance in cloned generations is desired. Tools for circuitry engineering may be DNA-, RNA- or protein based [9–16]. Usually, the core of a synthetic gene circuit will be a transcriptional module comprising a transcription factor protein which will activate or repress a cognate regulator protein. The most commonly used tools are summarized in Table 6.1. The application of synthetic gene circuits in the

Table 6.1 List of some common synthetic gene circuit components

Group	Regulator	Target	Effect
Transcriptional activators	LuxR	RNAP holoenzyme	Activation
	HrpRS (T7RNAP)	(RNAP orthogonal promoters)	
	TetR Transcription activator-like effectors (TALEs)	DNA	Repression
CRISPR-dCas9 transcriptional regulation	dCas9	Transcription initiation or elongation	Repression
RNA-mediated transcription control	STARs pT181 Riboswitches	Transcription termination	Repression Activation
Control of protein concentration	Toehold switches sRNAs	Translation inhibition	Repression Activation

development of novel, highly specific and selective cell-based sensors will be boosted by parallel progress in the fields of synthetic gene assembly and improved cell–sensor interface (in which bioelectrical principles may overtake optical ones, due to greater speed and spatial resolution).

6.3 Cell-based toxicity biosensors

By default, when it comes down to toxicology risk assessment, cells are the biorecognition elements of choice, since they represent the actual targets of any toxic compound. In spite of the differences between in vitro assay systems and in vivo approaches, cell-based biosensors are becoming increasingly popular as reliable, cost-efficient and high-throughput tools for a variety of toxicology tests, including acute toxicity testing, toxicokinetics, genotoxicity and mutagenicity. An excellent review on this topic is provided by Banerjee *et al.* [17].

In principle, cell-based biosensors utilize the principles of cell-based assays. Until recently, the overwhelming majority of CBBs for toxicity testing was based on either prokaryotic or eukaryotic cells engineered with luminescent or fluorescent reporters, in other words optical detection systems. Several reporters have been used in transfected bacterial, yeast and mammalian cells, whereas the choice of the optimal reporter is a determining factor for each cell type × application combination. The use of bacteria is advantageous in respect of increased specificity (due of cross-talk between screened analytes and other biochemical pathways), low cost (especially associated with the facile proliferation of microbial cells) and robustness (cell endurance under storage conditions, even extreme ones). Jouanneau *et al.* [18] have recently published a quite extensive review on this topic. On the other hand, mammalian cells are also attractive as biorecognition elements, since they constitute a more realistic target of toxic effects, in fact providing information more immediate to the analyst's needs [19,20].

A number of widely used commercial toxicity testing kits based on bioluminescent bacteria have been released over the last 20 years, with the most prominent one shown in Table 6.2. The wide distribution of these kits as well as the ongoing development of variants could be ascribed to the availability of an extended range of photoenzymes with high quantum yield emission efficiency, in particular luciferases which are derived from a plethora of species, including *Aliivibrio fischeri, Photinus pyralis, Pyrophorus plagiophthalamus, Photobacterium phosphoreum, Metridia longa* and *Oplophorus gracilirostris* [21–23]. Equally important and broadly utilized is the group of reporters expressing the green fluorescent protein primary derived from the jellyfish species *Aequorea victoria* [24]. Further improvements have been reported regarding miniaturization of read-out devices and increased control of operational parameters [25]. There exists wide variability in sensitivity and speed among the reported assays, depending on the analyte [26,27]. There is also a progress regarding the nature of the luminescent reporters themselves: For example, a cloned luciferase from the sea shrimp species *O. gracilirostris*, in combination with the novel imidazopyrazinone substrate furimazine, has been shown to produce a 150-fold stronger luminescence than that of firefly luciferase. This luciferase species offers the additional advantage of non-requirement of ATP as a substrate for the photochemical reaction. The novel reporter is commercially available under the name Nanoluc luciferase [28]. There are currently at least 11 commercially available luciferase species used in bioluminescence-mediated detection. Almost half of them use D-luciferin as a substrate, requiring ATP as a co-substrate. Other luciferases use another type of substrate. In addition to the aforementioned furimazine (catalysed by Nanoluc), coelenterazine is another light-emitting substrate-target of various luciferases. It is found in many aquatic species, including *A. victoria* and *Gaussia princeps* [29].

Similarly to bacterial cells, mammalian cell-based biosensors have been built upon genetically modified cells to express the receptor-like molecule, thus achieving the desired level of target-specific and selective interaction [30]. The textbook paradigm is the CANARY® (Cellular Analysis and Notification of Antigen Risks and Yields) system [31] in which specific antibodies are expressed on the surface of B cells by stable transfection together with the photoprotein aequorin. Following the binding of the antibody-specific antigen, an intracellular calcium concentration-dependent, aequorin-mediated light emission is observed. Equally famous (and more widely used) is the CALUX® (Chemical Activated Luciferase gene eXpression) system. As its name implies, it is based on the firefly luciferase (*luc*) gene, the expression of which is controlled by human receptors such as aryl hydrocarbon, oestrogen and androgen receptors, all of them being in turn regulated by endocrine-disruptors and dioxin-like molecules [32]. For this reason and having a fairly high sensitivity against oestrogenic compounds, CALUX® has been favourably adopted by several toxicological laboratories around the world. In yet another approach, Mandon *et al.* [33,34] have transfected the mammalian liver cell HepG2 line with the *Drosophila melanogaster* stress-induced *hsp22* promoter in order to construct a cell-based screening test against cadmium chloride and the dithiocarbamate fungicide thiram, with detection limits 0.1 and 0.01 μM, respectively. It is possible also to opt for a more generic approach to developing transfected cells with a broad responsiveness against xenobiotic compounds. Very recently, Evangelia Flampouri (personal communication) at the Laboratory of Cell Technology, AUA, constructed

Table 6.2 Representative commercial toxicology biosensors based on bioluminescent cells

Commercial name	Manufacturer	Reporter system	Target analyte(s)/pathogen(s) Target application(s)
Milliflex® Rapid Testing	Merck-Millipore	ATP-dependent luciferin-luciferase system (ALLS)	*Saccharomyces cerevisiae* *Pseudomonas aeruginosa* *Methylobacterium mesophilicum* *Escherichia coli* *Burkholderia cepacia* *Staphylococcus epidermidis* *Ralstonia pickettii*
DeltaTox® II MicroTox®M500	Modern Water	ALLS (*Vibrio fischeri*)	A range of 2,700 different simple and complex chemicals (including heavy metals, pathogenic toxins and chlorinated organic compounds)
Mutatox™	Modern Water	ALLS (*Photobacterium phosphoreum* M169)	Mutagenic agents
Clean-Trace™	3M	ALLS	A wide range of pathogenic microorganisms
EnSURE™ SystemSURE Plus™ AquaSnap™ WaterShot™ MicroSnap™	Hygiena	ALLS	A wide range of pathogenic microorganisms
TOXmini® i TOXc o n t r o l® BACTcontrol®	microLAN	ALLS (*Vibrio fischeri*)	General toxicity Wastewater toxicity Bacterial pollution monitoring
novaLUM II	Charm Sciences	ALLS	Allergens Water safety screening A wide range of pathogenic microorganisms
ToxAlert 10®	Merck	ALLS (*Vibrio fischeri*)	A range of ~1,400 different simple and complex chemicals
(Dioxin-Responsive) DR-CALUX®	BioDetection Systems	Rat (H4IIE) or mouse hepatoma cell lines, stably transfected with the luciferase reporter gene plasmid pGudLuc1.1	Dioxins and dioxin-like compounds
Accupoint®	Neogen	ALLS	A wide range of pathogenic microorganisms
MVP Icon®	BioControl Systems	ALLS	A wide range of pathogenic microorganisms

a cell-based biosensor based on the culture of mammalian cells on the surface of the biocompatible and conductive polymer poly(3,4-ethylenedioxythiophene) (PEDOT). Cells were subjected to post-translational modifications silencing transcription factor Nrf2 in a successful attempt to improve the sensitivity of the biosensor by controlling the expression of their detoxifying and antioxidant genes.

However, the use of genetically modified cells has a number of series of considerable limitations, such as

1. Transfected cells express receptor molecules able to interact with particular substances under determination in an unknown sample. In the majority of cases (especially when cell types other than B-cells are used, e.g. HEK-293), the expressed receptors are intracellular. This presents a challenge to the use of the transfected cells as detectors, due to the following limitations:
   i. The target compound must enter the cell, which may be not possible or may take considerable time (hours to days), necessitating the prior co-incubation of transfected cells with the sample.
   ii. The target compound may be metabolized inside the cell.
   iii. The concentration of the target compound that will reach the intracellular receptors is different from its actual concentration in the sample, due to diffusion, binding to cellular components etc.

2. Expression of receptors on the cell surface is not always feasible, whereas the co-expression of a membrane-spanning sequence is usually a requirement. This limits the scope of receptor-like molecules which can be located on the surface of genetically transfected cells. On the contrary, not every cell line is amenable to transfection. In fact, only a limited number of cells lines can be used for this purpose, e.g. HEK293T or COS7. These cells, while typically used as models, have limited relevance to physiological systems [35]. Moreover, in most transfected cell-related applications, it is always required to use an optimized luminescent reporter, which is not possible always.
3. Cell transfection is often transient.
4. Cell transfection often results in the over-expression of individual genes, with the probability of producing a biased cellular phenotype.

In spite of these limitations, recent progress in the development of methods to investigate signalling pathways and their phenotypic outputs in living cells is quite remarkable. For example, the concept of functional selectivity relies on G protein-coupled receptor (GPCR) ligands able to activate distinct but parallel signalling cascades (biased signalling) [36]. This approach is well suited for label-free assay techniques, mainly surface plasmon resonance (SPR) spectroscopy, whereas a number of commercial systems are already available [37]. Depending on the nature of the receptors, the measured reflectance value may be reduced (e.g. if cell apoptosis takes place) or increased (due to cell proliferation and/or migration, e.g. following activation of a growth factor receptor) (Figure 6.3). Naturally, a prerequisite for the success of this methodological approach is the prior availability of transfected cells stably expressing high levels of the receptor. In the same sense, it is preferable to establish a homogenous expression of the receptor

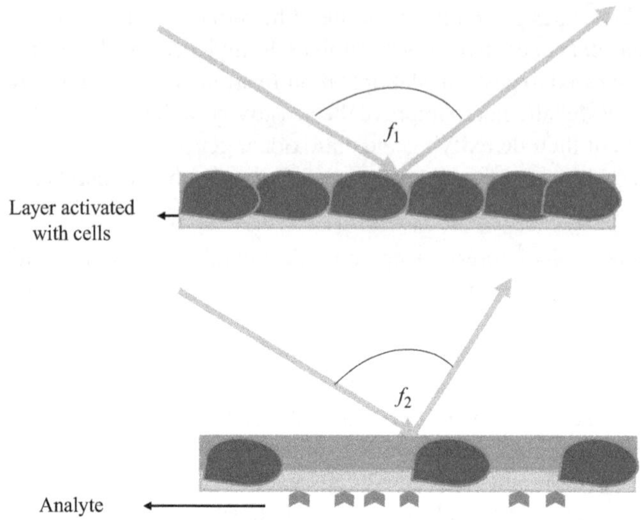

Figure 6.3 Application of SPR spectroscopy to the study of selective interaction between G protein-coupled receptor and agonistic/antagonistic ligands (analytes). In the example depicted in the figure, receptor activation leads to cell apoptosis, thus reducing the cell density on the reflectance layer and the value of the measured reflectance

in the assayed cell population; otherwise, the percentage of non-transfected cells will lower respectively the overall response to a ligand. The use of GPCR ligand-mediated signalling as a biosensor principle has been substantially upgraded by the recent of custom-triggered luminescent reporters [38], leading to the emergence of commercial platforms (e.g. GloSensor™, Promega) able to deliver responses in a few minutes.

In parallel, cell transfection protocols are being constantly improved. One of the latest developments in this field is the so-called reverse transfection, which allows the creation of multiple cell phenotypes based on a single cell line on the same chip [39]. Newest alternatives to cell transfection also include invasive dye reagents that measure changes in microbial metabolic activities as a result of redox reactions, in particular those involving cellular respiration components such as NADH and NADPH [40,41]. Representative examples of such dyes are potassium ferricyanide, 2-(*p*-iodophenyl)-3-(*p*-nitrophenyl)-5-phenyl tetrazolium chloride and resazurin-based compounds.

An emerging technique in cell-based electrochemical biosensors is the measurement of exocytotic events in response to selective cell–analyte interactions. Ground work on this topic has been conducted by Simonsson *et al.* [42], who developed an artificial secretory cell, containing secretory vehicles having target-*N*-ethylmaleimide-sensitive factor attachment protein receptor (t-SNARE) proteins substituted for target-cholesterol–DNA strands. Exocytosis ensued as a result of the hybridization of the DNA constructs in a zipper-like fashion to bring about docking of the artificial secretory vesicles, combined with the addition of Ca^{2+}. Vesicular release was measured amperometrically. More recently, Jiang *et al.* [43] reported the development of a novel biosensor based on rat basophilic leukaemia cells (RBL-2H3) transfected with lipidosome/SiO_2 shell-coated Fe_3O_4 magnetic nanoparticles tagged with fluorescein isothiocyanate, provisionally designated as cationic magnetic fluorescent nanoparticles (CMFNPs). RBL-2H3 cells express the FcεR receptor, which can selectively bind murine IgE antibodies and trigger a cascade of exocytotic responses, thus offering a suitable platform for allergen detection. Therefore, the presence of an allergen in the assayed sample will trigger the exudation of a percentage of CMFNPs via degranulation, a process which can be measured by the decrease in cell magnetism and absorption by a magnetic electrode. A variation of this approach was recently reported by the same research group [44] in which cell impedance changes were detected using a cell-based electrochemical assay on a microfluidic chip.

Compared to prokaryotes, kits based on mammalian cells are evolving into more flexible assay platforms, and frequently utilizing assay principles other than optical. For example, many cell-based methods are centred on the measurement of bioelectric properties. A very popular approach is the application of electric impedance spectroscopy (EIS) on cultured cells and the indirect assessment of toxicity through the measurement of impedance reduction, as a result of cells dying in response to the toxic effect. Although EIS has been mainly used for studying the toxicity of chemical residues (such as pesticides and hydrocarbons), it has been recently extended to cover biotoxins, for example mycotoxins such as deoxynivalenol and zearalenone using human hepatocellular carcinoma BEL-7402 cells as the biosensory elements [45]. Many commercial cell-based EIS systems are available nowadays, although the cost of their acquisition is rather considerable. A major advantage of EIS is its essentially label-free character, as well as its high-throughput capacity. On the downside, EIS does not represent a significantly faster assay type, since it is necessary to incubate the cells with the sample for at least several hours to (most usually) a whole day, thus rendering the assay time comparable to conventional in vitro toxicological protocols. A few reports exist on the electrochemical assessment of toxicity on mammalian cells. A now established method is the Bioelectric Recognition Assay (BERA), which is based on the measurement of changes of the cell membrane potential as a result of the cells interaction with bioactive compounds [46–49]. In the last years, the specificity and selectivity of BERA-based biosensors have been considerably increased by using membrane-engineered cells with electroinserted antibodies or other receptor-like molecules on their surface as biorecognition elements. This combinatory approach has been used for the detection of human viruses (hepatitis B and C viruses) [50], plant viruses [51,52], superoxide [53–55], prions, foot-and-mouth disease virus, blue-tongue virus, pesticide residues and 2,4,6-trichloroanisole [56,57]. Quite recently, Mavrikou *et al.* [58] have reported the development of a novel biosensor for the detection of aflatoxin B1 (AFB1), which is based on cyclic voltametric and chronoamperometric measurements on Vero cells membrane-engineered with anti-AFB1 antibodies reacting with AFB1 molecules on gold nanoparticle/screen printed electrodes. The biosensor displayed a high sensitivity (with a detection limit of 0.5 ng mL^{-1}).

While microorganisms are used as biosensory elements for screening more or less general toxicants, such as heavy metals and aromatic hydrocarbons, mammalian cells are preferably employed for the detection of specialized agents, including bacterial and fungal toxins. For example, Banerjee *et al.* [59] developed a murine hybridoma cell-based assay for listeriolysin O produced by the foodborne pathogen *Listeria monocytogenes*. The same research group extended the biosensor analytical scope to include several other toxins, including crude toxin preparations from *Bacillus cereus* A926, α-haemolysin from *Staphylococcus aureus*, phospholipase C from *Clostridium perfringens* and cytolysin from *Stoichactis helianthus* with a limit of detection (LOD) of 10 ng g^{-1} [60]. The only prerequisite was that the assayed toxins have a cytolytic and/or haemolytic property.

Miniaturized cell microarrays, in particular microarrays of reverse-transfected cells with plasmid DNA or siRNA spotted on the surface of a substrate, have opened exciting perspectives in high-throughput pharmacological and toxicological research (an excellent review on this subject is given by Berthuy *et al.* [39]). Doubtless, the development of operational cell microarrays presents the researcher with more challenges than their gene or protein counterparts.

As with many other classes of biosensor technologies and inventions, toxicity sensors based on engineered prokaryotic or eukaryotic cells have a long way to go before they become a part of everyday analytical practice. In fact, even though the level of technological sophistication increases with every new wave of scientific publications, it is rather doubtful than more than a handful of reported methods will ever turn into commercial products. There are many reasons for this [61], the main culprits being inadequate validation trials with real samples, lack of cost-efficiency considerations and a low Technology Readiness Level, which usually does not exceed four, that is too low to get attraction from corporate developers. Even so, one can only view as positive the fact that a number of commercial systems have been already in use, some of them for a relatively long time (e.g. DeltaTox® II, CALUX®) (Table 6.2) [62–66]. The majority of the products are based on ATP-dependent, luciferase-luciferin associated bioluminescence. Prices for read-out devices range from $1,500 to $5,000, whereas the consumable kits are fairly economic, usually at $2–$3 per test. The LOD for toxic chemic compounds and for pathogenic microorganisms can be as low as 0.001 ppm and 1 CFU mL^{-1}, respectively.

The use of mammalian cells in biosensor platforms, in particular those equipped with optical instrumentation, such as SPR and coupled with immunoassay and/or microfluidic configurations may provide some specific challenges. For example, binding kinetics between cells and immobilized antibodies is a complex due to the presence of the multiple antigens on the cell surface. Also, the control of cell localization in a microfluidic system is not always optimal. Having this restriction in mind, Schasfoort *et al.* [67] have developed a biosensor based on red blood cells (RBCs) immobilized via glycophorin antibodies on planar carboxy-modified gold layer SPR sensor. In this context, they took advantage of the abundance of glycophorin on the membrane of RBCs. In this way, they were able to build a label-free, reusable (up to 100 times) optical biosensor for RBC typing, an approach which could be used modified for other cell types, too, thus providing an interesting alternative to EIS.

Even simpler approaches can be found with bacteria-based sensors for assessing biochemical oxygen demand. For example, it is possible to measure directly and rapidly (10 min) the oxygen concentration in a sample, by using an electrochemical or optical probe as a result of O_2 consumption by gel-immobilized bacteria [68]. Although this technique has been known for several decades, its commercial application has been limited by the non-favourable growth of immobilized bacteria causing lack of test reproducibility and consistency over time [69].

Mammalian cell immobilization or encapsulation in an appropriate matrix remains the easiest and by far the most feasible approach to extend cell viability and, thereof, the storability of the consumable part of the associated cell-based biosensor system. The relationship between cellular physiology and the physicochemical properties of the immobilization matrix has been established and investigated in numerous reports (e.g. see Kintzios *et al.* [70] for the effect of gel shape on selected biochemical parameters of the immobilized cells). More recently, Nunes *et al.* [71] and Miklas *et al.* [72] developed an advanced biosensor platform using human pluripotent stem cell-derived cardiomyocytes. Following electric stimulation, 3D aligned cardiac tissues (biowires) with markedly increased myofibril ultrastructural organization, elevated conduction velocity and improved both electrophysiological and Ca^{2+} handling properties were formed. The advantages of encapsulated cell-based systems have been demonstrated in very recent studies, as shown, for example, by Schukur *et al.* [73] for designer white blood cells immobilized in an alginate-(poly-L-lysine)-alginate gel using a commercial encapsulation device. Successful encapsulation techniques have been reported with bacteria, as well: recently, Jouanneau *et al.* [22] reported the development of a bioluminescent biosensor for naphthalene screening based on the wild marine bioluminescent bacterial strain *A. (Vibrio) fischeri* ATCC®49387™ entrapped in an agarose matrix within a disposable card. With the advent of bioprinting technology [74], 3D cell cultures can be constructed with high special precision and using a variety of materials resembling the cellular microenvironment in vitro. For example, components of the extracellular matrix can be combined with synthetic nanomaterials, such as carbon nanotubules and peptide hydrogels to provide a cell growth substrate with improved biosensing capacities due to its increased conductivity and controllable porosity [75]. At the same time, advances in 3D printing accuracy have allowed the reproducible fabrication of immobilized cell arrays with a special resolution of 50 μM or less [76]. Miniaturization of the actual detector unit, comprising the cells and the immobilization matrix, should reduce the unfavourable effects associated with the limitations in mass transfer as well as matrix erosion over time, leading to reduced reproducibility of measurements [70,77]. Newest versions of 3D culture-based biosensor platforms are built upon so-called organoids comprising cell layers of different origin and type, as well as semi-artificial vascular networks, thus closely resembling morphological and functional characteristics of the actual target organ. A recent example is reported by Rennert *et al.* [78] with the development of a liver organoid comprising non-parenchymal cells, endothelial cells, tissue macrophages and stellate cells co-cultured with hepatocytes. The organoid was incorporated in a microfluidically perfused biochip (ChipShop, Germany) equipped with oxygen sensors and compatible with various fluorescence assays.

Further progress in cell encapsulation may arise from the use of novel biomaterials such as silk proteins [79] and thermoresponsive biodegradable polyurethane hydrogels [80]. Advances in immobilization techniques may provide a solution to the long-standing issue of limited storability of cells as biorecognition elements in commercial biosensor settings. Currently, the use of cell-based devices as portable point-of-use analytical tools is limited due to the requirement of special cell preservation methods and dedicated cell culture infrastructure [20]. This is particularly the case with mammalian cells, whereas prokaryotes are eligible for a wider variety of cell storage techniques, for example freeze-drying [81]. Other approaches towards increasing cell storability include the use of customized culture media [82] or even the employment of novel cell lines with increased robustness, such as rainbow trout gill epithelial cells (RTgill-W1) [83,84].

References

[1] Wang B., Buck M. 'Customizing cell signalling using engineered genetic logic circuits'. *Trends in Microbiology*. 2012;**20**(8):376–84.

[2] Bonnet J., Yin P., Ortiz M.E., Subsoontorn P., Endy D. 'Amplifying genetic logic gates'. *Science*. 2013;**340**:599.

[3] Ausländer D., Ausländer S., Charpin-El Hamri G., *et al*. 'A synthetic multifunctional mammalian pH sensor and CO_2 transgene-control device'. *Molecular Cell*. 2014;**55**(3):397–408.

[4] Ausländer S., Fussenegger M. 'Synthetic biology: dynamic genome engineering in living cells'. *Science*. 2014;**346**(6211):813–14.

[5] Ausländer S., Fussenegger M. 'Engineering gene circuits for mammalian cell-based applications'. *Cold Spring Harbor Perspectives in Biology*. 2016; pii: a023895 (advanced online article).

[6] Weber W., Schoenmakers R., Keller B., *et al*. 'A synthetic mammalian gene circuit reveals antituberculosis compounds'. *Proceedings of the National Academy of Sciences of the United States of America*. 2008;**105**(29):9994–8.

[7] Saxena P., Heng B.C., Bai P., Folcher M., Zulewski H., Fussenegger M. 'A programmable synthetic lineage-control network that differentiates human IPSCs into glucose-sensitive insulin-secreting beta-like cells'. *Nature Communications*. 2016;**7**:11247.

[8] Kotula J.W., Kerns S.J., Shaket L.A., *et al*. 'Programmable bacteria detect and record an environmental signal in the mammalian gut'. *Proceedings of the National Academy of Sciences of the United States of America*. 2014;**111**:4838–43.

[9] Bikard D., Jiang W., Samai P., Hochschild A., Zhang F., Marraffini L.A. 'Programmable repression and activation of bacterial gene expression using an engineered CRISPRC as system'. *Nucleic Acids Research*. 2013;**41**:7429–37.

[10] Na D., Yoo S.M., Chung H., Park H., Park J.H., Lee S.Y. 'Metabolic engineering of *Escherichia coli* using synthetic small regulatory RNAs'. *Nature Biotechnology*. 2013;**31**:170–4.

[11] Politz M.C., Copeland M.F., Pfleger B.F. 'Artificial repressors for controlling gene expression in bacteria'. *Chemical Communications*. 2013;**49**:4325–7.

[12] Green A.A., Silver P.A., Collins J.J., Yin P. 'Toehold switches: de-novo-designed regulators of gene expression'. *Cell*. 2014;**159**:925–39.

[13] Krajewski S.S., Narberhaus F. 'Temperature-driven differential gene expression by RNA thermosensors'. *Biochimica et Biophysica Acta*. 2014;**1839**(10):978–88.

[14] Bradley R.W., Wang B. 'Designer cell signal processing circuits for biotechnology'. *New Biotechnology*. 2015;**32**(6):635–43.

[15] Chappell J., Takahashi M.K., Lucks J.B. 'Creating small transcription activating RNAs'. *Nature Chemical Biology*. 2015;**11**:1–9.

[16] Bradley R.W., Buck M., Wang B. 'Tools and principles for microbial gene circuit engineering'. *Journal of Molecular Biology*. 2016;**428**(5 Pt B):862–88.

[17] Banerjee P., Kintzios S., Prabhakarpandian B. 'Toxin detection using cell-based sensors'. *Toxins*. 2013;**5**:2366–83.

[18] Jouanneau S., Recoules L., Durand M.J., *et al*. 'Methods for assessing biochemical oxygen demand (BOD): a review'. *Water Research*. 2014; **49**:62–82.

[19] Roda A., Mirasoli M., Michelini E., *et al*. 'Progress in chemical luminescence-based biosensors: a critical review'. *Biosensors and Bioelectronics*. 2016;**76**:164–79.

[20] Kintzios S., Banerjee P. 'Mammalian cell-based sensors in food safety'. In: Bhunja A.K., Kim M.S., Taitt C.R. (eds.). *High Throughput Screening for Food Safety Assessment: Biosensor Technologies, Hyperspectral Imaging and Practical Applications*. Amsterdam: Woodhead; 2014. pp. 123–46.

[21] Sakaguchi T., Morioka Y., Yamasaki M., *et al*. 'Rapid and onsite BOD sensing system using luminous bacterial cells-immobilized chip'. *Biosensors and Bioelectronics*. 2007;**22**(7):1345–50.

[22] Jouanneau S., Durand-Thouand M.J., Thouand G. 'Design of a toxicity biosensor based on *Aliivibrio fischeri* entrapped in a disposable card'. *Environmental Science and Pollution Research International*. 2016; **23**:4340–5.

[23] Kokkali V., van Delft W. 'Overview of commercially available bioassays for assessing chemical toxicity in aqueous samples'. *TrAC Trends in Analytical Chemistry*. 2014;**61**:133–55.

[24] Walker C.L., Lukyanov K.A., Yampolsky I.V., *et al*. 'Fluorescence imaging using synthetic GFP chromophores'. *Current Opinion in Chemical Biology*. 2015;**27**:64–74.

[25] Tsai H.F., Tsai Y.C., Yagur-Kroll S., Palevsky N., Belkin S., Cheng J.Y. 'Water pollutant monitoring by a whole cell array through lens-free detection on CCD'. *Lab on a Chip*. 2015;**15**:1472–80.

[26] Charrier T., Durand M.J., Jouanneau S., *et al.* 'A multi-channel bioluminescent bacterial biosensor for the on-line detection of metals and toxicity. Part I: design and optimization of bioluminescent bacterial strains'. *Analytical and Bioanalytical Chemistry*. 2011;**400**:1051–60.

[27] Charrier T., Chapeau C., Bendria L., Picart P., Daniel P., Thouand G. 'A multi-channel bioluminescent bacterial biosensor for the on-line detection of metals and toxicity. Part II: technical development and proof of concept of the biosensor'. *Analytical and Bioanalytical Chemistry*. 2011;**400**:1061–70.

[28] Hall M.P., Unch J., Binkowski B.F., *et al.* 'Engineered luciferase reporter from a deep sea shrimp utilizing a novel imidazopyrazinone substrate'. *ACS Chemical Biology*. 2012;**7**(11):1848–57.

[29] Vece V., Vuocolo G. 'Multicomponent synthesis of novel coelenterazine derivatives substituted at the C-3 position'. *Tetrahedron*. 2015;**71**:8781–5.

[30] Yoshikawa T., Uchimura E., Kishi M., Funeriu D.P., Miyake M., Miyake J. 'Transfection microarray of human mesenchymal stem cells and on-chip siRNA gene knockdown'. *Journal of Controlled Release*. 2004;**96**(2): 227–32.

[31] Rider T.H., Petrovick M.S., Nargi F.E., *et al.* 'A B cell-based sensor for rapid identification of pathogens'. *Science*. 2003;**301**:213–15.

[32] Sonneveld E., Jansen H.J., Riteco J.A., Brouwer A., van der Burg B. 'Development of androgen- and estrogen-responsive bioassays, members of a panel of human cell line-based highly selective steroid-responsive bioassays'. *Toxicological Sciences*. 2005;**83**:136–48.

[33] Mandon C.A., Diaz C., Arrigo A.P., Blum L.J. 'Chemical stress sensitive luminescent human cells: molecular biology approach using inducible *Drosophila melanogaster* hsp22 promoter'. *Biochemical and Biophysical Research Communications*. 2005;**335**(2):536–44.

[34] Mandon C.A., Diaz-Latoud C., Arrigo A.P., Blum L.J. 'Dithiocarbamate fungicide thiram detection: comparison of bioluminescent and fluorescent whole-cell bioassays based on hsp22 stress promoter induction'. *Journal of Biotechnology*. 2006;**124**:392–402.

[35] Michiels F., van Es H., van Rompaey L., *et al.* 'Arrayed adenoviral expression libraries for functional screening'. *Nature Biotechnology*. 2002;**20**(11):1154–7.

[36] Urban J.D., Clarke W.P., von Zastrow M., *et al.* 'Functional selectivity and classical concepts of quantitative pharmacology'. *The Journal of Pharmacology and Experimental Therapeutics*. 2007;**320**(1):1–13.

[37] Bourassa P., Söllradl T., Maltais J.S., Charette P.G., Gendron L., Grandbois M. 'Surface plasmon resonance to study cell signaling and GPCR functional selectivity in live cells'. In: Fang Y. (ed.). *Label-Free Biosensor Methods in Drug Discovery*. New York: Humana Press-Springer; 2015. pp. 183–95.

[38] Robinson K.H., Yang J.R., Zhang J. 'FRET and BRET-based biosensors in live cell compound screens'. *Methods in Molecular Biology*. 2014;**1071**:217–25.

[39] Berthuy O.I., Blum L.J., Marquette C.A. 'Cells on chip for multiplex screening'. *Biosensors and Bioelectronics*. 2016;**76**:29–37.

[40] Potapova I., Eglin D., Laschke M.W., Bischoff M., Geoff Richards R., Fintan Moriarty T. 'Two-step labelling of *Staphylococcus aureus* with Lysostaphin-Azide and DIBO-Alexa using chemistry'. *Journal of Microbiological Methods*. 2013;**92**(1):90–8.

[41] Cregut M., Jouanneau S., Brillet F., *et al*. 'High throughput and miniaturised systems for biodegradability assessments'. *Environmental Science and Pollution Research International*. 2014;**21**:9545–52.

[42] Simonsson L., Kurczy M.E., Trouillon R., Hook F., Cans A.S. 'A functioning artificial secretory cell'. *Scientific Reports*. 2012;**2**:824.

[43] Jiang D., Zhu P., Jiang H., *et al*. 'Fluorescent magnetic bead-based mast cell biosensor for electrochemical detection of allergens in foodstuffs'. *Biosensors and Bioelectronics*. 2015;**70**:482–90.

[44] Jiang H., Jiang D., Zhu P., *et al*. 'A novel mast cell co-culture microfluidic chip for the electrochemical evaluation of food allergen'. *Biosensors and Bioelectronics*. 2016;**83**:126–33.

[45] Gu W., Zhu P., Jiang D., *et al*. 'A novel and simple cell-based electrochemical impedance biosensor for evaluating the combined toxicity of DON and ZEN'. *Biosensors and Bioelectronics*. 2015;**70**:447–54.

[46] Kintzios S., Pistola E., Konstas J., *et al*. 'Application of the bioelectric recognition assay (BERA) for the detection of human and plant viruses: definition of operational parameters'. *Biosensors and Bioelectronics*. 2001;**16**:467–80.

[47] Kintzios S., Pistola E., Panagiotopoulos P., *et al*. 'Bioelectric recognition assay (BERA)'. *Biosensors and Bioelectronics*. 2001;**16**:325–36.

[48] Kintzios S., Bem F., Mangana O., *et al*. 'Study on the mechanism of bioelectric recognition assay: evidence for immobilized cell membrane interactions with viral fragments'. *Biosensors and Bioelectronics*. 2004;**20**:907–16.

[49] Kintzios S., Papageorgiou K., Yiakoumettis I., Baričevič D., Kušar A. 'Evaluation of the antioxidants activities of four Slovene medicinal plant species by traditional and novel biosensory assays'. *Journal of Pharmaceutical and Biomedical Analysis*. 2010;**53**:773–6.

[50] Perdikaris A., Alexandropoulos N., Kintzios S. 'Development of a novel, ultra-rapid biosensor for the qualitative detection of hepatitis B virus-associated antigens and anti-HBV, based on "Membrane-engineered" fibroblast cells with virus-specific antibodies and antigens'. *Sensors*. 2009;**9**:2176–86.

[51] Moschopoulou G., Vitsa K., Bem F., *et al*. 'Engineering of the membrane of fibroblast cells with virus-specific antibodies: a novel biosensor tool for virus detection'. *Biosensors and Bioelectronics*. 2008;**24**:1033–6.

[52] Perdikaris A., Vassilakos N., Yiakoumettis I., Kektsidou O., Kintzios S. 'Development of a portable, high throughput biosensor system for rapid plant virus detection'. *Journal of Virological Methods*. 2011;**177**:94–9.

[53] Moschopoulou G., Kintzios S. 'Application of "membrane-engineering" to bioelectric recognition cell sensors for the detection of picomole concentrations of superoxide radical: a novel biosensor principle'. *Analytica Chimica Acta*. 2006;**573–574**:90–6.

[54] Moschopoulou G., Valero T., Kintzios S. 'Superoxide determination using membrane-engineered cells: an example of a novel concept for the construction of cell sensors with customized target recognition properties'. *Sensors and Actuators B: Chemical*. 2012;**175**:88–94.

[55] Moschopoulou G., Kintzios S. 'Noninvasive superoxide monitoring of in vitro neuronal differentiation using a cell-based biosensor'. *Journal of Sensors*. 2015;**5**:768352.

[56] Varelas V., Sanvicens N., Marco M.P., Kintzios S. 'Development of a cellular biosensor for the detection of 2,4,6-trichloroanisole (TCA)'. *Talanta*. 2010;**84**:936–40.

[57] Apostolou T., Pascual N., Marco M.P., *et al*. 'Extraction-less, rapid assay for the direct detection of 2,4,6-trichloroanisole (TCA) in cork samples'. *Talanta*. 2014;**125**:336–40.

[58] Mavrikou S., Flampouri E., Iconomou D., Kintzios S. 'Development of a cellular biosensor for the detection of aflatoxin B_1, based on the interaction of membrane engineered Vero cells with anti-AFB_1 antibodies on the surface of gold nanoparticle screen printed electrodes'. *Food Control*. 2016;**73**:64–70.

[59] Banerjee P., Morgan M.T., Rickus J.L., *et al*. 'Hybridoma Ped-2E9 cells cultured under modified conditions can sensitively detect *Listeria monocytogenes* and *Bacillus cereus*'. *Applied Microbiology and Biotechnology*. 2007;**73**:1423–34.

[60] Banerjee P., Bhunia A.K. 'Cell-based biosensor for rapid screening of pathogens and toxins'. *Biosensors and Bioelectronics*. 2010;**26**:99–106.

[61] Kintzios S. 'Biosensors in food safety control: an update'. In: Columbus F. (ed.). *Biosensors: Properties, Materials and Applications*. Hauppauge, NY: Nova Science Publishers; 2009. pp. 53–86.

[62] Kwan K.K., Dutka B.J., Rao S.S., Liu D. 'Mutatox test: a new test for monitoring environmental genotoxic agents'. *Environmental Pollution*. 1990;**65**(4):323–32.

[63] Murk A.J., Legler J., Denison M.S., Giesy J.P., van de Guchte C., Brouwer A. 'Chemical-activated luciferase gene expression (CALUX): a novel in vitro bioassay for Ah receptor active compounds in sediments and pore water'. *Fundamental and Applied Toxicology*. 1996;**33**(1):149–60.

[64] Jennings V.L., Rayner-Brandes M.H., Bird D.J. 'Assessing chemical toxicity with the bioluminescent photobacterium (*Vibrio fischeri*): a comparison of three commercial systems'. *Water Research*. 2001;**35**(14):3448–56.

[65] Moore G., Smyth D., Singleton J., Wilson P. 'The use of adenosine triphosphate bioluminescence to assess the efficacy of a modified cleaning program implemented within an intensive care setting'. *American Journal of Infection Control*. 2010;**38**(8):617–22.

[66] Dudziak M. 'MICROTOX as a tool to evaluate unfavourable phenomenon occurrences during micropollutants decompositions in AOPS'. *Architecture Civil Engineering Environment*. 2015;**2**:85–90.

[67] Schasfoort R.B.M., Bentlage A.E.H., Stojanovic I., *et al.* 'Label-free cell profiling'. *Analytical Biochemistry*. 2013;**439**:4–6.

[68] Karube I. 'Microbial electrode BOD sensors'. *Biotechnology and Bioengineering*. 1977;**19**(10):1535–1547.

[69] Jouanneau S., Durand M.J., Thouand G. 'Online detection of metals in environmental samples: comparing two concepts of bioluminescent bacterial biosensors'. *Environmental Science and Technology*. 2012;**46**(21):11979–87.

[70] Kintzios S., Yiakoumetis I., Moschopoulou G., Mangana O., Nomikou K., Simonian A. 'Differential effect of the shape of calcium alginate matrices on the physiology of immobilized neuroblastoma N2a and Vero cells: a comparative study'. *Biosensors and Bioelectronics*. 2007;**23**:543–8.

[71] Nunes S.S., Miklas J.W., Liu J., *et al.* 'Biowire: a platform for maturation of human pluripotent stem cell-derived cardiomyocytes'. *Nature Methods*. 2013;**10**(8):781–7.

[72] Miklas J.W., Nunes S.S., Zhang B., Radisic M. 'Design and fabrication of biological wires'. *Methods in Molecular Biology*. 2014;**1181**:157–65.

[73] Schukur L., Geering B., Fussenegger M. 'Human whole-blood culture system for ex vivo characterization of designer-cell function'. *Biotechnology and Bioengineering*. 2016;**113**:588–97.

[74] Pati F., Ha D.H., Jang J., Han H.H., Rhie J.W., Cho D.W. 'Biomimetic 3D tissue printing for soft tissue regeneration'. *Biomaterials*. 2015;**62**: 164–75.

[75] Pérez R.A., Won J.E., Knowles J.C., Kim H.W. 'Naturally and synthetic smart composite biomaterials for tissue regeneration'. *Advanced Drug Delivery Reviews*. 2013;**65**:471–96.

[76] Fujita S., Onuki Nagasaki R., Fukuda J., Enomoto J., Yamaguchi S., Miyake M. 'Development of super-dense transfected cell microarrays generated by piezoelectric inkjet printing'. *Lab on a Chip*. 2013;**13**(1):77–80.

[77] Affi M, Solliec C, Legentillomme P, Comiti J., Legrand J., Thouand G. 'Numerical design of a card and related physicochemical phenomena occurring inside agarose-immobilized bacteria: a valuable tool for increasing our knowledge of biosensors'. *Sensors and Actuators B: Chemical*. 2009;**138**:310–17.

[78] Rennert K., Steinborn S., Gröger M., *et al.* 'A microfluidically perfused three dimensional human liver model'. *Biomaterials*. 2015;**71**:119–31.

[79] Thurber A.E., Omenetto F.G., Kaplan D.L. 'In vivo bioresponses to silk proteins'. *Biomaterials*. 2015;**71**:145–57.

[80] Hsieh F.Y., Lin H.H., Hsu S.H. '3D bioprinting of neural stem cell-laden thermoresponsive biodegradable polyurethane hydrogel and potential in central nervous system repair'. *Biomaterials*. 2015;**71**:48–57.

[81] Jouanneau S., Durand M.J., Lahmar A., Thouand G. 'Main technological advancements in bacterial bioluminescent biosensors over the last two decades'. *Advances in Biochemical Engineering/Biotechnology*. 2015;**154**: 101–16.

[82] Harper J.D., Nargi F.E., Petrovick M.S., Rider T.H., Schwoebel E.D. 'Pathogen-detecting cell preservation systems'. *US Patent 20090221056*, 2009.

[83] Lee L.E.J., Dayeh V.R., Schirmer K., Bols N.C. 'Applications and potential uses of fish gill cell lines: examples with RTgill-W1'. *In Vitro Cellular Developmental Biology-Animal* 2009;**45**:127–34.

[84] Brennan L.M., Widder M.W., Lee L.E.J., Van Der Schalie W.H. 'Long-term storage and impedance-based water toxicity testing capabilities of fluidic biochips seeded with RTgill-W1 cells'. *Toxicology In Vitro*. 2012;**26**(5): 736–45.

Part II

Sub-component design and optimization

Chapter 7

Novel nanocomposite materials for miniaturized biosensor fabrication

G. Roussos[1] and N. Chaniotakis[1]

7.1 Introduction

Since the development of the first biosensor by Clark and Lyons [1], these devices have come to be one of the most important analytical tools for the detection and monitoring of chemical and biochemical substances in the biological world. According to International Union of Pure and Applied Chemistry (IUPAC) [2], biosensor is *A device that uses specific biochemical reactions mediated by isolated enzymes, immunosystems, tissues, organelles or whole cells to detect chemical compounds usually by electrical, thermal or optical signals*. Biosensors provide analytical data on many biochemical substances. Such data is very valuable as it can aid significantly in dealing with many human activity scientific issues related to health, environment, food quality and safety. In general, biosensors are indispensable tools in all areas where direct and selective quantitative chemical or biochemical analysis and monitoring are required. Their high selectivity, simplicity in their use, very low detection limits and continuous or online monitoring are indispensable characteristics that no other analytical methodology can provide [3]. Since the initial appearance of biosensor technology, and due to the fact that their potential as detection and monitoring devices was very early realized, there is a great effort on developing reliable biosensing devices. There are indeed a number of biosensors that have succeeded and are on the market, especially for clinical and environmental analyses. Despite though, the great scientific effort devoted for more than 50 years now to a number of analytical and physical issues that prevent their widespread application and commercial success.

A successful biosensor must possess specific analytical features. It must be stable under harsh environmental conditions and independent of physical parameters such as temperature and stirring. Its response should be rapid, accurate, linear and reproducible over the required analytical range. The whole device should be relatively inexpensive to manufacture, portable, user friendly and biocompatible, especially if it is to be used in clinical situations. To this end, it is envisaged

[1]Department of Chemistry, University of Crete, Heraklion, Crete Greece

that nanotechnology can play a vital role in dealing with some of the technical issues for biosensor fabrication.

Nanotechnology allows modern science to utilize nanomaterials, and their unique characteristics to face up with problems and optimize application development. Nanostructures have tunable exceptional chemical and physical properties that are unique and thus they can improve many of the required analytical and physical properties of the modern biosensors. Novel nanostructures can be adjusted to the requirements of each specific biosensor application, which allows the improvement of a specific problem. There are already successful examples in the literature in which nanowires [4], nanorods [5], nanoparticles [6], carbon nanotubes (CNTs) [7,8], hybrid materials [9] and nanocomposites [10,11] have been used in biosensor development.

This chapter deals specifically with the nanocomposite materials that are or can be used in the biosensing technology, since they are highly controllable and custom fabricated. We will emphasize particularly the application of nanocomposite materials as a solution to miniaturization process. The need of miniature devices is becoming very important, as they are the main detection and monitoring components in implantable and point-of-care systems for real-time analysis and diagnosis. The ability to design biosensors at micro- or nanoscale allows for medical real-time continuous monitoring of certain vital health-related conditions.

7.2 Principles of biosensors

The art of biochemical sensing is based on two fundamental scientific principles, that of biochemical recognition, and the transduction. It is a system that from its early days was able to fuse biochemistry and physics for the development of bioanalytical devices. In general, a biosensor is an integrated device which is capable of providing information continuously on concentration and type of analyte using a biological recognition system. Although the recognition or sensing element is responsible for probing the test solution for the analyte, it requires a second part in order to relay this information to the user. For this to be achieved, the sensing element is placed in close proximity to a transduction element or a transducer. The transducer, as used in biosensors, is the device that converts the chemical energy into electrical energy (Figure 7.1).

Since the early stages of biosensor development, there have been two main transduction methods used, namely electrochemical and optical. When an electrochemical transducer is used, the signal is one of the three parameters of Ohm's law ($V = I \times R$). Thus, we have potentiometric, amperometric, conductometric and impedometric sensors. When optical transduction is used, the signal obtained is either radiation intensity decrease (absorption), increase (fluorescence, chemiluminescence) or wavelength shift (change of colour).

Nanomaterials and nanocomposites can play a fundamental role in biosensor development [12–14]. They have high surface-to-volume ratios, unique catalytic activities and high electron-transfer rates at relatively low overpotentials. These characteristics make them ideal platforms for the design of electrochemical biosensors. They can be used either as transducers [15] or in other biosensor development

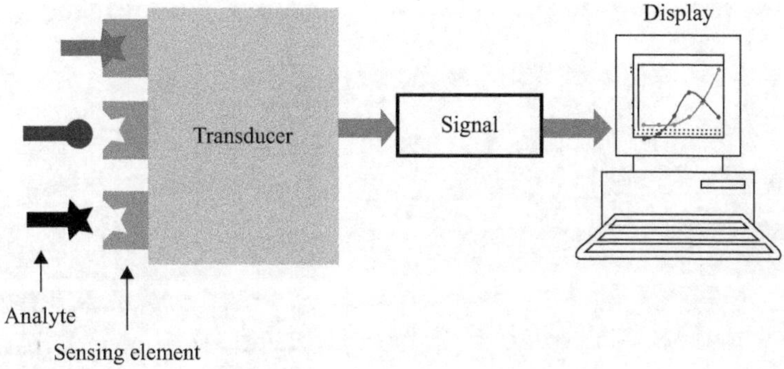

Figure 7.1 Biosensor fabrication

stages, such as biological element immobilization [16], enzyme stabilization [17,18] and amplification or mediation components [19].

7.3 Miniaturization of biosensors

The implementation of biosensor systems in specific areas require that the whole integrated system is of small size. For example, an implantable biosensor must have some very specific characteristics. To start with, it should not require large amounts of power, or better yet, can be self-sufficient. It must be implanted for a long period; a fact that mandates long operational stability and biocompatibility. Finally, it should be able to self-calibrate if possible. It is well known that the most critical issues in the application of in vivo biosensor systems are the stability of the biological sensing element (protein, enzyme, peptide etc.) and the efficient and facile signal mediation and transduction for improved sensitivity and elimination of interferences.

When considering miniaturized implantable biosensor requirements, one must pay attention to the possible causes of the issues, as well as possible solutions to the problems. The issue of stability is a major one. The operation of a biosensor is based on the activity of the biological element. In the case when an enzyme is used, it must be active, with a turnover rate which must remain constant over time, or at least predictable. Another major barrier in development of electrochemical systems is the electrical communication between the signal transducer and the sensing element such as redox proteins or enzymes. For example, the active centre of an enzyme is usually hidden by its prosthetic groups and the direct electron transfer (DET) is prohibitively slow resulting in low sensitivity [20]. To achieve direct electrochemical operation, researchers have turned their attention to the nanoscale materials. The most efficient procedure for DET is achieved when the sensing element is closely and efficiently immobilized in nanomaterials. Due to the high electron conductivity of the nanomaterials used [21], the charge transfer increases and thus the analytical characteristics of the sensors can be improved [22].

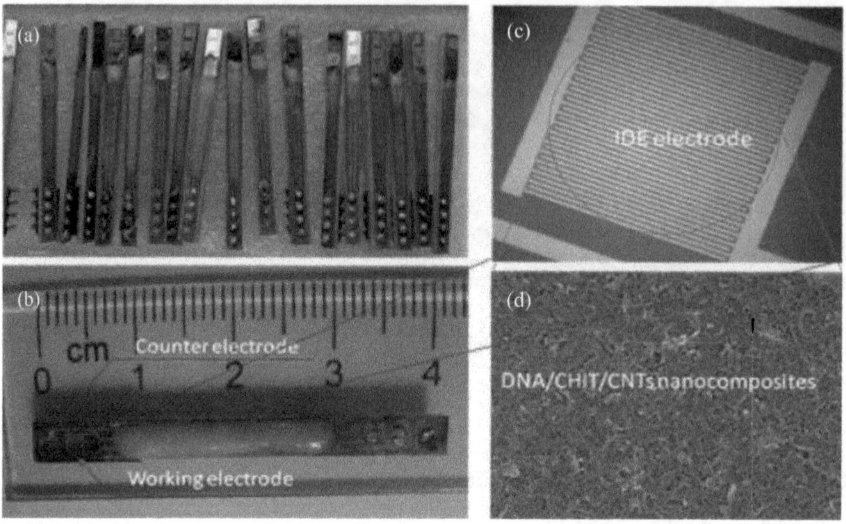

Figure 7.2 Images of miniaturized electrodes (a), one electrode (b), working electrode (c), and density of CNTs on electrode surface (d). Adopted, with permission, from Reference 23

We now have at our disposal nanocomposite materials, which theoretically can undertake some of the issues raised above, and have already wide used, especially for the miniaturization process of biosensors. A typical example of miniaturized biosensors that uses nanocomposite technology it is presented in Figure 7.2.

7.4 Nanocomposite materials

Nanocomposite materials are well-controlled physical mixtures of different materials, in which at least one of them is at its initial shape at dimensions in the nanometre range (1 nm = 10^{-9} m). Nanocomposite materials are promising alternatives to simple nanomaterials, since their chemical and physical characteristics can be precisely control and thus provide unique characteristics. The conductivity, reactive sites, length-to-diameter ratios, optical characteristics etc. are some of the parameters that are easily controlled. They can be adjusted to the requirements of each specific application and allow for the improvement of a specific biosensor problem. The control of elemental composition of the material allows them to possess unique property combinations that are not found in plain neat materials. As new nanomaterials are developed, together with the tools to control and characterize the final nanocomposite become available, new nanocomposites will also be available and their usefulness will be realized for biosensor and other scientific applications.

The main reason behind the success of nanomaterials in the biosensor development is the fact that, below a critical size, there are dramatic changes and effects on its chemical and physical properties. At the nanolevel, the interfacial interactions become very important, which in turn enhance the chemical properties

Table 7.1 Critical sizes for rapid changes in properties reported in nanocomposite systems

Properties	**Feature size (nm) at which changes might be expected**
Catalytic activity	<8
Making hard magnetic materials soft	<20
Modifying hardness and plasticity	<100
Producing strengthening and toughening	<100
Producing electromagnetic phenomena	<100

of the materials. For example, the surface-to-volume ratio of nanomaterials is much larger than that of microstructured ones, and it is crucial to the improvement of reactivity, stability, strength etc. Some of the properties that are affected by the size of the nanocomposite materials are listed in Table 7.1.

These effects are very evident in the case of carbon-based nanocomposite materials. For this reason, we are going to analyze them further.

7.4.1 Application of carbon nanocomposites in biosensor optimization

Carbon nanostructures are some of the most promising nanomaterials that have been already used successfully for the development of a variety of biosensors. Carbon nanofibres (CNFs), CNTs, graphene, fullerenes and carbon dots offer many advantages with their properties, such as large active surface area, chemical stability, biocompatibility, electrical conductivity and mechanical strength. One of the first nanomaterials to be used in biosensor development is that of CNTs. CNTs are composed of carbon atoms with an sp^2 hybridism arranged in hexagons forming a layer of graphite which in turn folds and produces a cylindrical sheet. Nanotubes are divided depending on their structure into single-wall nanotubes or multi-wall nanotubes. Side walls of CNTs are combined at their ends with pentagons and create caps (end caps) which resemble fullerene hemispheres. As a result both ends of CNTs are closed [24]. They present tensile strength hundred times greater than steel and thermal conductivity much higher than diamond. At the same time, CNTs have electrical conductivity similar to copper with the difference that CNTs can withstand greater currents (1012 A/cm^2 as opposed to 106 A/cm^2 for copper) [25]. CNTs have been used in biosensor design since early 2000. In these early works, CNTs were used as immobilization matrix and as a mediator for the development of an amperometric biosensor with good overall analytical characteristics [7].

Another carbon nanomaterial that can be used for biosensor development is CNFs. CNFs are cylindrical nanostructures of graphite layers stacked around the fibre's axis in different variations giving distinct well-organized structures [25]. This material can be used for stabilization of the sensing element of a biosensor system, like in the example of Figure 7.3, in which the enzyme glucose oxidase was immobilized on different CNFs and nanotubes [26].

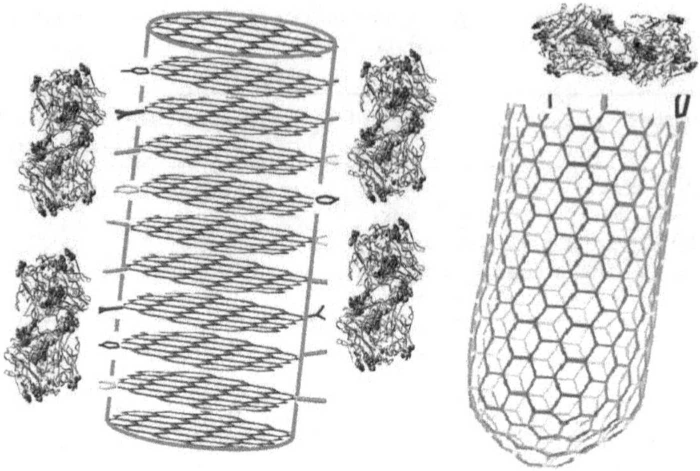

Figure 7.3 Schematic picture of the immobilization of the model enzyme GOx on carbon nanofibres and single-walled carbon nanotubes. Adopted, with permission, from Reference 26

It has been shown in this work that CNFs are the best matrixes for the immobilization of proteins and enzymes for biosensor development. CNF-based glucose biosensors showed higher sensitivity, longer lifetimes and excellent measurement-to-measurement reproducibility over a prolonged period of time. The very high surface area of the nanofibres, together with their large number of active sites, provides the grounds for the adsorption of enzymes. In addition, they allow for both the DET and increased stabilization of the enzymatic activity. These CNF materials are thus very promising substrates for the development of a series of highly stable and novel biosensors [26].

The characteristics on the CNFs can also be combined with other nanostructured materials for the development of nanocomposites which will possess innovative characteristics of both materials. Such an example is the work by Vamvakaki *et al.* [10] in which biomimetically synthesized silica–CNF architectures were used in biosensor design. This Si–CNF nanocomposite provided two fundamental advantages to the biosensor designed: stability through biosilica encapsulation and electrochemical transduction through the highly conductive CNFs. The unstable enzyme acetylcholine esterase was encapsulated in a silica/nanofibre nanocomposite architecture as shown in Figure 7.4, and the corresponding biosensor had both drastically improved lifetime and good transduction of the electrochemical signal.

Since then, carbon nanocomposite materials have been considerably used for the fabrication of many electrochemical biosensors with large lifetime by immobilizing the unstable sensing elements in their matrix. Carbon nanostructures are often combined with conductive polymers, metal nanoparticles or semiconductor dots with the aim of improving the characteristics of the developed sensors.

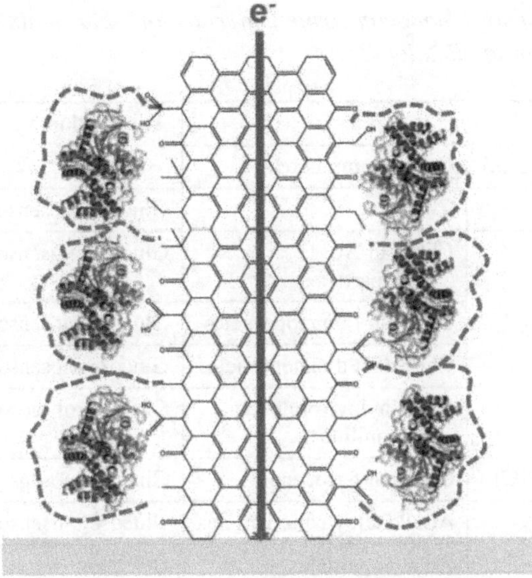

Figure 7.4 *Schematic representation of the electrochemical silica/nanofiber-based biosensor setup. The dashed line represents the shield provided by the biomimetically synthesized silica nanostructures. The electron transfer is achieved via the conductive carbon nanofibres, onto which the active enzyme is immobilized. Adopted, with permission, from Reference 10*

Consequently, different nanomaterials and their nanocomposites have been used in many applications (Table 7.2).

7.5 Future trends

Nanocomposites are very unique and scientifically interesting materials. In the near future, they are expected to generate a great scientific and industrial impact based on the variety of their properties, which are suited for a variety of applications. Although packaging, coating and automotive sectors are the ones to be the main sectors that will utilize these new materials, biosensor and in general detection technologies have also a lot to gain with their use. Of particular, interest to the biosensor sector is the use of electroconductive polymers, polyelectrolyte-composites and nanofibre-biomaterial composites. The research of nanocomposites is very interdisciplinary. Knowledge and tools originating from physics, chemistry, biology, materials science and engineering are required. For this reason, any new knowledge in this area will create new science and in particular new materials, with far reaching technological implications. Basic research activities aiming at examining the structure-property correlations will lead the development of new screening evaluation and synthesis methodologies. The understanding of their interactions at the

Table 7.2 List of some nanocomposite materials and their main characteristics to certain applications

Nanocomposite		Application	Reference
Carbon nanomaterial	**Other composite(s)**		
	Chitosan	Glucose biosensor	[27]
Graphene	Chitosan/Au nanoparticles	Glucose biosensor	[28]
	Chitosan/Pt nanoparticles	Glucose biosensor	[29]
	Chitosan/Pd nanoparticles	Glucose biosensor	[30]
	Polyvinylpyrrolidone/ polyaniline	Cholesterol biosensor	[31]
Graphene oxide (GO)	Chitosan/ferrocene	Glucose biosensor	[32]
Reduced GO	Au–Pd alloy nanoparticles	Glucose biosensor	[33]
	Au nanoparticles	Detection of pesticides	[34]
Single-walled carbon nanotubes	Chitosan/graphite	Indicator-free DNA hybridization for monitoring Hepatitis B virus	[35]
Multi-walled carbon nanotubes	Chitosan	Lactate biosensor	[36]
	Chitosan/Au–Pt alloy nanoparticles	Glucose biosensor for human blood and urine analysis	[37]
	Poly*(o-anisidine) (POAS)*	Biosensor development	[38]
	Chitosan/DNA	DNA biosensor for pathogenic virus detection	[23]
	Chitosan/glutaraldehyde	Label-free DNA biosensor	[39]

nanoscale size will aid in improving the interaction between the sensing elements and the signal transduction, allowing for the development of highly sophisticated integrated biosensing devices. Healthcare and prognosis based on biosensing devices built with new nanocomposites are expected to be of high impact in the improvement of our quality of life in the coming years. As the properties of nanostructured composites are highly complex and environment dependent, significant effort must be devoted to basic research in this area to provide a better understanding of the structure–property relationship in nanostructure-based systems. In conclusion, successful implantable biosensor systems are highly dependent on the advancement of nanocomposite materials. Such materials with novel properties and/or improved performance are still required to meet the emerging demands arising from the need to have reliable miniature detection and monitoring systems.

References

[1] Clark L.C. Jr, Lyons C. 'Electrode systems for continuous monitoring in cardiovascular surgery'. *Annals of the New York Academy of Sciences*. 1962;**102**:29–45.

[2] IUPAC. *Compendium of Chemical Terminology*, 2nd ed. (the "Gold Book"). Compiled by A.D. McNaught, A. Wilkinson. Blackwell Scientific Publications, Oxford (1997). XML on-line corrected version: http://goldbook.iupac.org (2006-) created by M. Nic, J. Jirat, B. Kosata; updates compiled by A. Jenkins. ISBN 0-9678550-9-8. doi:10.1351/goldbook.

[3] Cristea C, Harceaga V., Săndulescu R. 'Electrochemical sensor and biosensors'. In: Moretto LM, Kalcher K, editors. *Environmental Analysis by Electrochemical Sensors and Biosensors*. Vol. 1 Fundamentals. 1st ed. New York: Springer; 2014. pp. 155–65.

[4] Kulkarni A., Xu Y., Ahn C., *et al*. 'The label free DNA sensor using a silicon nanowire array'. *Journal of Biotechnology*. 2012;**160**(3–4):91–6.

[5] Byun K.M., Kim N.H., Ko Y.H., Yu J.S. 'Enhanced surface plasmon resonance detection of DNA hybridization based on ZnO nanorod arrays'. *Sensors and Actuators B: Chemical*. 2011;**155**(1):375–9.

[6] Jayakumar K., Rajesh R., Dharuman V., Venkatasan R., Hahn J.H., Pandian S.K. 'Gold nano particle decorated graphene core first generation PAMAM dendrimer for label free electrochemical DNA hybridization sensing'. *Biosensors and Bioelectronics*. 2012;**31**(1):406–12.

[7] Sotiropoulou S., Chaniotakis N.A. 'Carbon nanotube array-based biosensor'. *Analytical and Bioanalytical Chemistry*. 2003;**375**(1):103–5.

[8] Tam P.D., Van Hieu N., Chien N.D., Le A.T., Tuan M.A. 'DNA sensor development based on multi-wall carbon nanotubes for label-free influenza virus (type A) detection'. *Journal of Immunological Methods*. 2009;**350**(1–2):118–24.

[9] Lv W., Jin F.M., Guo Q., Yang Q.H., Kang F. 'DNA-dispersed graphene/NiO hybrid materials for highly sensitive non-enzymatic glucose sensor'. *Electrochimica Acta*. 2012;**73**:129–35.

[10] Vamvakaki V., Hatzimarinaki M., Chaniotakis N. 'Biomimetically synthesized silica–carbon nanofiber architectures for the development of highly stable electrochemical biosensor systems'. *Analytical Chemistry*. 2008;**80**(15):5970–5.

[11] Wu Y.L., Lin J.J., Hsu P.Y., Hsu C.P. 'Highly sensitive polysilicon wire sensor for DNA detection using silica nanoparticles/γ-APTES nanocomposite for surface modification'. *Sensors and Actuators B: Chemical*. 2011;**155**(2):709–15.

[12] Chaniotakis N., Buiculescu R. 'Semiconductor quantum dots in chemical sensors and biosensors' *Nanosensors for Chemical and Biological Applications: Sensing with Nanotubes, Nanowires and Nanoparticles*. Woodhead Publishing; 2014. p. 267.

[13] Chen A., Chatterjee S. 'Nanomaterials based electrochemical sensors for biomedical applications'. *Chemical Society Reviews*. 2013;**42**(12):5425–38.

[14] Zhu C., Yang G., Li H., Du D., Lin Y. 'Electrochemical sensors and biosensors based on nanomaterials and nanostructures'. *Analytical Chemistry*. 2014;**87**(1):230–49.

[15] Li H., Liu S., Dai Z., Bao J., Yang X. 'Applications of nanomaterials in electrochemical enzyme biosensors'. *Sensors (Basel)*. 2009;**9**(11):8547–61.

[16] Putzbach W., Ronkainen N.J. 'Immobilization techniques in the fabrication of nanomaterial-based electrochemical biosensors: a review'. *Sensors (Basel)*. 2013;**13**(4):4811–40.

[17] Kim J., Grate J.W., Wang P. 'Nanostructures for enzyme stabilization'. *Chemical Engineering Science*. 2006;**61**(3):1017–26.

[18] Hwang E.T., Gu M.B. 'Enzyme stabilization by nano/microsized hybrid materials'. *Engineering in Life Sciences*. 2013;**13**(1):49–61.

[19] Wang J. 'Nanomaterial-based amplified transduction of biomolecular interactions'. *Small*. 2005;**1**(11):1036–43.

[20] Zhang W., Li G. 'Third-generation biosensors based on the direct electron transfer of proteins'. *Analytical Sciences*. 2004;**20**(4):603–9.

[21] Wang J. 'Nanomaterial-based electrochemical biosensors'. *Analyst*. 2005;**130**(4):421–6.

[22] Sanzó G., Tortolini C., Antiochia R., Favero G., Mazzei F. 'Development of carbon-based nano-composite materials for direct electron transfer based biosensors'. *Journal of Nanoscience and Nanotechnology*. 2015;**15**(5):3423–8.

[23] Van Thu V., Dung P.T., Tam L.T., Tam P.D. 'Biosensor based on nanocomposite material for pathogenic virus detection'. *Colloids and Surfaces B: Biointerfaces*. 2014;**115**:176–81.

[24] Dai H. 'Carbon nanotubes: opportunities and challenges'. *Surface Science*. 2002;**500**(1):218–41.

[25] Frasco M.F., Buiculescu R., Vamvakaki V., Chaniotakis N.A. 'The applications of carbon nanostructures and semiconductor materials in the development of biosensors'. In *2009 International Semiconductor Conference*. IEEE; 2009: **1**:99–106.

[26] Vamvakaki V., Tsagaraki K., Chaniotakis N. 'Carbon nanofiber-based glucose biosensor'. *Analytical Chemistry*. 2006;**78**(15):5538–42.

[27] Kang X., Wang J., Wu H., Aksay I.A., Liu J., Lin Y. 'Glucose oxidase–graphene–chitosan modified electrode for direct electrochemistry and glucose sensing'. *Biosensors and Bioelectronics*. 2009;**25**(4):901–5.

[28] Shan C., Yang H., Han D., Zhang Q., Ivaska A., Niu L. 'Graphene/AuNPs/chitosan nanocomposites film for glucose biosensing'. *Biosensors and Bioelectronics*. 2010;**25**(5):1070–4.

[29] Wu H., Wang J., Kang X., *et al.* 'Glucose biosensor based on immobilization of glucose oxidase in platinum nanoparticles/graphene/chitosan nanocomposite film'. *Talanta*. 2009;**80**(1):403–6.

[30] Zeng Q., Cheng J.S., Liu X.F., Bai H.T., Jiang J.H. 'Palladium nanoparticle/chitosan-grafted graphene nanocomposites for construction of a glucose biosensor'. *Biosensors and Bioelectronics*. 2011;**26**(8):3456–63.

[31] Ruecha N., Rangkupan R., Rodthongkum N., Chailapakul O. 'Novel paper-based cholesterol biosensor using graphene/polyvinylpyrrolidone/polyaniline nanocomposite'. *Biosensors and Bioelectronics*. 2014;**52**:13–19.

[32] Qiu J.D., Huang J., Liang R.P. 'Nanocomposite film based on graphene oxide for high performance flexible glucose biosensor'. *Sensors and Actuators B: Chemical*. 2011;**160**(1):287–94.

[33] Yang J., Deng S., Lei J., Ju H., Gunasekaran S. 'Electrochemical synthesis of reduced graphene sheet–AuPd alloy nanoparticle composites for enzymatic biosensing'. *Biosensors and Bioelectronics*. 2011;**29**(1):159–66.

[34] Liu T., Su H., Qu X., Ju P., Cui L., Ai S. 'Acetylcholinesterase biosensor based on 3-carboxyphenylboronic acid/reduced graphene oxide–gold nanocomposites modified electrode for amperometric detection of organophosphorus and carbamate pesticides'. *Sensors and Actuators B: Chemical*. 2011;**160**(1):1255–61.

[35] Erdem A., Muti M., Karadeniz H., Congur G., Canavar E. 'Electrochemical monitoring of indicator-free DNA hybridization by carbon nanotubes–chitosan modified disposable graphite sensors'. *Colloids and Surfaces B: Biointerfaces*. 2012;**95**:222–8.

[36] Tsai Y.C., Chen S.Y., Liaw H.W. 'Immobilization of lactate dehydrogenase within multiwalled carbon nanotube-chitosan nanocomposite for application to lactate biosensors'. *Sensors and Actuators B: Chemical*. 2007;**125**(2):474–81.

[37] Kang X., Mai Z., Zou X., Cai P., Mo J. 'A novel glucose biosensor based on immobilization of glucose oxidase in chitosan on a glassy carbon electrode modified with gold–platinum alloy nanoparticles/multiwall carbon nanotubes'. *Analytical Biochemistry*. 2007;**369**(1):71–9.

[38] Carrara S., Bavastrello V., Ricci D., Stura E., Nicolini C. 'Improved nanocomposite materials for biosensor applications investigated by electrochemical impedance spectroscopy'. *Sensors and Actuators B: Chemical*. 2005;**109**(2):221–6.

[39] Wang Q., Zhang B., Lin X., Weng W. 'Hybridization biosensor based on the covalent immobilization of probe DNA on chitosan-multiwalled carbon nanotubes nanocomposite by using glutaraldehyde as an arm linker'. *Sensors and Actuators B: Chemical*. 2011;**156**(2):599–605.

Chapter 8

Monolithically integrated optoelectronic biosensors for point-of-need applications

Panagiota Petrou[1], Eleni Makarona[1], Ioannis Raptis[1], Konstantinos Misiakos[1] and Sotirios Kakabakos[1]

8.1 Introduction

Biosensors emerged approximately 50 years ago, if one considers the electrode made by Clark and Lyons [1] as the first sensor demonstrated. The research effort put forward since then resulted in a great variety of transducers as an answer to demand for compact, low-cost systems that could provide results in short time preferably for more than one analyte in a single sample with high sensitivity, specificity and stability. Despite the variations of transduction principles that have been employed, all of them could be grouped in a few basic categories termed mass-sensitive, temperature-sensitive, electrochemical and optical.

Another general classification of the various biosensors so far developed lies on the use or not of labels (label-free sensors) in order to obtain a measurable signal. The first ones most commonly employing fluorescent substances or nanoparticles usually demonstrate higher detection sensitivities and in some cases have even succeeded single-molecule detection [2]. In addition, the availability of different types of labels, some of which have been developed specifically for a particular sensor category, renders these sensors more versatile and more easily adaptable to the requirements of specific applications. Nevertheless, the label-free sensors are gaining ground against the labelled ones taking advantage of the fast-paced progress of nanotechnology. Therefore, during the past decade, label-free sensors have also exhibited high sensitivities with the additional advantage of being unburdened of the need for labels, decreasing thus the cost and time of sample preparation [3].

Optical sensors are considered particularly powerful in terms of detection sensitivity, especially when label-free detection is pursuit. Moreover, they are donned with additional advantages over other classes of sensors, such as the increased potential for miniaturization and incorporation to lab-on-a-chip systems, multiplexing capabilities, which automatically translate into synchronous

[1]Institute of Microelectronics, National Center for Scientific Research “Demokritos” Ag. Paraskevi, Athens Greece

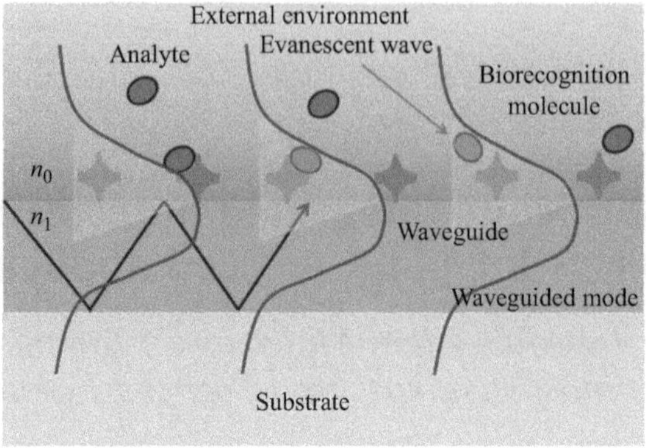

Figure 8.1 Schematic showing the evanescent field sensing principle. The binding of analyte by the immobilized onto the waveguide biorecognition molecule taking place within the evanescent field region results in a change in the effective refractive index of the transmitted light mode that is proportional of the number of bound analyte molecules

multi-analyte detection schemes, a great potential for integration with conventional electronic components, and last but not least electromagnetic immunity. The detection principle of most optical biosensors relies on evanescent field optics (Figure 8.1). The evanescent field is part of the electromagnetic field of the waveguided light and decays exponentially in the medium above the sensors surface with a depth of tens to hundreds of nanometres depending on the constituent materials and particular geometry of the waveguide. This electromagnetic field is very sensitive to changes in the refractive index taking place on the transducer surface. Thus, when a biomolecule that has been immobilized onto the optical transducer surface interacts with its counterpart molecule, the density/thickness of the biomolecular adlayer is increased modifying the effective refractive index on top of the optical transducer. This effective refractive index change in turn affects the light guiding properties of the transducer and, as a result, leads to modulations of the output beam in terms of intensity, polarization, phase etc. Due to the fact that the evanescent field intensity is higher in short distances from the sensor surface comparable to the thickness of the biolayer formed, real-time monitoring in label-free mode becomes feasible.

Evanescent-wave-based optical biosensors could acquire different formats and realized using different materials and technologies. The most popular evanescent-wave-based optical sensors are surface plasmon resonance (SPR), fibre optic, photonic crystal, ring resonator and interferometric sensors. SPR sensors are the most widely used sensors mainly due to the availability of commercial instruments based on this detection principle. They rely on charges in the reflectivity on a metallic layer (mostly gold) deposited on a dielectric material layer. A light beam is guided through a coupling prism on the metal–dielectric interface, and it is totally reflected at the interface generating an evanescent wave field which penetrates the metal layer. For a given angle or wavelength of the incident light, the propagation constant of the evanescent filed matches that of the surface plasmon wave that is the oscillation of the charge density at the interface of the materials with opposite signs of dielectric constants. Since this resonant angle depends on the surrounding medium refractive index, biomolecular reactions that modify the refractive index on the sensor surface can be monitored and quantified. The SPR biosensors have been extensively exploited in various applications and have demonstrated really outstanding performance with detection limits suitable for real-life applications [4]. Nevertheless, SPR devices have relatively large size and their miniaturization is not a straightforward task.

On the other hand, sensors based on integrated optics can be down-sized more easily and integrated with extra components, such as fluidic and electrical ones rendering to the most promising candidates for viable lab-on-chip and point-of-need applications. Additional advantages of these sensors are the high versatility of materials and technologies that could be applied for their fabrication, as well as the ability to detect biomolecules at very low concentrations [5]. With regards to the materials employed for the fabrication of integrated optics devices silicon, in the form of Si, SiO_2, Si_3N_4 or SiON, and polymers are the most widely used, combined by standard microfabrication techniques such as ion chemical vapour deposition, diffusion in glass, photo- and electron-beam lithography, etc. Moreover, the employment of silicon as a substrate and hence of the well-established semiconductor industry fabrication processes provides the means for reliability, robustness and reduced production cost of the final devices.

Another advantage of integrated optics devices is the ability to create arrays of sensors with the same operational characteristics suitable for multiplex analyses through modification of each array element with a different biorecognition molecule. In addition, one can maximize the detection sensitivity by optimizing the waveguide structure, material and working wavelength. One should not, however, oversee the fact that despite the inherent performance of the sensor as optical waveguide and detector of minute refractive index changes, the performance of a biosensor is by a great degree determined by the functionality of immobilized biomolecules. Similarly, the fluidic design plays an essential role since it defines the reaction kinetics, the sample and reagents volume as well as the size and complexity of the final device.

In the following section, the main categories of integrated optical sensors that is grating-couplers, interferometers, photonic crystals, microring resonators, slot waveguides or silicon wires will be discussed. Emphasis will be given to the analytical performance of these sensors upon their application in clinical diagnostics, food analysis schemes or environmental monitoring. Finally, the attempts to build up lab-on-chips or portable devices for application of these sensors at the point-of-need will be mentioned. The comparison of the various types of sensors will be performed using the detection sensitivity and in particular the limit of detection (LOD) as figures of merit. The LOD is referred to the lowest amount of analyte that the sensor can detect as a signal distinguishable from the measurement noise. Usually, the LOD is expressed as refractive index units (RIU) or as surface mass density (g mm^{-2}). It can be also expressed as analyte concentration, although this value could complicate the comparison between different sensors which have been evaluated using different analytes or recognition molecules with different affinity constants. Therefore, whenever applicable, the first two expressions of LOD are preferable to be employed as a more objective performance criterion.

8.2 Integrated optical sensors

8.2.1 Grating-coupled waveguide sensors

Grating-coupled waveguide sensors are amongst the first optical transducers exploited as biosensors [6]. They are based on planar waveguides with periodic patterns on top of the waveguides, the grating that permits the excitation of a guided mode in the waveguide by the incident light when the following condition is satisfied:

$$n_{\text{eff}} = n_{\text{air}}\sin\alpha + l\left(\frac{\Lambda}{\lambda}\right) \tag{8.1}$$

where n_{eff} is the effective refractive index of the waveguide, n_{air} is the refractive index of air, α is the angle of the incident light, l is the diffraction order, λ is the wavelength and Λ is the grating period (Figure 8.2). As is obvious by (8.1) the

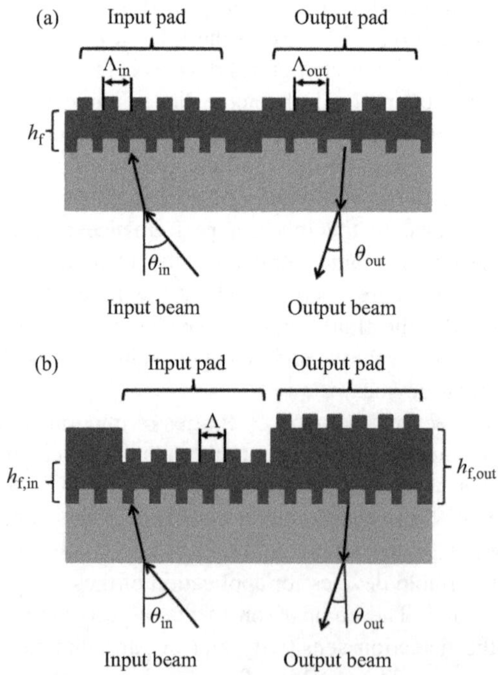

Figure 8.2 Optical sensor based on in- and output-grating couplers of two different configurations: (a) grating with different periods for in and output and (b) a single grating with a different thickness layer

incoupling angle depends on the effective refractive index at the waveguide surface (n_{eff}) in the region of the evanescent field. Thus, by monitoring the changes in the incoupling angle one can monitor biomolecular reactions that affect the refractive index near the waveguide surface. Similarly, the changes in the outcoupling angle can be used for biosensing. To achieve coupling, polarized laser beams are guided at different angles to the gratings. Thus, precise mechanical movements are required to bring the sensor in alignment with the light source and the photodetector [7,8]. This requirement is alleviated when the outcoupling angle is monitored; thus, a simpler set-up can be used and the sensor performance is improved [9,10].

The first grating coupling set-ups developed during the early 1990s [7–10] exhibited sensitivities in the range of 10^{-6} RIU employing model binding assay systems such as IgG/anti-IgG or biotin/streptavidin. A simplified set-up based on reflection-mode operation that avoided any moving parts provided similar LOD values in terms of RIU [11], whereas it was used for both competitive immunoassays [12] and DNA hybridization experiments [13]. To improve the performance of grating coupler sensors, Grego *et al.* [14] combined waveguides made of silicon oxynitride (deposited by plasma-enhanced vapour deposition) with colloidal self-assembly and imprint lithography for the fabrication of gratings. These technologies led to two-dimensional (2D) grating structures resulting in enhancement of detection sensitivity by a factor of at least 2 as compared to one-dimensional (1D) grating sensors [15]. Another group developed a self-referenced sensor by employing microfluidics to divide the grating into two-halves [16]. The sensor operated in reflection mode and the responses were collected by a CCD and the recorded patterns analysed. Another interesting configuration of grating coupler sensors developed was the so-called wavelength interrogated optical sensor [17,18]. The device was based on a single-mode waveguide with two gratings, one to incouple and the other to outcouple the light. The two gratings were designed in such a way that the in and out coupling angles were different. By keeping fixed the angle of incident light, it was possible to monitor changes in the refractive index over the waveguide as a result of biomolecular reactions by scanning the resonance peak using a tunable laser diode. A multimode fibre was used to collect the emitted light and guided it to photodiodes. The device was featured with four channels and self-referencing pads that could allow for multi-analyte determinations (up to 24 different sensing areas). The device was initially evaluated using low molecular weight analytes, for example biotin, as well as large biomolecules, exhibiting detection limits of 0.3 pg mm^{-2} and ability to detect molecules as small as 200 Da. The analytical capabilities of the device were also exploited for the detection of sulphonamides, a family of widely used veterinary antibiotics following a competitive immunoassay format [19]; according to that the antigen was covalently coupled onto the sensor surface which was activated with a photopolymerizable dextran layer. To improve the detection limit a signal enhancement step was employed using a secondary antibody. Thus, an antigen concentration as low as 0.5 ng mL^{-1} could be detected. The same principle was employed in a multiplexed sensor for the simultaneous detection of four different classes of antibiotics (34 in total) in milk samples with detection similar to that of the sulphonamides sensor [20]. The device was further improved to the direction of a lab-on-a-chip for on-site determinations through redesign of the microfluidics cartridge [21], which was then applied to the semi-quantitative detection of three families of veterinary antibiotics. Moreover, by applying a non-competitive immunoassay format, three different cytokines were simultaneously detected and quantified in cell cultures [22]. The device was also used to evaluate the hybridization efficiency achieved upon covalent immobilization of double-stranded oligonucleotides followed by sequential denaturation [23].

8.2.2 Microring resonators

Ring-resonators are a very popular class of optical transducers due to the high sensitivity and increased multiplexed analysis potential. In the basic ring-resonator configuration, the light is coupled through the evanescent field from an input waveguide to a circular one and propagates through the loop in the form of whispering-gallery modes (Figure 8.3). The spectral position of these modes is affected by changes in the refractive index in the vicinity of the ring and resonance is achieved according to (8.2) at wavelength λ according to the following equation:

$$\lambda = 2\pi n_{\text{eff}} \frac{r}{m} \tag{8.2}$$

where m is an integer describing the whispering-gallery modes angular momentum, r is the radius of the ring, and n_{eff} is the effective refractive index. Thus, when a bioreaction takes place onto the ring surface, the change in the effective refractive index causes a shift in the resonance spectrum which is then monitored either by scanning the output spectrum or by measuring the intensity at a fixed wavelength.

A basic difference between ring resonators and straight waveguides is that the interaction of the waveguided light with the environment is independent of the waveguide length and is determined by the number of revolutions within the ring, which is in turn expressed by the resonator quality factor (Q factor). High values of Q (in the order of 10^6) are indicative of low optical losses and long photon lifetimes, leading to narrow linewidths and high peak resolution and therefore to high

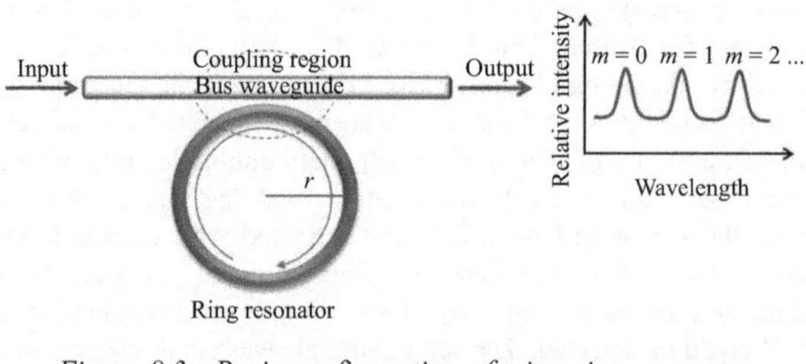

Figure 8.3 Basic configuration of microring resonator

sensitivity. Such Q values can be achieved with resonators of 50–200 μm that correspond to straight waveguides of several centimetres in length. Thus, higher sensitivities can be achieved with devices of smaller size, facilitating the fabrication of arrays of resonators in a relatively small surface area. With respect to resonator configurations so far developed and evaluated as biosensors, these include planar microring resonators based on microdisks [24–27], microrings [28–30] or microtoroids [31–34]. Such structures made either of glass [35], Si_xN_y–SiO_2 [36] or polymers [37,38] have been evaluated with respect to detection of proteins, DNA or bacteria [35]; whereas ring-resonators based on silicon-on-insulator (SOI) are also widely used [39]. SOI microring resonators with a bulk refractive index LOD of 10^{-5} RIU [28] have been reported to provide detection limits around 10 ng mL^{-1} in terms of avidin binding to a biotinylated surface [28,40]; comparable with that of other label-free sensors. In an advanced version of the same sensor, an array composed of three series of four rings connected to a single-input waveguide (one for each series) and each one with individual output waveguides [41] was fabricated. Each one of the four resonators of a single series had a distinct circumference ratio and thus, each one had a unique non-overlapping resonance spectrum. A grating was used to couple the incoming and outgoing light, whereas the output signal of the twelve resonators was imaged by an infrared camera. To allow for multiplexed detection, the sensor chip was combined with an appropriate microfluidic compartment for the delivery of the reagents, while one resonator was used as reference [40]. The surface mass LOD was calculated to be 3.4 pg mm^{-2}. A different sensor design [29] of two concentric ring structures was realized to increase the notch depth and the sensing area at the same time. The sensor was fabricated over a SOI wafer, and the estimated Q was around 5.1×10^4 resulting in bulk sensitivity of about 683 nm RIU^{-1}. Another research group integrated 32 ring resonators accessed via a bus waveguide; 24 of them, having an etched window on the surface for sample interaction, were used as sensing resonators, whereas the remaining eight, covered with the cladding layer, were used as reference in order to compensate for temperature-induced drifts [30]. A microfluidic compartment featuring fluidic ports, channels and reservoirs for fluid delivery to the 32 sensors and using gaskets to separate the individual sensors allowed for the simultaneous monitoring of 24 interactions. A bulk refractive index LOD of 7.6×10^{-7} RIU was determined, which corresponded to a mass LOD of 1.5 pg mm^{-2} [30]. The device was initially evaluated using model binding assays such as anti-IgG/IgG and biotin/streptavidin reactions as well as real-time multiplexed analysis of DNA hybridization reactions [30]. Apart from these proof-of-concept experiments, the device has applied to detect cytokines at very low concentrations (<0.1 ng mL^{-1} in buffer) by immobilizing specific antibodies onto different sensors [42] and implementing a secondary antibody as signal enhancement step [43]. Moreover, a direct assay was applied for the detection of carcinoembryonic antigen (CEA) in serum with a detection limit of 25 ng mL^{-1}, which is close to clinically relevant concentrations [44]. The multiplexed determination of five protein analytes, namely prostate-specific antigen, α-fetoprotein CEA, α-tumour necrosis factor, and interleukin-8 (IL-8) was achieved by immobilizing six different antibodies (five specific and one non-specific as control), each one in quadruplicates [45]. Similarly, the multiplexed detection of four different micro-RNAs using microrings modified with the respective single-strand DNA probes onto the microring array in about ten minutes and LOD in the sub-nM range was reported [46].

Comparable sensitivities have been also reported the toroidal-shaped microcavities. These structures exhibit ultrahigh Q factors (approximately 10^8) [47,48] and could be fabricated and integrated into arrays as easily as the planar ring resonators. The toroids usually have dimensions ranging from 30 to 150 μm for the major diameter to 2.5–6 μm for the minor diameter, which can be fabricated on silicon wafer by standard lithography techniques. Such sensors have been evaluated as biosensors for the detection of cytokines through immobilization of specific antibodies onto protein G functionalized surface. An LOD of about 5 aM (5×10^{-18} M) in buffer was achieved combined with a 12 orders of magnitude wide working range and with the ability to resolve single-molecule detection [31]. The same group fabricated toroids on low-loss polymer [32] or polymer-silica hybrid materials [49], which although have lower Q factors (10^5–10^7) as compared to silicon based toroids, their manufacturers claim detection sensitivities comparable to that of other ring resonators. Nevertheless, these devices have not been yet evaluated as biosensors.

8.2.3 Photonic crystal waveguides

Photonic crystal sensors constitute another class of integrated optical sensors. A photonic crystal is a nanostructure characterized by a periodicity in the dielectric constant of the order of a wavelength in one, two or three dimensions (1D, 2D or three-dimensional (3D) photonic crystals, respectively), which generates photonic bandgaps in which light cannot propagate in the crystal. By inducing a defect in the periodic structure one can introduce a 'defect mode' within the bandgap. Thus, when the incident light is in resonance with the defect mode, it can propagate in the photonic crystal. Moreover, the spectral position of the defect mode is sensitive to the local variation of the defect or to its surrounding environment, in other words it can be influenced by the refractive-index changes over the photonic crystal providing thus the means for monitoring of biomolecular interactions. Thus, by tuning both the photonic bandgap and the defect mode wavelength through appropriate design of the nanostructure, one can also tune the response of the device to refractive index changes and fabricate sensitive biosensors [50]. Another characteristic of the photonic crystal sensors is their compact size resulting from the strong confinement of the light in the periodic pattern. The most widely used structures are linear or 2D gratings on which discrete or line defects are introduced to create the transduction sites (Figure 8.4).

The first photonic crystal biosensors [51–54] were based in a 1D polymer grating covered with a high refractive index layer of TiO_2. The detection of biomolecular reactions was based on monitoring the shift of light reflected from the sensor surface when was probed with perpendicularly incident white light. This detection principle allowed the individual interrogation of different areas of the sensor surface without any cross-talk effects, leading thus to highly multiplexed

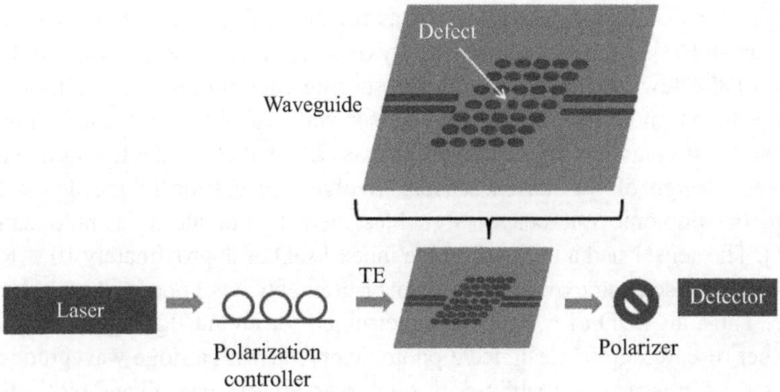

Figure 8.4 Schematic presentation of a photonic crystal and the measurement instrumentation set-up

detection capability [55]. Moreover, the sensor could be incorporated on the bottom of microtiter wells or on microarray slides [55], or integrated with microfluidics on the same substrate, providing for shorter analysis time and higher detection sensitivity as compared to static assays [56]. The particular sensors have been applied in a wide range of studies [57] including amongst others cell-based assays, cells attachment to surfaces, protein and virus detection, drug screening and protein–protein interactions. Moreover, the technology was commercialized by SRU Biosystems (www.srubiosystems.com).

A lot of work has been devoted to the design and fabrication of 1D and 2D planar photonic crystal structures. Most of them are based on waveguides with hole-arrays on which the guided wave is generated by missing holes or defects in the bulk photonic-crystal structure [58]. Following this concept, 1D SOI photonic-crystal waveguide consisting of a hexagonal lattice of holes with a single missing row line defect and point cavities in different arrangements was fabricated [59]. The sensor featured a bulk refractive index LOD of 10^{-3} RIU, a mass sensitivity of 24.7 nm pg^{-1} and an LOD of 500 pg mm^{-2} for direct adsorption of bovine serum albumin (BSA) or avidin. Another group prepared a photonic crystal with a length of 20 μm, a lattice constant and hole radius of 390 and 111 nm, respectively, and used tapered fibres for the in- and outcoupling of transverse electric (TE) -polarized light [60,61]. Combined with appropriate microfluidic, this sensor provided a mass LOD of 2.1 pg mm^{-2} for an anti-BSA antibody binding to BSA modified sensor surface [60]; and an LOD in the nM range in DNA hybridization experiments [61]. Taking advantage of the fact that the resonant wavelength can be tuned by changing the defect cavity spacing, Erickson's group developed the nanoscale optofluidic sensor array that was based on a microcavity structure evanescently coupled to an adjacent single-mode silicon waveguide [62]. By adjusting the cavity spacing, parallel resonator arrays of different Q factors could be realized on a single waveguide, which combined with a flow channel, allowed for multiplexed detection with a bulk refractive index LOD of 7×10^{-5} RIU [62]. The sensor was used in proof-of-concept hybridization experiments, as well as for the multiplexed determination of ILs-4, -6, and -8 [63]. Although the sensitivity of this particular sensor was not high, this is one of the few examples showing biosensing capabilities in a multi-analyte format. A SOI 2D photonic crystal-based sensor was also fabricated and evaluated using the biotin–streptavidin interaction and a mass LOD of about 2.5 fg was detected [64]. Another design of SOI-based sensors involved generation of the defect line adjacent to the photonic microcavities enables them to operate as a multichannel sensor [65]. The sensor had a bulk refractive index LOD of approximately 10^{-2} RIU, whereas for the biosensing experiments the measurements were obtained after drying the device. Thus, an LOD of 67 nM was determined for an anti-IgG/IgG interaction [65]. Another research group designed a photonic crystal with a ridge waveguide and demonstrated its biosensing capabilities through reaction of immobilized biotin-BSA with anti-biotin antibodies. An LOD of 20 pM was reported, which is translated to a mass sensitivity of about 4.5 fg [66]. Taking into account the performances reported, photonic crystal sensors have lower detection sensitivities as compared to other classes of integrated optical sensors. In order to improve their analytical performance, the immobilization of biorecognition molecules mainly onto the holes in which the light confinement is maximum resulting in highest resonant shifts was investigated. Following this approach, detection of single particles, virus or pathogens has been demonstrated [67,68].

8.2.4 Integrated interferometers

Interferometric sensors are the most developed class of integrated optical sensors due to the high detection sensitivity they offer. Three are the main interferometer configurations that have been employed for biosensing: Mach–Zehnder interferometer (MZI), Young interferometer (YI), Hartman and bi-modal interferometers (Figure 8.5).

Integrated MZIs are composed by a waveguide that at some point splits into two arms – the sensing and the reference arm, which recombine again after a certain distance across the length of which the biochemical interaction takes place. The reference arm is usually covered by a cladding layer, whereas the sensing arm is both properly modified biochemically for detecting specific analytes and exposed to the analyte solution. Detection is achieved through the interaction of the waveguided light with the environment through the evanescent field along the sensing arm. When a biomolecular reaction takes place on the sensing arm exposed area, the induced change in the effective refractive index results in a phase difference between the light propagating in the sensing and the reference arm. Thus, the output light intensity I and the phase difference $\Delta\Phi$ are modulated in a manner described by (8.3) and (8.4), respectively:

$$I = \frac{I_0}{2\left(E_S^2 + E_R^2 + 2E_sE_R\cos\Delta\Phi\right)} \tag{8.3}$$

$$\Delta\Phi = \frac{2\pi L}{\lambda\left(n_{\text{eff},S} - n_{\text{eff},R}\right)} \tag{8.4}$$

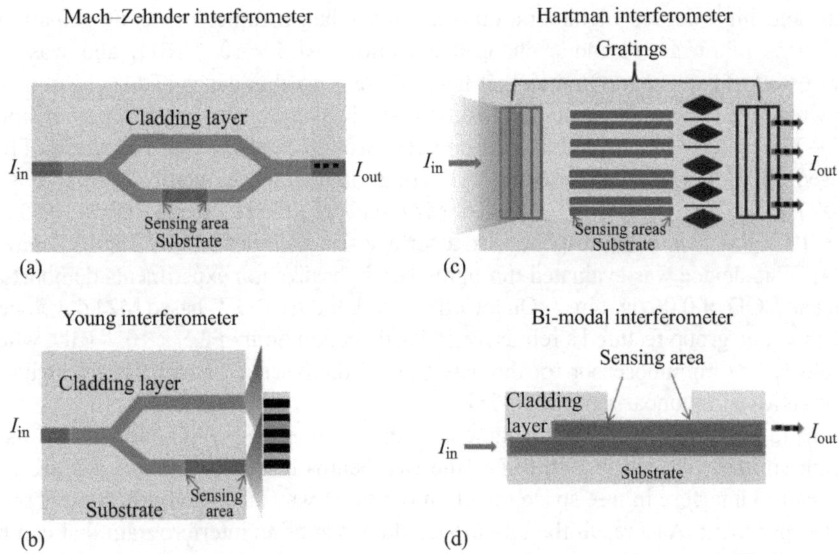

Figure 8.5 Schematic presentations of basic configurations of integrated (a) Mach–Zehnder, (b) Young, (c) Hartman and (d) bi-modal interferometers

where E is the electric field propagating along the waveguide, n_{eff} is the effective refractive index, Φ is the phase, L is the detection length, λ is the light wavelength, and I_0 and I are the light intensity at the input and output, respectively. S and R refer to the sensing and the reference arm, respectively. From the equations, it is obvious that (a) by increasing the interaction length the sensitivity can be also increased, and (b) the output light intensity has a cosine dependency from the phase difference that leads to higher sensitivities at the quadrature points but is considerably reduced in the vicinity of the local extrema. Moreover, for interferometers to be employed as biosensors, there exists a stringent prerequisite: The waveguides should be monomodal. If multiple modes are employed, due to the fact that each mode has a different propagation velocity each mode will produce a different output intensity modulation, and the final signal will appear "scrambled". Therefore, suitable photonic design must be performed prior to the fabrication of any interferometric sensor.

Apart from the sensing arm length, there are two more key parameters that affect the sensor performance: the geometrical characteristics of the waveguide and the refractive index difference between the waveguide and the cladding layers – directly related to the resulting effective refractive indices and hence the sensitivity of the sensor – according to (8.3) and (8.4). With respect to the materials usually employed for waveguide fabrication, these include glass, SiO_2, Si_3N_4, as well as polymers. A 3D MZI fabricated on fused silica using femtosecond laser writing is the most recent glass interferometer reported in the literature [69]. Despite its small size and high degree of integration, the sensor had a rather moderate sensitivity in terms of refractive index changes detection (~1.5×10^{-4} RIU) and was not evaluated as biosensor. On the other hand, there is an abundance of MZI biosensors in which the waveguide is made of silicon nitride and the cladding layer from silica [70–73]. Silicon-based MZIs could operate through total internal reflection (TIR-based MZI) [74] or anti-resonant reflecting optical waveguide (ARROW-based MZI) [75]. The reported TIR-based MZI had a bulk refractive index LOD of 8×10^{-6} RIU, which corresponded to a surface sensitivity of around 2×10^{-4} nm^{-1} [74]. This device was evaluated through DNA hybridization experiments demonstrating an LOD of 0.06 pg mm^{-2}. On the other hand, the ARROW-based MZI developed by the same group featured a refractive-index detection limit of 2.5×10^{-6} RIU, when evaluated as immunosensor for the detection of the insecticide carbaryl following a competitive immunoassay format [76].

YIs are also based on waveguides with an integrated Y-junction acting as a beam splitter, but contrary to MZIs, the two beams do not recombine but are left instead to interfere in free space, much in the same way as in Young's famous two-slit experiment. As a result the output signal consist of an interferogram that can be onto a CCD camera The change in the effective refractive index on the sensing arm due to biomolecular reaction induces a phase difference between the two interfering beams that is directly translated into a spatial shift of the interference fringes as expressed by the following equation:

$$\begin{aligned} I(\lambda) &= I_1 + I_2 + 2\sqrt{I_1 I_2}\cos\left(\frac{2\pi b}{\lambda L}x + \Delta\varphi\right) \\ &= I_1 + I_2 + 2\sqrt{I_1 I_2}\cos\left(\frac{bk}{L}x + \Delta\varphi\right) \end{aligned} \tag{8.5}$$

where I_1, I_2 are the light intensities through the two arms, b is the distance between the two arms, L is the distance between the waveguide end and the readout camera, and x is the position of the fringe pattern on the camera. The phase difference $\Delta\varphi$ between the two arms is given by the same equation (8.4) as in the case of MZI-based biosensors.

YIs have simpler read-out than MZIs and are less sensitive to temperature and wavelength drifts. One of the first YIs developed [77,78] was based on a monomodal Ta_2O_5 waveguide and had two reaction channels. Two separate beams were coupled into the sensor via a grating coupler and after propagation in the channels, the light was coupled out by a second grating and diffracted by a double slit to a CCD camera. An effective refractive index LOD of 9×10^{-8} RIU was achieved [77]. The device was used to detect serum proteins after modification with antibodies as well as pharmaceutical substances (methotrexate) in serum [79], to monitor the production of a recombinant in cell lysates [80] and to detect tuberculosis-specific antibodies in serum samples from infected patients [81]. Another research group developed a four-channel integrated YI that allowed independent measurements in each sensing area, one of which was used as reference [82]. The device was completed with a microfluidic module for minimum sample consumption. The refractive index LOD was about 8.5×10^{-8} RIU, corresponding to one of the most low LODs in terms of mass coverage resolution (20 fg mm^{-2}) [83]. In addition, by immobilizing specific antibodies against herpes simplex virus type 1, detection of about 850 particle mL^{-1} was achieved in buffer, whereas the LOD was increased a bit when the analysis was performed in serum [84,85].

Hartman interferometers are based on coupling a linearly polarized light in an array of interferometers by means of an integrated on the chip grating [86]. The first sensor was used to detect hormones in serum and blood samples at concentrations as low as 5 ng mL^{-1} [87,88]. Another more robust sensor that integrated the laser diode, the waveguides, the flow cell and the CCD camera provided an LOD of 10^{-6} RIU and was further evaluated for the detection of avian influenza [89].

A completely different design of interferometers was developed more recently, the so-called bimodal waveguide [90]. The device consists of a single waveguide with two different zones; a first one with single-mode behaviour and a second one supporting two modes (zero- and first-order modes) that propagate at different velocities depending on the refractive index of the cladding layer. The interference pattern at the waveguide output changes when the refractive index changes as a result of a biomolecular interaction. This type of sensor exhibited bulk refractive sensitivities comparable to those of more standard interferometric configurations (LOD 2.5×10^{-7} RIU [90]). It has been already applied to immunodetection of thyroid stimulating hormone via an indirect immunoassay demonstrating detection limit within the clinically relevant range [91], as well as for the immunodetection of bacteria with LOD of a few bacteria per millilitre [92].

8.2.5 Silicon nanowires, slot waveguides and other sensor configurations

Silicon nanowires and slot waveguides belong to a new class of sensing devices based on nanophotonic structures that are characterized by high detection sensitivity (due to the confinement of the electromagnetic field) and increased miniaturization and integration potential. Their viability as biosensors, however, remains to be proved since, up to now, very few devices have been evaluated in real bioanalytical applications.

Silicon photonic nanowires are fabricated by electron-beam lithography and reactive ion etching on SOI wafers. Due to the high index contrast between the silicon core and silicon oxide cladding, the waveguides can bend without significant optical losses thus allowing realization of very compact structures on small surface area. The first such sensor developed was an MZI based on SOI photonic wire waveguides and a folded double spiral millimetre-long resonator [93]. An updated version of the sensor included reference arms for suppression of wavelength and temperature drifts [94], integration of a fluidic channel patterned on an epoxy-based negative photoresist marketed as SU-8 [95] or use of four resonators for multiplex analyses [96]. The MZI sensor was evaluated using IgG/anti-IgG antibody reactions with an LOD in terms of surface coverage of about 0.3 pg mm^{-2} [95]. The resonator was also evaluated using similar reaction and demonstrated an LOD of about 20 pM or 40 ag in resolvable mass [96].

Slot waveguides consist of two slabs of high refractive index materials separated by a nanometre-scale low refractive index slot region and surrounded by low refractive index cladding materials [97]. Thus, the light is strongly confined in the slot region, and a stronger interaction with the sample is expected compared to more standard rib or planar waveguides. One of the first configurations developed was a slot-waveguide ring resonator structure [98]. This sensor had a bulk sensitivity of 212 nm RIU^{-1}, and a surface coverage LOD of 16 pg mm^{-2} when functionalized with an anti-BSA antibody for detection of BSA [99]. An updated version of this sensor included an array of eight slot-waveguide ring resonators, integrated microfluidics and alignment with the readout instrumentation for simultaneous detection [100]. The bulk refractive index LOD was 5×10^{-6} RIU, and a surface mass LOD of 0.9 pg mm^{-2} was reported that was significantly lower than that of the initial sensor. Horizontally slotted ring microdisc resonators were also exploited as biosensors after modification with biotin groups and probing with streptavidin demonstrating a moderate LOD in the order of 30 ng mL^{-1} [25,26]. Slot waveguides were also combined with planar photonic crystals [101]. A bulk refractive index of 7×10^{-6} RIU, which is well below those achieved by conventional waveguide photonic crystals, was achieved; however, the sensor has not been evaluated as a biosensor. It should be noted that in order to take advantage of the increased sensitivity of slot waveguides compared to standard ones, optimization of fluidics and sensor surface biofunctionalization is required to ensure uniform delivery of the sample and efficient capture of the analyte of interest.

8.3 Monolithically integrated optoelectronic transducers

All the optical sensors mentioned above rely on external optical components either for the in- or out-coupling of light. Thus, independently of the integration level of the sensing structure, the final device performance, volume and complexity are by a great extent defined by the implementation of external components like light sources and detectors, whereas the coupling of light is achieved mainly by gratings. Such dependence on external optical components increases considerably the cost, size and operational complexity of the final biosensing system, rendering it ill-suited for practical point-of-need applications. The only way out this bottleneck is the monolithic integration of both the light source and/or the detector on the same substrate with the integrated waveguide. Preferably, this should be performed following mainstream semiconductor industry fabrication procedures in order to ensure compatibility with the waveguides fabrication and reliable mass production.

To this direction, a technology developed by Misiakos *et al.* [102,103] allowed the monolithic integration of light-emitting diodes (LEDs), thin silicon nitride optical waveguides and photodetectors on the same silicon wafer substrate. A schematic of the monolithically integrated silicon transducer is presented in Figure 8.6(a). The LEDs are silicon avalanche diodes that emit light when biased beyond their breakdown point, a phenomenon known since the mid-1950s [104]. Nevertheless, the light intensity is very low when compared to lasers usually

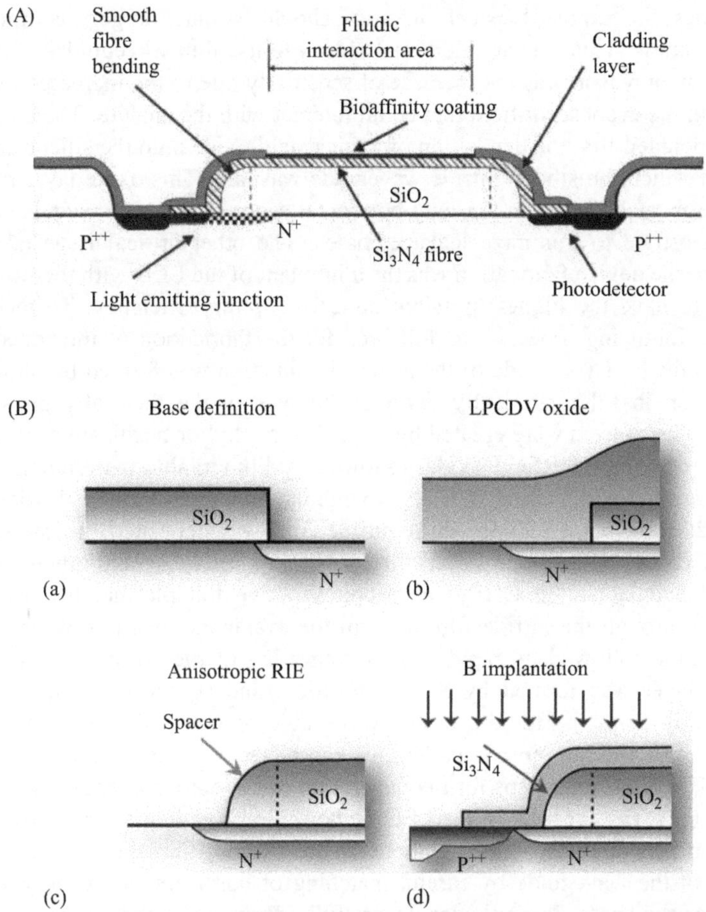

Figure 8.6 (A) Schematic of the monolithic transducer. (B) The basic steps of the monolithic optocoupler fabrication: (a) avalanche diode base formation, (b,c) spacer formation through dioxide deposition and plasma etching, and (d) self-aligned emitter formation by implantation through the waveguide nitride film (d)

employed as light sources in integrated sensors and in addition the emitted light is polychromatic covering all the visible and near infrared (IR) spectrum [103]. Moreover, two other critical issues had to be resolved in order to accomplish the integration of light source with the waveguide, the bending of the waveguide from the horizontal direction to the vertical towards both the light source and the detector and the alignment of the up-going waveguide segment with the integrated LED. The first issue was resolved by creating appropriate SiO_2 spacers that allowed the bending of waveguide towards the LED and detector with minimal optical losses. In particular, it was found that in order to assure the smooth bending of the waveguides, the radius of spacer curvature should be much larger than the waveguide thickness. This was an additional reason to use thin waveguides (~100–150 nm); the main reason was the increase of sensitivity due to the increased available photons in the evanescent field that could interact with the analyte. The creation of spacers included first the deposition of a thick field oxide onto the silicon substrate on top of which the silicon nitride waveguide was laid. This oxide layer provided also enough space between the long horizontal segment of the waveguide and the silicon substrate to minimize leakage losses. The other critical issue taken into account in the device fabrication was the alignment of the LED with the waveguide in order to have the highest possible optical coupling efficiency. To tackle this issue, the following process was followed for the fabrication of integrated transducer: (a) the base (N^+) side of the avalanche junction was formed by phosphorus implantation in lithographically defined windows in the thermally grown field oxide, (b) the spacers were created by a 2-μm tetraethyl orthosilicate deposition of SiO_2, over the thermal field oxide, followed by lithographic patterning, and anisotropic reactive ion etching in CHF_3, to end up with an overall field oxide thickness of 2.5 μm, (c) a 150-nm thick nitride film was deposited by low pressure chemical vapour deposition in a mixture of NH_3 and SiH_2Cl_2, and lithographically patterned and etched in CHF_3 anisotropic reactive ion plasma, (d) boron was implanted through the nitride film to form the avalanche junction by phosphorus compensation followed by rapid thermal annealing of the boron implant, (e) the cladding layer was formed by SiO_2 deposition, and (f) the cladding layer was etched away from the middle of the waveguide to create the sensing window. Contact hole and aluminium patterning completed the optocoupler fabrication process. The four basic steps for the fabrication of the optocoupler are presented in Figure 8.6(a). The creation of the LED by boron implantation after the waveguide deposition permitted the self-alignment of the emitting junction to the up-going segment of the waveguide by careful matching of boron implant energy with the waveguide thickness. In particular, to get full advantage of the built-in high coupling efficiency, the misalignment of the avalanche diode emitter to the base should be restricted to less than 1 μm.

Using 1,000-μm long waveguide and a 10-mA excitation current on the emitter, photocurrents of nearly 800 pA have been obtained; thus, taken into account that the overall external quantum efficiency of the LED is 2×10^{-7}, a 40 per cent external optical coupling efficiency between emitter and detector has been achieved, which is excellent if compared to the coupling efficiencies obtained on thicker waveguides when integrating compound semiconductor LEDs on silicon by microbonding techniques [105]. In addition, the small leakage current and capacitance value of the detector enabled the accurate current reading, whereas the stability of the LED allowed for the application of the monolithic optoelectronic transducer to various biosensing applications. For this purpose, the transducer was coupled to appropriately designed fluidic devices [106,107] that allowed either the simultaneous interrogation of a large number of optocouplers with the sample or the selective coating of adjacent optocouplers with different capture molecules. Signal transduction is based on the modulation of the optical coupling efficiency by the analyte molecules binding on the recognition biomolecules immobilized on the waveguide. The potential of the device to perform as a sensitive real-time affinity sensor was initially tested in single-analyte assay mode using the biotin-streptavidin model assay [103]. Taking into account that, the spectral response of the LEDs extends from blue (450 nm) to near infrared (850 nm) with a peak at 600 nm, colloidal gold nanoparticles were employed as labels due to their strong SPR absorption in the green-yellow spectral region. Furthermore, the use of gold nanoparticles provided the ability to increase considerably the detection sensitivity by silver-plating of the surface bound gold nanoparticles, which resulted in increase of their volume and of the associated light transmission losses. The LOD achieved was 3.8 pM in terms of gold-labelled gold nanoparticles, which reached 20 fM after 20 min of silver plating of gold nanoparticles corresponding to a surface density of 0.6×10^7 cm^{-2} indicating the great analytical potential of the proposed sensor. Moreover, using specially engineered liposomes bearing biotin moieties in their membrane and fluorescent labels in their core, the detection of single binding events were demonstrated [108]. The multiplexed real-time detection capabilities of the device was exemplified through the detection of seven deleterious mutations in BRCA1 gene related to predisposition to hereditary breast/ovarian cancer using chips featuring arrays of ten monolithic integrated transducers each one functionalized with mutation-specific oligonucleotides. Detection was based on waveguided photons elimination through interaction with fluorescently labelled PCR products. The sensor was packaged in a compact form and a portable device was built for fluidic management and signal acquisition [109,110]. The transducer can also perform label-free detection of biomolecular reactions after engineering of the optical fibre surface with gold nanoparticles [109] or through creation of a latent photonic crystal [110]. In the first case, the dominant signal generation mechanism is SPR of the surface-immobilized metal nanoparticles, whereas in the second one a grating of recognition biomolecules was created at the waveguide surface through introduction of 1D pattern of protein-adsorbing and non-adsorbing areas, the latent photonic crystal thereby formed, was further developed during analyte binding resulting in additional filtering out of resonant photons.

Although feasible, both label-free approaches lacked the required sensitivity for application in real sample analysis. Therefore, the integrated waveguide was patterned as MZI [111,112] and the capability of the transducers developed for label-free determinations was investigated. Due to the polychromatic nature of the integrated LED, a new detection concept that of frequency-resolved MZI was introduced [112]. This new detection principle overcame the phase ambiguity and signal fading disadvantages of standard single wavelength MZI transducers, since the phase change is different for each wavelength, and thus, it is easier to deduce from the full transmission spectrum, the change in the effective refractive index due to the biomolecular reaction taking place on the sensing arm of the MZI. Two different transducer configurations were realized. In the first one, termed 'fully-integrated', the sensing and reference arm effective indices are selected so that the phase difference of the light travelling through them is more or less wavelength independent in the emission spectral range of the LEDs [113]. This was achieved by choosing the right core thicknesses of the two arms. The integrated MZI output was recorded by an integrated photodetector as photocurrent intensity. In the second mode of operation, termed 'semi-integrated', the sensing arm was thinned with respect to the reference, so that the observable phase shifts to be a bell-curved function of the wavelength [114]. In the semi-integrated case, the output spectrum is recorded through a spectrophotometer and the spectral shifts over the entire LED spectrum are monitored in real time. As a result of the appropriate photonic circuit engineering, the observable spectral shifts were at least two-order of magnitude larger than those obtained with ring resonators. These large spectral shifts recorded by the external spectrometer were subjected to discrete Fourier transform (DFT) analysis, which leads to a dominant peak in the wavenumber domain for each of the two polarization modes (transverse electric and transverse magnetic or TE and TM, respectively), and the phase is identified with the argument of the complex DFT value at the main peak, whereas noise is filtered by ignoring non resonant frequencies in the spectrum. From the fabrication point-of-view, the steps for the LED and SiO_2 spacers' formation were the same as in the original transducer. It should be noted that an important element of the photonic circuits in both approaches is the existence of a so-called mode filter introduced prior to the MZI structure of the waveguide to ensure that the light travelling is monomodal.

From an analytical performance point-of-view, a bulk refractive index LOD of 5×10^{-6} was calculated for the fully integrated version by assessing its performance with solutions of known refractive indices, whereas its multi-analyte detection capabilities were demonstrated through the simultaneous detection of anti-mouse IgG-mouse IgG and biotin–streptavidin binding with LODs of 10 nM and 1 nM, respectively [114]. For the semi-integrated MZI, bulk refractive index LODs 7.4×10^{-7} and 3.1×10^{-7} have been predicted for the TE and TM modes, respectively, for a device with sensing arm length of 2 mm [115]. These LODs were translated to 32 and 5 pM, respectively, when the device was applied for detection of anti-mouse IgG-mouse IgG and biotin–streptavidin binding reactions. The analytical capabilities of the device were also demonstrated through the detection of bovine milk in goat milk at levels as low as 0.04 per cent (v/v) and with a dynamic range that varied between 0.1 and 1.0 per cent (v/v), showing the potential of the device for application to goat milk adulteration control [116]. Furthermore, the device was applied for the detection of mycotoxin ochratoxin A in beer samples with a detection limit of 2.0 ng mL^{-1} and a dynamic range 4.0–100 ng mL^{-1}, demonstrating its application potential to another real-life analysis [117].

8.4 Conclusion and outlook

Despite the enormous number of biosensors developed during the past few years, a very limited number of them managed to attain the excellent performance demonstrated by integrated optical sensors. In addition, integrated optics technology seems to be the only viable solution towards the development of truly portable lab-on-a-chip platforms that could be used for analysis outside a laboratory environment. Although few working solutions are already available, due to existing limitations in the integration of all the components in a small-sized microsystem, the most recent advances of integrated optical platforms are in the process of forming a solid basis for viable, truly portable biosensor systems suitable for a wide spectrum of applications like food and beverage safety, environmental pollution monitoring or identification and quantification of disease markers during emergencies at the doctor's office or whenever and wherever a rapid diagnosis is needed. This future view is not only supported by the intense research work in the field, but also from the attested interest shown from both public institutions and private companies to invest on the development of such devices, due to the expected imminent impact in our everyday lives. The future remains to be seen, and it will certainly be exciting sooner than later and full of novel, radical optical systems that will transform our abilities to safeguard our health and well-being.

References

[1] Clark Jr L., Lyons C. 'Electrode systems for continuous monitoring in cardiovascular surgery'. *Annals of the New York Academy of Sciences*. 1962;**102**:29–45.

[2] Walt D.R. 'Optical methods for single molecule detection and analysis'. *Analytical Chemistry*. 2013;**85**(3):1258–63.

[3] Estevez M.-C., Alvarez M., Lechuga L.M. 'Integrated optical devices for lab-on-a-chip biosensing applications'. *Laser and Photonics Reviews*. 2011;**6**(4):463–87.

[4] Homola J. 'Surface plasmon resonance sensors for detection of chemical and biological species'. *Chemical Reviews*. 2008;**108**(2):462–93.

[5] Estevez M.-C., Alvarez M., Lechuga L.M. 'Integrated optical devices for lab-on-a-chip biosensing applications'. *Laser and Photonics Reviews*. 2011;**6**(4):1–25.

[6] Voros J., Ramsden J.J., Csucs G., *et al*. 'Optical grating coupler biosensors'. *Biomaterials*. 2002;**23**(17):3699–710.

[7] Nellen P.M., Tiefenthaler K., Lukosz W. 'Integrated optical input grating couplers as biochemical sensors'. *Sensors and Actuators*. 1988;**15**(3):285–95.

[8] Nellen P.M., Lukosz W. 'Integrated optical input grating couplers as chemo- and immunosensors'. *Sensors and Actuators B: Chemical*. 1990;**1**(1–6):592–5.

[9] Lukosz W., Clerc D., Nellen P.M., Stamm C., Weiss P. 'Output grating couplers on planar optical waveguides as direct immunosensors'. *Biosensors and Bioelectronics*. 1991;**6**(3):227–32.

[10] Lukosz W., Nellen P.M., Stamm C., Weiss P. 'Output grating couplers on planar waveguides as integrated optical chemical sensors'. *Sensors and Actuators B: Chemical*. 1990;**1**(1–6):585–8.

[11] Brandenburg A., Polzius R., Bier F., Bitewski U., Wagner E. 'Direct observation of affinity reactions by reflected-mode operation of integrated optical grating coupler'. *Sensors and Actuators B: Chemical*. 1996:**30**(1):55–9.

[12] Piehler J., Brandenburg A., Brecht A., Wagner E., Gauglitz G. 'Characterization of grating couplers for affinity-based pesticide sensing'. *Applied Optics*. 1997;**36**(25):6554–62.

[13] Bier F.F., Kleinjung F., Scheller F.W. 'Real-time measurement of nucleic-acid hybridization using evanescent-wave sensors: steps towards the genosensor'. *Sensors and Actuators B: Chemical*. 1997;**38**(1–3):78–82.

[14] Grego S., Naskar S., Patel A.M., Huffman A., Bower C.A., Stoner B.R. 'Novel optical-waveguide sensing platform based on input grating coupler'. *Proceedings of SPIE—The International Society for Optical Engineering*. 2006;**6123**:art.61230D (11 pages).

[15] Grego S., McDaniel J.R., Stoner B.R. 'Wavelength interrogation of grating-based optical biosensors in the input coupler configuration'. *Sensors and Actuators B: Chemical*. 2008;**131**(2):347–55.

[16] Yuen P.K., Fontaine N.H., Quesada M.A., Mazumder P., Bergman R., Mozdy E.J. 'Self-referencing a single waveguide grating sensor in a micron-sized deep flow chamber for label-free biomolecular binding assays'. *Lab on a Chip*. 2005;**5**(9):959–65.

[17] Wiki M., Kunz R.E. 'Wavelength-interrogated optical sensor for biochemical applications'. *Optics Letters*. 2000;**25**(7):463–5.

[18] Cottier K., Wiki M., Voirin G., Gao H., Kunz R.E. 'Label-free highly sensitive detection of (small) molecules by wavelength interrogation of integrated optical chips'. *Sensors and Actuators B: Chemical*. 2003:**91**(1–3): 241–51.

[19] Adrian J., Pasche S., Diserens J.M., *et al*. 'Waveguide interrogated optical immunosensor (WIOS) for detection of sulfonamide antibiotics in milk'. *Biosensors and Bioelectronics*. 2009;**24**(11):3340–6.

[20] Adrian J., Pasche S., Pinacho D.G., *et al*. 'Wavelength-interrogated optical biosensor for multi-analyte screening of sulfonamide, fluoroquinolone, β-lactam and tetracycline antibiotics in milk'. *TrAC Trends in Analytical Chemistry*. 2009;**28**(6):769–77.

[21] Suarez G., Jin Y.H., Auerswald J., *et al*. 'Lab-on-a-chip for multiplexed biosensing of residual antibiotics in milk'. *Lab on a Chip*. 2009;**9**(11): 1625–30.

[22] Pasche S., Giazzon M., Wenger B., *et al*. 'Monitoring of cellular immune responses with an optical biosensor: a new tool to assess nanoparticle toxicity'. *Procedia Chemistry*. 2009;**1**(1):738–41.

[23] Razumovitch J., de Franca K., Kehl F., Wiki M., Meier W., Vebert C. 'Optimal hybridization efficiency upon immobilization of oligonucleotide double helices'. *Journal of Physical Chemistry B*. 2009;**113**(24):8383–90.

[24] Boyd R.W., Heebner J.E. 'Sensitive disk resonator photonic biosensor'. *Applied Optics*. 2001;**40**(31):5742–7.

[25] Lee S., Eom S.C., Chang J.S., Huh C., Sung G.Y., Shin J.H. 'Label-free optical biosensing using a horizontal air-slot SiNx microdisk resonator'. *Optics Express*. 2010;**18**(20):20638–44.

[26] Lee S., Eom S.C., Chang J.S., Huh C., Sung G.Y., Shin J.H. 'A silicon nitride microdisk resonator with a 40-nm-thin horizontal air slot'. *Optics Express*. 2010;**18**(11):11209–15.

[27] Schweinsberg A., Hocde S., Lepeshkin N.N., Boyd R.W., Chase C., Fajardo J.E. 'An environmental sensor based on an integrated optical whispering gallery mode disk resonator'. *Sensors and Actuators B: Chemical*. 2007; **123**(2):727–32.

[28] De Vos K., Bartolozzi I., Schacht E., Bienstman P., Baets R. 'Silicon-on-insulator microring resonator for sensitive and label-free biosensing'. *Optics Express*. 2007;**15**(12):7610–15.

[29] Li X., Zhang Z., Qin S., *et al*. 'Sensitive label-free and compact biosensor based on concentric silicon-on-insulator microring resonators'. *Applied Optics*. 2009;**48**(25):F90–4.

[30] Luchansky M.S., Washburn A.L., Martin T.A., Iqbal M., Gunn L.C., Bailey R.C. 'Characterization of the evanescent field profile and bound mass sensitivity of a label-free silicon photonic microring resonator biosensing platform'. *Biosensors and Bioelectronics*. 2010;**26**(4):1283–91.

[31] Armani A.M., Kulkarni R.P., Fraser S.E., Flagan R.C., Vahala K.J. 'Label-free, single-molecule detection with optical microcavities'. *Science*. 2007;**317**(5839):783–6.

[32] Armani A.M., Srinivasan A., Vahala K.J. 'Soft lithographic fabrication of high Q polymer microcavity arrays'. *Nano Letters*. 2007;**7**(6):1823–6.

[33] Hunt H.K., Soteropulos C., Armani A.M. 'Bioconjugation strategies for microtoroidal optical resonators'. *Sensors*. 2010;**10**(10):9317–36.

[34] Zhang X.M., Choi H.S., Armani A.M. 'Ultimate quality factor of silicamicrotoroid resonant cavities'. *Applied Physics Letters*. 2010;**96**(15):153304.

[35] Ramachandran A., Wang S., Clarke J., *et al*. 'A universal biosensing platform based on optical micro-ring resonators'. *Biosensors and Bioelectronics*. 2008;**23**(7):939–44.

[36] Ksendzov A., Lin Y. 'Integrated optics ring-resonator sensors for protein detection'. *Optics Letters*. 2005;**30**(24):3344–6.

[37] Chen A., Dalton L., Sherwood T., *et al*. 'All-organic and organic-silicon photonic ring micro-resonators'. *Proceedings of SPIE—The International Society for Optical Engineering*. 2005;**5708**:187–92.

[38] Chao C.-Y., Guo L.J. 'Biochemical sensors based on polymer microrings with sharp asymmetrical resonance'. *Applied Physics Letters*. 2003;**83**(8): 1527.

[39] Dekker R., Usechak N., Forst M., Driessen A. 'Ultrafast nonlinear all-optical processes in silicon-on-insulator waveguides'. *Journal of Physics D: Applied Physics*. 2007;**40**(14):R249–71.

[40] De Vos K., Girones J., Popelka S., Schacht E., Baets R., Bienstman P. 'SOI optical microring resonator with poly(ethylene glycol) polymer brush for label-free biosensor applications'. *Biosensors and Bioelectronics*. 2009;**24**(8): 2528–33.

[41] De Vos K., Girones J., Claes T., *et al.* 'Multiplexed antibody detection with an array of silicon-on-insulator microring resonators'. *IEEE Photonics Journal*. 2009;**1**(4):225–35.
[42] Byeon J.Y., Limpoco F.T., Bailey R.C. 'Efficient bioconjugation of protein capture agents to biosensor surfaces using aniline-catalyzed hydrazone ligation'. *Langmuir*. 2010;**26**(19):15430–5.
[43] Luchansky M.S., Bailey R.C. 'Silicon photonic microring resonators for quantitative cytokine detection and T-cell secretion analysis'. *Analytical Chemistry*. 2010;**82**(5):1975–81.
[44] Washburn A.L., Gunn L.C., Bailey R.C. 'Label-free quantitation of a cancer biomarker in complex media using silicon photonic microring resonators'. *Analytical Chemistry*. 2009;**81**(22):9499–506.
[45] Washburn A.L., Luchansky M.S., Bowman A.L., Bailey R.C. 'Quantitative, label-free detection of five protein biomarkers using multiplexed arrays of silicon photonic microring resonators'. *Analytical Chemistry*. 2010;**82**(1): 69–72.
[46] Qavi A.J., Bailey R.C. 'Multiplexed detection and label-free quantitation of microRNAs using arrays of silicon photonic microring resonators'. *Angewandte Chemie International Edition*. 2010;**49**(27):4608–11.
[47] Armani D.K., Kippenberg T.J., Spillane S.M., Vahala K.J. 'Ultra-high-Q toroid microcavity on a chip'. *Nature*. 2003;**421**(6926):925–8.
[48] Armani A.M., Armani D.K., Min B., Vahala K.J., Spillane S.M. 'Ultra-high-QQ microcavity operation in H_2O and D_2O'. *Applied Physics Letters*. 2005;**87**(15):151118.
[49] Choi H.S., Zhang X.M., Armani A.M. 'Hybrid silica-polymer ultra-high-Q microresonators', *Optics Letters*. 2010;**35**(4):459–61.
[50] Block I.D., Ganesh N., Lu M., Cunningham B.T. 'A sensitivity model for predicting photonic crystal biosensor performance'. *IEEE Sensors Journal*. 2008;**8**(3–4):274–80.
[51] Cunningham B., Li P., Lin B., Pepper J. 'Colorimetric resonant reflection as a direct biochemical assay technique'. *Sensors and Actuators B: Chemical*. 2002;**81**(2–3):316–28.
[52] Cunningham B., Lin B., Qiu J., Li P., Pepper J., Hugh B. 'A plastic colorimetric resonant optical biosensor for multiparallel detection of label-free biochemical interactions'. *Sensors and Actuators B: Chemical*. 2002;**85**(3): 219–26.
[53] Cunningham B., Qiu J., Li P., Lin B. 'Enhancing the surface sensitivity of colorimetric resonant optical biosensors'. *Sensors and Actuators B: Chemical*. 2002;**87**(2):365–70.
[54] Block I.D., Chan L.L., Cunningham B.T. 'Photonic crystal optical biosensor incorporating structured low-index porous dielectric'. *Sensors and Actuators B: Chemical*. 2006;**120**(1):187–93.
[55] Cunningham B.T., Li P., Schulz S., *et al.* 'Label-free assays on the BIND system'. *Journal of Biomolecular Screening*. 2004;**9**(6):481–9.

[56] Choi C.J., Cunningham B.T. 'A 96-well microplate incorporating a replica molded microfluidic network integrated with photonic crystal biosensors for high throughput kinetic biomolecular interaction analysis'. *Lab on a Chip*. 2007;**7**(5):550–6.

[57] Cunningham B.T., Laing L. 'Microplate-based, label-free detection of biomolecular interactions: applications in proteomics'. *Expert Review of Proteomics*. 2006;**3**(3):271–81.

[58] Foresi J.S., Villeneuve P.R., Ferrera, J., *et al*. 'Photonic-bandgap microcavities in optical waveguides'. *Nature*. 1997;**390**:143–5.

[59] Dorfner D., Zabel T., Hurlimann T., *et al*. 'Photonic crystal nanostructures for optical biosensing applications'. *Biosensors and Bioelectronics*. 2009; **24**(12):3688–92.

[60] Garcıa-Ruperez J., Toccafondo V., Banuls M.J., *et al*. 'Label-free antibody detection using band edge fringes in SOI planar photonic crystal waveguides in the slow-light regime'. *Optics Express*. 2010;**18**(23):24276–86.

[61] Toccafondo V., Garcıa-Ruperez J., Banuls M.J., *et al*. 'Single-strand DNA detection using a planar photonic-crystal-waveguide-based sensor'. *Optics Letters*. 2010;**35**(21):3673–5.

[62] Mandal S., Erickson D. 'Nanoscale optofluidic sensor arrays'. *Optics Express*. 2008;**16**(3):1623–31.

[63] Mandal S., Goddard J.M., Erickson D. 'A multiplexed optofluidic biomolecular sensor for low mass detection'. *Lab on a Chip*. 2009;**9**(20):2924–32.

[64] Lee M., Fauchet P.M. 'Two-dimensional silicon photonic crystal based biosensing platform for protein detection'. *Optics Express*. 2007;**15**(8): 4530–5.

[65] Pal S., Guillermain E., Sriram R., Miller B.L., Fauchet P.M. 'Silicon photonic crystal nanocavity-coupled waveguides for error-corrected optical biosensing'. *Biosensors and Bioelectronics*. 2011;**26**(10):4024–31.

[66] Zlatanovic S., Mirkarimi L.W., Sigalas M.M., *et al*. 'Photonic crystal microcavity sensor for ultracompact monitoring of reaction kinetics and protein concentration'. *Sensors and Actuators B: Chemical*. 2009;**141**(1):13–19.

[67] Mandal S., Serey X., Erickson D. 'Nanomanipulation using silicon photonic crystal resonators'. *Nano Letters*. 2010;**10**(1):99–104.

[68] Erickson D., Serey X., Chen Y.-F., Mandal S. 'Nanomanipulation using near field photonics'. *Lab on a Chip*. 2011;**11**(6):995–1009.

[69] Crespi A., Gu Y., Ngamsom B., *et al*. 'Three-dimensional Mach–Zehnder interferometer in a microfluidic chip for spatially-resolved label-free detection'. *Lab on a Chip*. 2010;**10**(9):1167–73.

[70] Schipper E.F., Brugman A.M., Domínguez C., Lechuga L.M., Kooyman R.P.H., Greve J. 'The realization of an integrated Mach–Zehnder waveguide immunosensor in silicon technology'. *Sensors and Actuators B: Chemical*. 1997;**40**(2–3):147–53.

[71] Heideman R.G., Lambeck P.V. 'Remote opto-chemical sensing with extreme sensitivity: design, fabrication and performance of a pigtailed

integrated optical phase-modulated Mach–Zehnder interferometer system'. *Sensors and Actuators B: Chemical*. 1999;**61**(1–3):100–27.

[72] Brosinger F., Freimuth H., Lacher M., *et al*. 'A label-free affinity sensor with compensation of unspecific protein interaction by a highly sensitive integrated optical Mach–Zehnder interferometer on silicon'. *Sensors and Actuators B: Chemical*. 1997;**44**(1–3):350–5.

[73] Weisser M., Tovar G., Mittler-Neher S., *et al*. 'Specific bio-recognition reactions observed with an integrated Mach–Zehnder interferometer' *Biosensors and Bioelectronics*. 1999;**14**(4):405–11.

[74] Prieto F., Sepulveda B., Calle A., *et al*. 'An integrated optical interferometric nanodevice based on silicon technology for biosensor applications'. *Nanotechnology*. 2003;**14**(8):907–12.

[75] Prieto F., Llobera A., Jimenez D., Domínguez C., Calle A., Lechuga L.M. 'Design and analysis of silicon anti-resonant reflecting optical waveguides for evanescent field sensor'. *Journal of Lightwave Technology*. 2000;**18**(7): 966–72.

[76] Prieto F., Sepulveda B., Calle A., Llobera A., Domínguez C., Lechuga L.M. 'Integrated Mach–Zehnder interferometer based on ARROW structures for biosensor applications'. *Sensors and Actuators B: Chemical*. 2003;**92**(1–2): 151–8.

[77] Brandenburg A., Krauter R., Kunzel C., Stefan M., Schulte H. 'Interferometric sensor for detection of surface-bound bioreactions'. *Applied Optics*. 2000;**39**(34):6396–405.

[78] Schmitt K., Schirmer B., Hoffmann C., Brandenburg A., Meyrueis P. 'Interferometric biosensor based on planar optical waveguide sensor chips for label-free detection of surface bound bioreactions'. *Biosensors and Bioelectronics*. 2007;**22**(11):2591–7.

[79] Brynda E., Houska M., Brandenburg A., Wikerstal A. 'Optical biosensors for real-time measurement of analytes in blood plasma'. *Biosensors and Bioelectronics*. 2002;**17**(8):665–75.

[80] Hoffmann C., Schmitt K., Brandenburg A., Hartmann S. 'Rapid protein expression analysis with an interferometric biosensor for monitoring protein production'. *Analytical and Bioanalytical Chemistry*. 2007;**387**(5): 1921–32.

[81] Nagel T., Ehrentreich-Forster E., Singh M., *et al*. 'Direct detection of tuberculosis infection in blood serum using three optical label-free approaches'. *Sensors and Actuators B: Chemical*. 2008;**129**(2):934–40.

[82] Ymeti A., Kanger J.S., Greve J., Lambeck P.V., Wijn R., Heideman R.G. 'Realization of a multichannel integrated Young interferometer chemical sensor'. *Applied Optics*. 2003;**42**(28):5649–60.

[83] Ymeti A., Greve J., Lambeck P.V., Wijn R., Heideman R.G., Kanger J.S. 'Drift correction in a multichannel integrated optical Young interferometer'. *Applied Optics*. 2005;**44**(17):3409–12.

[84] Ymeti A., Kanger J.S., Greve J., *et al.* 'Integration of microfluidics with a four-channel integrated optical Young interferometer immunosensor'. *Biosensors and Bioelectronics*. 2005;**20**(7):1417–21.

[85] Ymeti A., Greve J., Lambeck P.V., *et al.* 'Fast, ultrasensitive virus detection using a Young interferometer sensor'. *Nano Letters*. 2007;**7**(2):394–7.

[86] Schneider B.H., Edwards J.G., Hartman N.F. 'Hartman interferometer: versatile integrated optic sensor for label-free, real-time quantification of nucleic acids, proteins, and pathogens'. *Clinical Chemistry*. 1997;**43**(9):1757–63.

[87] Schneider B.H., Dickinson E.L., Vach M.D., Hoijer J.V., Howard L.V. 'Highly sensitive optical chip immunoassays in human serum'. *Biosensors and Bioelectronics*. 2000;**15**(1–2):13–20.

[88] Schneider B.H., Dickinson E.L., Vach M.D., Hoijer J.V., Howard L.V. 'Optical chip immunoassay for hCG in human whole blood'. *Biosensors and Bioelectronics*. 2000;**15**(11–12):597–604.

[89] Xu J., Suarez D., Gottfried D. 'Detection of avian influenza virus using an interferometric biosensor'. *Analytical and Bioanalytical Chemistry*. 2007;**389**(4):1193–9.

[90] Zinoviev K.E., González-Guerrero A.B., Domínguez C., Lechuga L.M. 'Integrated bimodal waveguide interferometric biosensor for label-free analysis'. *Journal of Lightwave Technology*. 2011;**29**(13):1926–30.

[91] Duval D., González-Guerrero A.B., Dante S., *et al.* 'Nanophotonic lab-on-a-chip platforms including novel bimodal interferometers, microfluidics and grating couplers'. *Lab on a Chip*. 2012;**12**:1987–94.

[92] Maldonado J., González-Guerrero A.B., Domínguez C., Lechuga L.M. 'Label-free bimodal waveguide immunosensor for rapid diagnosis of bacterial infections in cirrhotic patients'. *Biosensors and Bioelectronics*. 2016;**85**:310–16.

[93] Densmore A., Xu D.X., Janz S., *et al.* 'Spiral-path high-sensitivity silicon photonic wire molecular sensor with temperature-independent response'. *Optics Letters*. 2008;**33**(6):596–8.

[94] Xu D.X., Vachon M., Densmore A., *et al.* 'Real-time cancellation of temperature induced resonance shifts in SOI wire waveguide ring resonator label-free biosensor arrays'. *Optics Express*. 2010;**18**(22):22867–79.

[95] Densmore A., Vachon M., Xu D.X., *et al.* 'Silicon photonic wire biosensor array for multiplexed real-time and label-free molecular detection'. *Optics Letters*. 2009;**34**(23):3598–600.

[96] Xu D.X., Vachon M., Densmore A., *et al.* 'Label-free biosensor array based on silicon-on-insulator ring resonators addressed using a WDM approach'. *Optics Letters*. 2010;**35**(16):2771–3.

[97] Dell'Olio F., Passaro V.M.N. 'Optical sensing by optimized silicon slot waveguides'. *Optics Express*. 2007;**15**(8):4977–93.

[98] Barrios C.A., Gylfason K.B., Sanchez B., *et al.* 'Slot-waveguide biochemical sensor'. *Optics Letters*. 2007;**32**(21):3080–2.

[99] Barrios C.A., Banuls M.J., González-Pedro V., *et al.* 'Label-free optical biosensing with slot-waveguides'. *Optics Letters*. 2008;**33**(7):708–10.

[100] Carlborg C.F., Gylfason K.B., Kazmierczak A., *et al.* 'A packaged optical slot-waveguide ring resonator sensor array for multiplex label-free assays in labs-on-chips'. *Lab on a Chip*. 2010;**10**(3):281–90.

[101] Di Falco A., O'Faolain L., Krauss T.F. 'Chemical sensing in slotted photonic crystal heterostructure cavities'. *Applied Physics Letters*. 2009;**94**(6):063503.

[102] Misiakos K., Kakabakos S.E., Douvas A., Argitis P. 'Monolithic silicon optoelectronic biochips'. *Technical Digest—International Electron Devices Meeting*. 2001;359–62.

[103] Misiakos K., Kakabakos S.E., Petrou P.S., Ruf H.H. 'A monolithic silicon optoelectronic transducer as a real-time affinity biosensor'. *Analytical Chemistry*. 2004;**76**(5):1366–73.

[104] Chynoweth A.G., McKay K.G. 'Photon emission from avalanche breakdown in silicon'. *Physical Review*. 1956;**102**:369–76.

[105] Nagata T., Namba T., Kuroda Y., *et al.* 'Single-chip integration of light-emitting diode, waveguide and micromirrors', *Japanese Journal of Applied Physics*. 1995;**34**(2B):1282–5.

[106] Misiakos K., Petrou P.S., Kakabakos S.E., *et al.* 'Monolithic silicon optoelectronic transducers and elastomeric fluidic modules for bio-spotting and bio-assay experiments'. *Microelectronic Engineering*. 2006;**83**(4–9): 1605–8.

[107] Ruf H.H., Knoll T., Misiakos K., *et al.* 'Biochip-compatible packaging and micro-fluidics for a silicon opto-electronic biosensor'. *Microelectronic Engineering*. 2006;**83**(4–9):1677–80.

[108] Mavrogiannopoulou E., Petrou P.S., Kakabakos S.E., Misiakos K. 'Real-time detection of BRCA1 gene mutations using a monolithic silicon optocoupler array'. *Biosensors and Bioelectronics*. 2009;**24**(5): 1341–7.

[109] Petrou P.S., Kakabakos S.E., Misiakos K. 'Silicon optocouplers for biosensing'. *International Journal of Nanotechnology*. 2009;**6**(1–2):4–17.

[110] Misiakos K., Petrou P.S., Kakabakos S.E., *et al.* 'Fully integrated monolithic optoelectronic transducer for real-time protein and DNA detection: the NEMOSLAB approach'. *Biosensors and Bioelectronics*. 2010;**26**(4): 1528–35.

[111] Misiakos K., Makarona E., Kitsara M., Raptis I. 'Monolithic silicon optocoupler engineering based on tapered waveguides'. *Microelectronic Engineering*. 2008;**85**(5–6):1074–6.

[112] Kitsara M., Raptis I., Misiakos K., Makarona E. 'Broad-band Mach–Zehnder interferometry as a detection principle for label-free biochemical sensing'. *Proceedings IEEE Sensor*. 2008;934–7, art.no. 4716594.

[113] Misiakos K., Raptis I., Salapatas A., *et al.* 'Broad-band Mach–Zehnder interferometers as high performance refractive index sensors: theory and monolithic implementation'. *Optics Express*. 2014;**22**(8): 8856–70.

[114] Misiakos K., Raptis I., Makarona E., *et al.* 'All-silicon monolithic Mach–Zehnder interferometer as a refractive index and biochemical sensor'. *Optics Express*. 2014;**22**(22):26803–13.

[115] Psarouli A., Salapatas A., Botsialas A., *et al.* 'Monolithically integrated broad-band Mach–Zehnder interferometers for highly sensitive label-free detection of biomolecules through dual polarization optics'. *Scientific Reports*. 2015;**5**:17600.

[116] Angelopoulou M., Botsialas A., Salapatas A., *et al.* 'Assessment of goat milk adulteration with a label-free monolithically integrated optoelectronic biosensor'. *Analytical and Bioanalytical Chemistry*. 2015;**407**(14):3995–4004.

[117] Pagkali V., Petrou P.S., Salapatas A., *et al.* 'Detection of ochratoxin A in beer samples with a label-free monolithically integrated optoelectronic biosensor' *Journal of Hazardous Materials*. 2016; in press, doi:10.1016/j.jhazmat.2016.03.019.

Chapter 9

Time-series processing for portable biosensors and mobile platforms for automated pattern recognition

C.P. Yialouris[1] and K.P. Ferentinos[1]

9.1 Introduction

A *time-series* is an ordered by time set of data over a predefined continuous time interval. Each data item of a time-series represents a measurement that was obtained within a certain period of time (e.g. a second, a minute, etc.) or it is a value that results as a statistical value (average, standard deviation, etc.) of a predefined number of different values obtained within a predefined time slice of the period of the time-series. Two representative examples are shown in the tables below. The first one presents a time-series comprising actual measurement values in Table 9.1; the second one presents a time-series that each item has resulted from average value of group of four consecutive measurement values in Table 9.2.

Data contained in time-series are often acquired by monitoring various phenomena, such as physical, biological, environmental etc. Information extracted

Table 9.1 Illustration of time-series with actual measurements as values

Seconds	1	2	3	4	5	6	7	8	9	10
Time-series value	0.25	0.27	0.29	0.3	0.31	0.32	0.37	0.39	0.4	0.4

Table 9.2 Illustration of time-series with values resulted from actual measurements

Seconds	1				2				3			
Time-series value	0.25				0.27				0.29			
Measurement	0.25	0.25	0.26	0.24	0.28	0.27	0.25	0.28	0.30	0.29	0.29	0.28

[1]Informatics Laboratory, Department of Agricultural Economics and Development, Agricultural University of Athens, Athens, Greece

from a time-series processing is mainly used to understand the phenomenon as well as to predict a future value or to classify the phenomenon monitored. So, the goal of time-series processing is the interpretation of the phenomenon which produced the time-series data under processing.

Often, different types of time-series provide a system with different types of information. By combining the plethora of information, a complicated conclusion can be drawn. Most time-series are captured using special sensors for monitoring the phenomenon.

In this chapter, techniques concerning processing of time-series produced by portable biosensors will be presented. A biosensor can be defined as a 'compact analytical device or unit incorporating a biological or biologically derived sensitive recognition element integrated or associated with a physio-chemical transducer' [1,2]. There are three main parts of a biosensor: (i) the bio-receptor which is the sensitive biological element that reacts with a specific type of chemicals or biological agents, (ii) a transducer that converts the biorecognition reaction into a measurable signal, and (iii) a signal processing system that reads the measurement signal, providing the user with proper information or advice.

Advanced requirements of a wide range of current sensing applications demand for portable biosensors and corresponding mobile platforms instead of large, laboratory-based devices. The construction of such mobile, portable platforms can be achieved with the reduction of the cost of the chips and several energy-conservation related technological advancements of recent years. The sensors and electronics that monitor the output of such portable systems must be compact and low power, whereas any existing mechanical interface must be robust to resist vibrations and shocks. Some years ago restrictions concerning power, size and speed limit the amount of signal processing available [3]. Although most of the above limitations can be overcome using powerful small and portable devices mainly on processing power and energy consumption, measured data analysis methods should be tailor-made and optimized towards limited computation power and high-energy conservation.

9.2 Time-series analysis

The two basic purposes of time-series analysis are prediction and pattern recognition. The former concerns prediction of future values of the measured variables included in the time-series, whereas the latter concerns the classification of time-series values into specific, predefined patterns. This work focuses on the pattern recognition task concerning time-series analysis.

A crucial step in time-series analysis for pattern recognition is time-series representation [4]. Various algorithms for such representation have been proposed, such as sampling, averaging and exponential smoothing [5–7], symbolic mappings [8,9] and Fourier transformations [10,11].

Computational intelligence techniques, such as artificial neural networks (ANNs), support vector machines (SVMs) and genetic algorithms (GAs), have become increasingly popular and successful in time-series analysis in recent years, mainly because of the development of adequate hardware capable of effectively simulating such models. ANNs constitute cost-effective methods for achieving good results with time-series while utilizing meta-heuristic methods [12–14]. They can also effectively be combined with other methodologies to construct novel hybrid models, which include combinations of auto regressive integrated moving average models and ANNs [15], rough set-back propagation [16], recursive structured set of multi-layered perceptrons or auto-deterministic networks [17]. On the other hand, SVMs offer a good alternative to ANNs, improving the generalization performance while achieving, at specific cases, better overall results, especially when combined with genetically driven pre-processing [18,19]. Several modified SVM models have also been developed, such as least-squares SVMs, normal [20] and recurrent [21], and minimum class variance SVMs [22]. Finally, GAs and variations of evolutionary programming have been effectively used in order to utilize time-series segmentation and facilitate effective representation [23–25].

Piecewise linear representation (PLR) models constitute a powerful approach to time-series segmentation. Based on those models, a time-series of length n is approximated by K straight lines, where K is typically much smaller than n. The algorithms that perform this type of time-series segmentation use an arbitrary number of segments, whereas the maximum and average error of all segments is less than an arbitrary threshold. Several variations of PLR algorithms have been proposed, towards the enhancement of the resulting representation. Ding *et al.* [26] used segment radian errors in order to enhance data compression and time-series representation. Guerrero *et al.* [27] tackled the segmentation process as a multi-objective optimization problem. Along with the obtained final approximation error, they introduced the number of segments in the final representation as another quality metric over the final results. Tseng *et al.* [28] proposed a segmentation approach which combined the clustering technique, discrete wavelet transformation and GAs in a process to find the segmentation points for deriving appropriate patterns. Finally, Glezakos *et al.* [29] developed an evolutionary computation-based PLR model using GAs for pre-processing the initial time-series information, and thus producing fitter secondary data sets used in the supervised training and testing of ANN and SVM classifiers.

Although there are a lot of approaches in time-series analysis, most of which may have problems to be implemented in order to install in a mobile device. In the next section, we will concentrate on three time series processing techniques, which can be used for the development of a software application that can be used by a mobile device for pattern recognition.

9.3 Extracting features from time-series created by biosensors for pattern recognition

In recent years, a significant increase in the application of diagnostic systems using biosensors is observed. The majority of biosensors are based on the conversion of a biological function of the biological component (bio-probe) of the biosensor to an electrical signal which can be measured by a suitable device. The continuous measurements of a phenomenon create a time-series. The processing of this time-series can have two main objectives:

1. To estimate the value of a biological agent, for example the value of glucose in the human blood or the concentration of a pesticide in fresh fruits. This is a regression problem.
2. To classify the factor that reacted with the bio-probe, caused the biological function of the biosensor and created the specific time-series, to a specific class. This is a classification problem.

The above can be achieved using one or more pattern recognition algorithms and tools.

9.3.1 Pattern recognition

Pattern recognition is a branch of artificial intelligence and machine learning, which mainly focuses on the recognition of patterns and regularities in data. A pattern recognition system is usually trained by examples using a set of data, namely the *training data set*. During training, the system is trained to associate inputs with output patterns. A training data set contains a significant number or records. In classification problems, the data to be processed are divided into groups called samples, and each one of them belongs to one of a pre-defined set of possible classes. In regression problems, each record of the training data set is assigned to a known value.

Once a pattern recognition system is trained, it is able to be used in order for a given input data to be classified to a member of a known predefined class (classification), or to be assigned to some value (regression).

There are many classification systems, based on several techniques. One of the most successful and widely used classification tools is based on ANNs. ANNs are inspired by the way a biological nervous system works, creating synapses in order to learn and to be used for future processing. ANNs are characterized by the ability to operate successfully using complicated, imprecise or even missing data. This ability can be used in classification problems as well as in regression ones.

In pattern recognition, through the training process, an ANN is configured for an application of a specific type, such as pattern recognition or data classification. During training, the network is trained to associate inputs with output patterns, using specialized training algorithms, like Backpropagation. Once the neural network is trained, it can be used for the identification of unknown samples, meaning data samples that have not been used during training.

A time-series usually contains a rather large number of values, for example 300 or more. These values cannot constitute inputs to a neural network model, because the training of networks with such a large number of inputs can be considerably slow and problematic because of the noise that the raw data contain. Thus, it is necessary that a set of features is extracted from each time-series, and these features are then used in the training process.

In the following subsections, three practical techniques are presented for the extraction of features from time-series, in order to create training data sets to train ANNs capable of identifying plant viruses and pesticides in food commodities.

9.3.2 Resampling

For a given time-series, we can apply a smoothing technique, like re-sampling, to extract the necessary features. The technique is simple. For a given time-series that contains N values and the re-sampling rate is K, the training set will contain values from the time-series taken every k time stamps. According to this smoothing technique, the number of the produced features and also the dimensionality of the problem are reduced and the number of the input nodes of the neural network is N/K. The final selection of the number K is depended on the problem and will decide after some test and trial procedures. This technique was used successfully for the development of a neural for plant virus identification [30].

9.3.3 Fixed segmentation

According to this technique, metadata are created form a time-series to be used to train a neural network.

The basic idea concerns the division of the time-series into a relative small (k) number of segments. For each segment, the average and the standard deviation is computed. The metadata used for the training can contain the following:

- The average of all values of the time-series.
- The standard deviation of all values of the time-series.
- The k average values of each of the k segments of the time-series.
- The k standard deviation values of each of the k segments of the time-series.
- The minimum value of the time-series.
- The maximum value of the time-series.

This technique was used in Ferentinos *et al.* [31], dividing the data samples into four equal-length segments; thus, each record of the training data set contained 12 items. The developed neural network was used for the identification of existence of specific pesticides residues in food commodities.

9.3.4 Feature extraction with genetic algorithms support

In this approach, metadata are created from time-series, to be used for training a neural network. The creation is based on a GA. GAs are inspired by biology and are based on the axiom *survival of the fittest*. GAs adopt terms of evolutionary biology, such as selection, crossover and mutation, to perform specific functions. The GA system can create the training data set, train the corresponding neural network, evaluate it and propose the best architecture of the neural network.

The basic idea of the metadata creation is the following. By randomly assigning a bit (0 or 1) to each item of the time-series, a record having the same length with the time-series is created, called chromosome genome. By mapping the chromosome genome to the time-series, virtual segments are created, according to 0 and 1 bits.

Based on the above approach, the system can combine

- Re-sampling techniques but with variable re-sampling rate.
- Variable length segmentation to extract different statistical indices.

The above approach has been used for identification of plant viruses using biosensors' time-series. Full description of the technique is presented in Glezakos *et al.* [29,32].

9.4 Portable biosensors using Smartphone capabilities

Today mobile devices such as Smartphones and tablets have features such as CPUs with 8 cores, 3+ GB of RAM memory and large data storage capacities with more than 64 GB and fully functional operating systems. These characteristics give mobile devices capabilities for high-level processing and satisfactory storage. Most of these devices can be connected to the internet in many ways, such as 4G or WiFi. For example, an Android app can use various internet protocols such as HTTP (HyperText Transfer Protocol), HTTPS (HyperText Transfer Protocol Secure), TCP/IP (Transmission Control Protocol/ Internet Protocol). The Android SDK (Software Development Kit) provides developers accessibility and use of these communication protocols.

Thus, a mobile device could be used as the transducer of a biosensor device. Although the development of a classifier, such as a neural network model, would be developed using a conventional computer, the developed software could be easily modified to create a suitable mobile app able to run on mobile devices. For the development of a biosensor data acquisition device, the use of a mobile device is necessary. Today manufacturers have developed data-specific acquisition devices having very small dimensions, which can be connected to a mobile device through the USB (Universal Serial Bus) port or wirelessly, using standard protocols such as Bluetooth.

We can distinguish three different uses of mobile devices as biosensors.

1. The mobile device works as stand-alone. An app installed in the Smartphone or tablet has embedded the software for the communication with the data acquisition module and one or more neural networks software for the pattern recognition process (Figure 9.1).
2. The mobile device works using either the WiFi or 3G/4G communication facilities for connection with a server in which the neural networks are stored. The mobile device is responsible for the communication with the data acquisition module. As soon as the measurement is acquired, the user can send the time-series to the server either by SMS or through a dedicated web application for uploading the data. The server processes the data and sends the answer to the user either by SMS or directly through the web application (Figure 9.2).
3. Portable biosensor-based web services. In the basic flow of the process, the sensor and data logger provide the raw data to the device (Figure 9.3).
   (i) In the mobile presentation layer, the mobile application serves as a thin client for Android/iOS operating systems, by implementing minimal functionality to the integrated system. Its core role is to prepare the XML

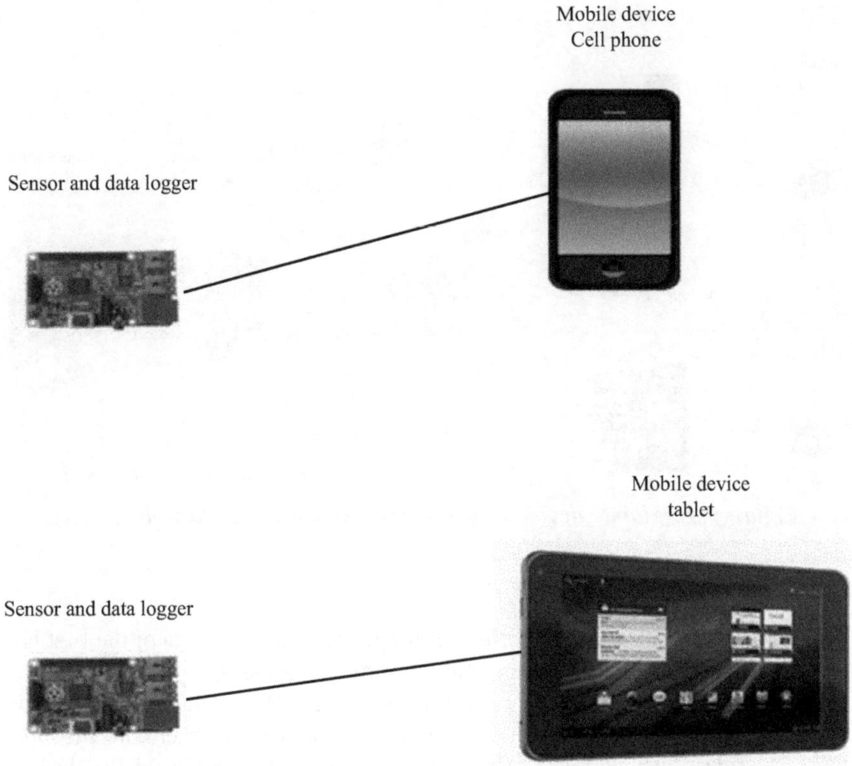

Figure 9.1 Stand-alone architecture of mobile device as a portable biosensor

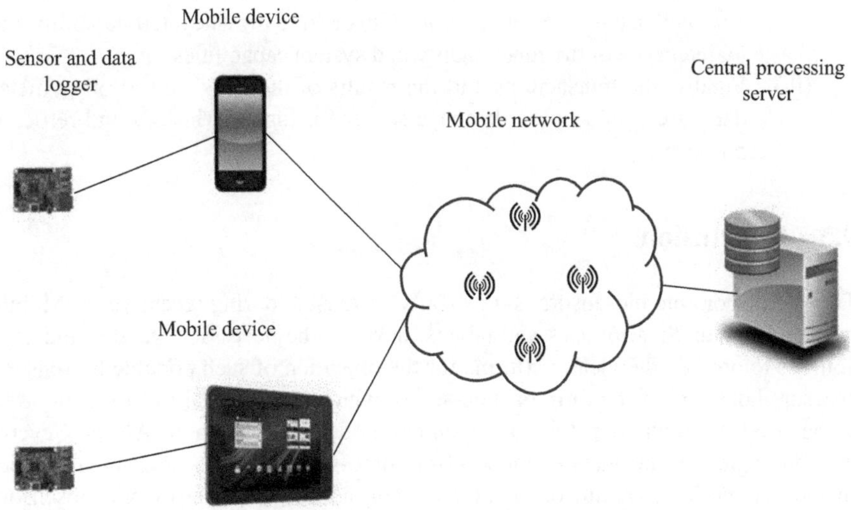

Figure 9.2 Mobile device as a portable biosensor based on mobile network

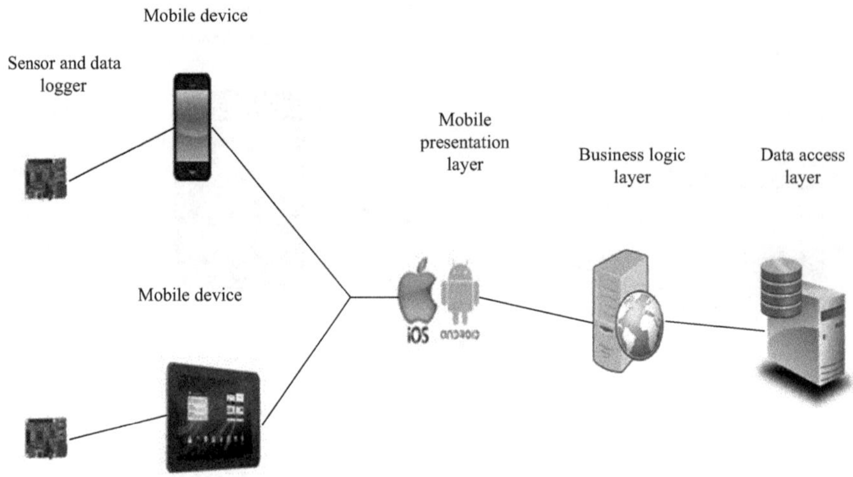

Figure 9.3 Mobile device as a portable biosensor-based web services

documents and their format, equipped with the raw data of the first layer and transmit to the back-end for further processing.

(ii) In the business logic layer, there is a server-side implementation of the data analysis algorithms that support the whole process. Furthermore, the server-side functionality is responsible to transmit the results of the decision-making process, back to the client (mobile device) and inform the user about the outcome of the process. It is worthwhile to mention that in the business logic layer, a web services implementation could be particularly designed in the scope of the project, in order to provide interoperability and maintenance of the functionality and system capabilities.

(iii) Finally, the transactions and the results of them, as well as other meta-data, are stored in the database server for logging, history and retrieval purposes.

9.5 Conclusion

The use of portable biosensors is drastically increasing during recent years. Mobile devices, such as Smartphones and tablets, have all the necessary features and capabilities to provide the ideal platforms for the operation of such portable biosensors. Measured data are of the form of time-series, which need special treatment in order to be used as input to pattern recognition models, for example ANNs. Several methodologies for metadata extraction from time-series are presented and analyzed. In addition, possible architectures of the uses of mobile devices as mobile biosensors are presented.

References

[1] Turner A.P. 'Biosensors—sense and sensitivity'. *Science*. 2000;**290**:1315–7.

[2] Koyun A., Ahlatcıoğlu E., Yeliz Koca I. 'Biosensors and their principles'. In: S. Kara (Ed.). *A Roadmap of Biomedical Engineers and Milestones*. 2012. Available from <http://www.intechopen.com/books/a-roadmap-of-biomedical-engineers-and-milestones> [accessed 1 August 2016].

[3] DeBusschere B.D., Borkholder D., Kovacs G. 'System requirements for a portable cell based sensor'. [online]. 2014. Available from <http://scholarworks.rit.edu/other/125> [accessed 1 August 2016].

[4] Fu T. 'A review on time series data mining'. *Engineering Applications of Artificial Intelligence*. 2011;**24**:164–81.

[5] Box G., Jenkins G. *Time Series Analysis: Forecasting and Control*, revised edition. Oakland, CA: Holden-Day; 1976. p. 575.

[6] Velleman P.W., Hoaglin D.C. *Applications, Basics, and Computing of Exploratory Data Analysis*. Boston, MA: Duxbury Press; 1981. p. 292.

[7] Frank R.J., Davey N., Hunt S.P. 'Time series prediction and neural networks'. *Journal of Intelligent and Robotic Systems*. 2001;**31**(1–3):91–103.

[8] Rajagopalan V., Ray A., Samsi R., Mayer J. 'Pattern identification in dynamical systems via symbolic time series analysis'. *Pattern Recognition*. 2007;**40**(11):2897–907.

[9] Lesher S., Guan L., Cohen A.H. 'Symbolic time-series analysis of neural data'. *Neurocomputing*. 2000;**32–33**:1073–81.

[10] Sharma K.K., Joshi S.D. 'Time delay estimation using fractional Fourier transform'. *Signal Process*. 2007;**87**:853–65.

[11] Giampaoli I., Ng W.L., Constantinou N. 'Analysis of ultrahigh-frequency financial data using advanced Fourier transforms'. *Finance Research Letters*. 2009;**6**:47–53.

[12] Prudencio R.B.C., Ludermir T.B. 'Meta-learning approaches to selecting time series models'. *Neurocomputing*. 2004;**61**:121–37.

[13] Elizondo D.A., Birkenhead R., Gongora M., Taillard E., Luyima P. 'Analysis and test of efficient methods for building recursive deterministic perceptron neural networks'. *Neural Networks*. 2007;**20**:1095–108.

[14] Sivagaminathan R.K., Ramakrishnan S. 'A hybrid approach for feature subset selection using neural networks and ant colony optimization'. *Expert Systems with Applications*. 2007;**33**:49–60.

[15] Diaz-Robles L.A., Ortega J.C., Fu J.S., *et al*. 'A hybrid ARIMA and artificial neural networks model to forecast particulate matter in urban areas: the case of Temuco, Chile'. *Atmospheric Environment*. 2008;**42**:8331–40.

[16] Xiao Z., Ye S.J., Zhong B., Sun C.X. 'BP neural network with rough set for short term load forecasting'. *Expert Systems with Applications*. 2009;**36**:273–9.

[17] Hamzacebi C. 'Improving artificial neural networks' performance in seasonal time series forecasting'. *Information Sciences*. 2008;**178**:4550–9.

[18] Lu W.Z., Wang W.J. 'Potential assessment of the "support vector machine" method in forecasting ambient air pollutant trends'. *Chemosphere*. 2005;**59**:693–701.

[19] Huang S.C., Wu T.K. 'Integrating GA-based time-scale feature extractions with SVMs for stock index forecasting'. *Expert Systems with Applications*. 2008;**35**:2080–8.

[20] Cao S.G., Liu Y.B., Wang, YP. 'A forecasting and forewarning model for methane hazard in working face of coal mine based on LS-SVM'. *Journal of China University of Mining and Technology*. 2008;**18**:0172–6.

[21] Sun J., Zheng C., Zhou Y., Bai Y., Luo J. 'Nonlinear noise reduction of chaotic time series based on multidimensional recurrent LS-SVM'. *Neurocomputing*. 2008;**71**:3675–9.

[22] Laskaris N.A., Zafeiriou S.P., Garefa L. 'Use of random time intervals (RTIs) generation for biometric verification'. *Pattern Recognition*. 2009;**42** (11):2787–96.

[23] Cao H., Recknagel F., Joo G.J., Kim D.K. 'Discovery of predictive rule sets for chlorophyll-a dynamics in the Nakdong River (Korea) by means of the hybrid evolutionary algorithm HEA'. *Ecological Informatics*. 2006;**19** (1):43–53.

[24] Gaur S., Deo M.C. 'Real-time wave forecasting using genetic programming'. *Ocean Engineering*. 2008;**35**(11–12):1166–72.

[25] Hung J.C. 'A genetic algorithm approach to the spectral estimation of time series with noise and missed observations'. *Information Sciences*. 2008; **178** (24):4632–43.

[26] Ding Y., Yang X., Kavs A., Li J. 'A novel piecewise linear segmentation for time series'. *Proceedings of the Second International Conference on Computer and Automation Engineering (ICCAE)*, Zhejiang University, Hangzhou, China, 2010. Vol. **4**. pp. 52–5.

[27] Guerrero J., Berlanga A., Garcia J., Molina M. 'Piecewise linear representation segmentation as a multiobjective optimization problem'. *Proceeding of the Seventh International Conference, Distributed Computing and Artificial Intelligence*: (Advances in Intelligent and Soft Computing, 79). Berlin: Springer; 2010. pp. 267–74.

[28] Tseng V.S., Chen C.H., Huang P.C., Hong T.P. 'Cluster-based genetic segmentation of time series with DWT'. *Pattern Recognition Letters*. 2009;**30**:1190–7.

[29] Glezakos T.J., Tsiligiridis T.A., Yialouris C.P. 'Piecewise evolutionary segmentation for feature extraction in time series models'. *Neural Computing and Applications*. 2012;**24**:243–57.

[30] Frossyniotis D., Anthopoulos Y., Kintzios S., Perdikaris A., Yialouris C.P. 'A multisensor fusion system for the detection of plant viruses by combining artificial neural networks'. In: S. Kollias *et al*. (Eds.): *ICANN 2006, Part II, LNCS 4132*. Berlin: Springer-Verlag; 2006. pp. 401–9.

[31] Ferentinos K.P., Yialouris C.P., Blouchos P., Moschopoulou G., Tsourou V., Kintzios, S. 'The use of artificial neural networks as a component of a

cell-based biosensor device for the detection of pesticides'. 2012;**47**: 989–92.

[32] Glezakos T.J., Moschopoulou G., Tsiligiridis T.A., Kintzios S., Yialouris C.P. 'Assessing the classification accuracy of an evolutionary neural network: the case of plant virus identification'. *Proceeding of the Fourth International Conference HAICTA 2008: Information and Communication Technology in Bio and Earth Sciences*. Athens, Sep 2008. pp. 370–8.

Part III

Applications

Chapter 10
Nanosensors in food safety

Preetam Sarkar[1]*, Shubham Subrot Panigrahi*[1]*,*
Emily Roy[2] *and Pratik Banerjee*[2]

10.1 Introduction

10.1.1 Food safety: global public health concern

Food safety is a critical and persistent public health issue. The concerns associated with food safety are further intensified by improper hygiene, poor food handling practices, and contaminated food supplies, leading to a financial burden of food-borne disease (FBD). The FBDs are often linked to consumer illness, which bears high medical costs and loss of productivity and sales [1,2]. To combat the threat of FBDs, an increased and comprehensive awareness of food safety is of paramount importance. The safety of the food enormously influences consumer health. There are several factors that ensure the safety of processed and packaged food commodities from pathogenic microorganisms, such as *Listeria monocytogenes, Escherichia coli* O157:H7, *Toxoplasma gondii, Campylobacter jejuni, Salmonella, Staphylococcus aureus, Campylobacter coli, Bacillus cereus, Norovirus*, and numerous others that can deleteriously impact human health [3,4]. Hence considering the roles that food safety plays in both health and development, relevant actions have been taken by various agencies in different nations to improve the safety of the food supplied to the consumers. In the United States, the US Food and Drug Administration (FDA) and the United States Department of Agriculture (USDA) are key agencies providing food safety guidelines, standards, and policies to US entities and many other nations. The European Food Safety Authority and the Food Standards Agency in the United Kingdom were authorized to survey the quality and safety of foods sold in stores. In addition, they are also assigned the responsibility of performing research in food safety and, therefore, plays crucial roles in uplifting the food safety scenario in the EU, the UK, and across the globe [5,6].

Various foodborne diseases cause significant morbidity and mortality worldwide. Historically, efforts to reduce the life-threatening consequences of food

[1]Department of Food Process Engineering, National Institute of Technology Rourkela, Rourkela, Odisha, India
[2]Division of Epidemiology, Biostatistics, and Environmental Health Science, School of Public Health, University of Memphis Robison Hall, Memphis, TN, USA

contamination have been made by innovation in food preservation. Sun drying and cooking were conceivably the first methods used; later more sophisticated technologies, such as fermentation and canning, came into existence. In recent times, advanced technologies in food preservation and packaging have made food safer. As the global population is increasing, scientists with innovative approaches towards science and technology are working efficiently to provide the best quality and safest foods to consumers. In between these approaches, nanotechnology provides an advanced and powerful platform utilizing unique properties of materials emerging from nanometric size (1–100 nm) that have the prospect of revolutionizing agriculture and food sectors, biomedicine, environment safety, energy conservation, and many other areas [7,8].

10.1.2 Food safety: a challenging field for nanotechnological innovations

The far-reaching implications of nanotechnology in the twenty-first century confer substantial impact on health, industry, and the global economy [9]. They possess the potential to maximize agricultural productivity and food quality through proper management of processing and packaging within food production facilities [10]. Apart from the food-related issues, this technology can also be implemented in the areas of toxic sensing and prevention by exploiting unique properties of nanomaterial. The integration of nanotechnology in electrochemistry, photosynthesis, and photo-electrochemical cells along with other modern technology to address the growing need for safe and renewable energy resources continues to provide high-impact solutions to these sectors [11]. Nanotechnology plays a major role in all stages of the food production chain starting from the initial agricultural production to food processing in which food ingredients can be encapsulated within nanomaterials to achieve enhanced functionality. In addition, nanotechnology has been explored tremendously in food packaging [12,13]. For example, the food-packaging sector has seen many transformations in barrier perfection with the advent of various nanoscale fillers, which have also resulted in reduced effects of accelerating factors due to contamination and spoilage. In addition to packaging, preservation of food materials also holds great potential in the application of nanotechnology. All these issues are regulated by different segments of nanotechnology, of which nanosensors often play a crucial role.

There are various factors that affect a country's economic conditions, of which agri-business is one of them. In today's world, many nations encounter pressing challenges, including environmental issues, such as the build-up of fertilizers and pesticides, loss of viable use of natural resources, urbanization, and climate change [14]. The field of nanoscience has the potential to help farmers to maintain their farms with precise control. Biomass conversion and material science technologies can be applied to develop proper handling of fertilizers and chemical pesticides. Through the use of nanomaterial-based nanosensors such as surface plasmon resonance (SPR) consisting of gold nanoparticle, quantum dots (QDs), and clay nanotube [15], farmers can detect pests, plant viruses, evidence of drought, and levels of soil nutrients, to name a few. Nano-encapsulated fertilizers that are released slowly can be used for the further consumption process and to minimize environmental pollution [16–18].

Smart packaging is categorized under an advanced generation of packaging, which incorporates nanosensors that react to the physical or chemical change in food samples within the packaging to delay spoilage or contamination. Various assets, such as mechanical, thermal, and optical features of nanosensor-possessing nanocomposites, are used to enhance the properties of foodstuffs to extend shelf life and maintain freshness and quality by regulating the route of unwanted gases and moisture through the packaging material [19,20]. Nanosensors in the form of an electronic nose or tongue can also be used for detecting chemicals released during spoilage of food, by placing them in direct contact with the packaging material so as to ensure better quality [21,22]. These nanosensors are in the form of "buttons" placed upon the packaging materials, which help in detecting the change in color. This technique can also be used to indicate whether or not packaged food is stored under optimum conditions [20].

Biosensors, which are increasingly integrating nanotechnology tools, is a portable sensor incorporating molecular recognition elements derived from biological materials, such as nucleic acid, antibodies, cells, or enzymes with suitable physico-chemical transducing mechanisms [23]. These biological sensing elements are either integrated within or associated with transducers, which transform physicochemical interaction into a decipherable form with the help of a transduction and electro-mechanical interpretation, depending upon target groups of analytes. A conventional biosensor consists of three components: bioreceptor, transducer, and detector [24–26]. The first component, bioreceptor, acts as a prototype for the detection of the analyte. The second component is the transducer system, which is an interface that determines the energy that arises with the reaction at the bioreceptor and then transforms that energy into the detectable electrical output. There are various elements of the transducer system, mostly optical, electrochemical, thermometric, piezoelectric, magnetic, and other ones [26–28]. The third component, the detector system, amplifies signals and analyses them in a microprocessor. Data are transformed into concentration units, which are then transferred to a data storage device [26].

Biosensors can be used to identify a broad array of targets, ranging from protein molecules to large pathogens [24]. Compared to the traditional methods, biosensors can provide results more rapidly than other culture-based methods, making them vital to the lab and real-world applications, promoting robustness, sensitivity, accuracy, real-time assay, and ease of use. The innovation of sensors is essential to the agro-food sector, as any incorrect result can lead to an extensive economic loss along with the loss of consumer confidence [29].

10.2 Nanosensors

The increasing application of sensors in different sectors, including medical diagnostics and therapeutics, food, air, and water quality monitoring, is evident. The word "nanosensors" is an integration of two segments: *nano*, which suggests a size range of 10^{-9} unit of the scale, and *sensors*, which is derived from a Greek word *Sentire* meaning to perceive [30]. Broadly, it is a device that ultimately converts a physical stimulus such as a thermal, luminescent, electric, or magnetic effect into an electrical signal. Sensitivity, selectivity, resolution, stability, and calibration characteristics are some of the crucial parameters to be considered during their functional operation in any field of the response.

General classification of nanosensors can be done in two major groups. One group has nanoscale dimensions that are further subgrouped into two types that is sensors whose three-dimensional properties are in nanometric proportions, like QDs resembling fluorescent nano-sized semiconductor crystals and thin films of any metal oxides for gas sensing having one nanoscale dimension. The other group is of nanoscale measurements but does not have any particular nanoscale dimensions. Further classification of nanosensors can be as active nanosensors that require a source of energy such as a thermistor (a thermally sensitive resister) and passive nanosensors that do not require the energy source like a piezoelectric sensor (a sensor that uses the piezoelectric effect to measure the strain, force and pressure). Another broad classification can be according to the signal detected in the form of energy, such as a physical, chemical, or biological effect. Physical nanosensors comprise mechanical, optical, and electromagnetic nanosensors [31], which are used for measuring properties such as force, pressure, density, viscosity, absorbance, fluorescence, luminescence, dielectric constants, and permeability. Although chemical nanosensors are meant to identify any foreign chemical particle, concentrations of certain substances, and pH, biological nanosensors (biosensors) have other essential purposes.

10.2.1 Optical nanosensors

Optical nanosensors use optical signals to transduce biological or chemical sequences for analytical purposes of all particles with dimensions less than 1000 nm. They include the option of measuring in small volume, high specificity, low toxicity, and other interferences similar to water on the sensors properties. The first optical nanosensor was based on fluorescein entrapped in a polyacrylamide nanoparticle, which was used for pH sensing [32]. The fluorescent molecules were composed of at least one photoactive element and a substrate unit. In the process, a light of particular wavelength was absorbed, which was followed by the emission of quanta of light possessing an energy equivalent to the difference between the excited state and the ground state, also known as luminescence phenomenon. This phenomenon is preferred for nanoscale chemical analysis and biological or environmental applications [33]. An optical receptor or a sensing element is integrated into the optical nanosensor consisting of an inert matrix for detecting the properties of the assigned analyte. Currently, fluorescence is the only transduction process due to the high relative ease of measurement and sensitivity [34].

10.2.1.1 Fiber-optic nanosensors

Optical nanosensors usually consist of a chemically or biologically sensitive layer, which can either be made up of chemical recognition element or biological recognition elements covalently attached to the optical transducer (Figure 10.1). This transduction mechanism holds a very attractive platform for chemical sensing application, among which minimal invasiveness is the most substantial technique for cell monitoring [35]. The small size, light weight, exemption to electromagnetic

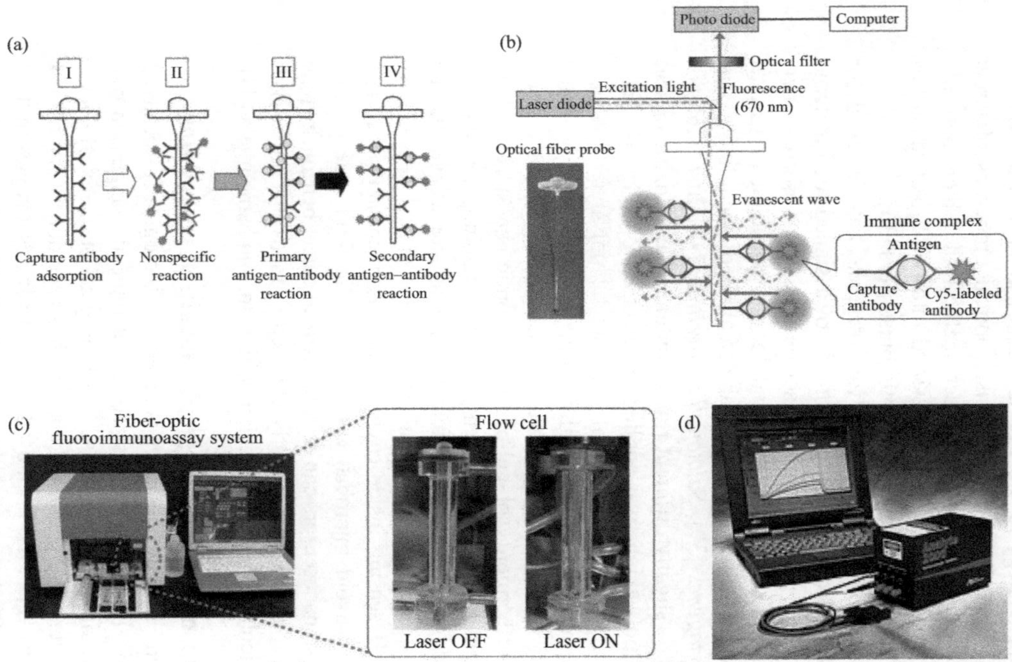

Figure 10.1 *Schematics of prototypes and models of fiber-optic sensors. (a) The fiber-optic probes are first immobilized with captured antibodies followed by interacting with samples with potential target analytes by a sandwich immunoassay. (b) and (c) The fluoroimmunoassay system developed by Miyajima et al. [36] consisting of fiber-optic probes, excitation light sources, optical filters, and detection devices (photo diode connected to a computer) is utilized to perform flow-through measurement analytes. The changes in the fluorescent intensities are measured to determine the concentration of the analyte. (d) A commercially available model of fiber-optic sensor. The Analyte 2000 is a 4-channel, single wavelength fluoroimmunoassay system developed in collaboration with the Naval Research Laboratory (courtesy: Research International, Inc., Monroe, WA, USA)*

interference (also called radiofrequency interference), and easy configuration of the optical fibers used in the nanosensors are some unique characteristics that enable sensing of biological and physiological processes in the corresponding fields. The interaction between the surface of a target analyte and the receptor is such that a physicochemical perturbation can be converted to a measurable electrical signal that is received by the optical probe, and then the signal is transmitted to the database. The sensing strategy for fiber optic nanosensors is based upon two methods; one is a direct method in which intrinsic optical properties of target analytes are measured. The other is an indirect method in which optically detectable bioprobes are monitored. It is also classified into two broad categories that fiber-optic chemical nanosensors and fiber optic biosensors [36].

The first fiber-optic sensor in the field of food science was designed for detection of *E. coli* O157:H7 from the complex matrix of assigned food material [37]. Analyte 2000 is a core type fiber-optic sensor designed by Research International (Monroe, WA, USA) and is widely used for the detection of various pathogens in a sandwich form using fluorophore-labeled antibody including *E. coli* O157:H7 [38], *L. monocytogenes* [39], *Salmonella enteritidis* [40], and staphylococcal enterotoxin B [41]. A slightly different fluoroimmunoassay utilizing optical fiber probes consisting of polystyrene was applied to detect *E. coli* O157:H7 [36]. During food packaging, a sensitive coating of syndiotactic polystyrene, a new crystalline engineering thermoplastic with a boiling point of 270 °C high heat, and chemical resistance with good dielectric properties, is used along with the fiber-optic nanosensors [42]. Furthermore, sol–gel-based fiber-optic nanosensors along with carbon dioxide, a sensing element that employs dual luminophore internal referencing, was developed [43]. Sensing of pH remains of great interest even though cross-sensitivity to ionic strength is the central problem for optical sensors. An emerging application of pH sensing inside cavities, such as microtiter plates and micro-bioreactors, can help avoid microbial contamination in foods due to non-contact measurement. These nanosensors can also be used to detect benzo[*a*]pyrene inside a single cell that is found to be in slightly elevated in foods that are cooked well on the barbecue, particularly in steaks, chicken with skin, and hamburgers [44].

10.2.1.2 Probes encapsulated by biologically localized embedding

PEBBLES is the acronym for “Probes encapsulated by biologically localized embedding.” It is a spherically optical nano-sized sensor typically ranging in size between 20 and 200 nm in diameter and transduces biological or chemical events into a measurable optical signal [45]. Limited analysis has been done using optical PEBBLE nanosensors; however, it has been used to detect various analytes in food additives in which Mg^{2+} PEBBLE sensor is one of them. It was designed by encapsulating a hydrophilic dye (cyanine which has an affinity for water molecules) and a commercial reference dye (Texas red which is a red fluorescent dye used for staining cell specimen) in polyacrylamide nanoparticle (a white polyamide related to acrylic acid which possesses a hydrophilic property). This sensor was designed to determine the role of Mg^{2+} inside human macrophage cells in the presence of attacking *Salmonella* (a Gram-negative rod-shaped bacteria which is a major cause for foodborne illness transmitted to individuals through ingestion of contaminated food of animal origin i.e., meat, poultry, mainly eggs, and milk) [46].

10.2.1.3 Recent developments in optical nanosensors

Surface-enhanced Raman scattering

In contrast to the detection strategy based on refractive index profile, surface-enhanced Raman scattering (SERS) is a unique vibrational spectroscopic method that generates vibrational signatures for small molecular analytes to distinguish between molecules such as structural isomers of fructose and glucose possessing certain similarities [47]. This was first observed on a silver surface electrode in 1974 [48]. The benefit of using Raman spectroscopy is its ability to display the molecular structure of a sample by transducing the data in the form of signal. With the progress being made in the fabrication of metal nanoscale and nanomaterial surfaces and the advancing technique of Raman scattering, SERS has become a delicate method, relevant for examining foods, and drugs [49].

SERS can be used for detecting quantitatively by picking an appropriate calibration data set, as its intensity varies between samples due to the substrate division. Real-time response, low detection limit, and quantitative and qualitative analytical capabilities are some of the advantages of this mechanism, which makes it applicable for identification and characterization of bacteria [50], pharmaceuticals [51], and other molecular species [52].

The pH sensor based on SERS is developed using a silver nanoparticle with a gold-silica core nanoparticle. The sensor is made up of a 50 to 80 nm diameter of silver nanoparticles mixed with *para*-mercaptobenzoic acid, which shows a characteristic SERS spectrum dependent on the pH of the adjoining solution. It is made sensitive to pH changes in the range between 6.0 and 8.0 [53] and has been successfully examined for the rapid detection of *B. subtilis* spore, which is a non-pathogenic surrogate for *B. anthracis* [54]. Ideally, the limit of detection (LOD) should be less than the threshold dose of a pathogen. The sensor was found to have an LOD of 2.1×10^{-14} M (2.6×10^3 spore in 0.2 μL, 0.02 M HNO_3) for *B. subtilis*, which was found to be less than anthrax infection dose of 10^4 spore [55]. Conventional Raman spectroscopy based sensors have also been used to examine the antifungal movement of ZNO nanoparticles against *Penicillium expansum* and *Botrytis cinerea*, as they play a major role in the economic loss during postharvest handling of fruits [56]. In various food industries, these nanosensors are used to monitor and detect chemical fungicide that is, thiabendazole (TBZ) against mold and blight formed on bananas and citrus fruits. With a portable Raman spectrometer, the estimated amount of TBZ in a 5 g of citrus sample (peel) is found to be 78 mg kg^{-1}, which is almost higher than the maximum allowed range of present-day regulations [57]. Another crucial development has emerged for fungicide detection at ultralow levels in solutions that is by using a dogbone-shaped gold nanoparticles in SERS sensors. Nanorods (aspect ratio of 2.19 and length of 38 nm) were used for the detection of dithiocarbamate fungicide *Thiram* in acetonitrile-water solutions in which LOD and limit of quantitation were estimated to be around 11 and 34.4 nM, respectively, which is much below the Environmental Protection

Agent tolerance value of 16.6 mM. But when dogbone-shaped gold nanoparticles were used, the LOD value further decreased to 44 nM [58]. Conventional high performance liquid chromatography (HPLC) analysis techniques can be integrated with the SERS-based detection system for rapid screening of food sample to eliminate false results. Along with the SERS mechanism, gold nanoparticles or nanorods (Figure 10.2) can be used as a nanomaterial for efficient detection of perchlorate (a food pollutant) at the nanometric level in contaminated water samples [59]. Silver substrates along with the SERS technique can rapidly screen *L. monocytogenes, E. coli*, and *Salmonella* [60]. Gold substrates have previously been used as a nanomaterial to detect viruses such as adenovirus, parvovirus, coronavirus, herpesvirus, norovirus, simian rotavirus, and Sendai virus with detection limit of 100 particles [61]. Application of magnetic nanoparticles in SERS-based detection of microbes renders a particular advantage. In this approach, first the magnetic nanoparticles that are covered with capture molecules (such as antibodies) participate in the initial capture and separation of microbes from sample matrices takes place. This is followed by direct detection of the captured species by Raman tags attached to the same or different nanoparticles in the same system. An example of such simultaneous capture and detection of *E. coli* O157:H7 was shown by Najafi *et al.* [62].

Surface plasmon resonance

SPR sensors utilize free electron oscillation at the interface of a metal surface with dielectric constants of opposite charges during the coupling of excited light of a specific wavelength. In 1982, this phenomenon was first implemented for gas detection and in biological sensors [63,64]; since then this technology has been extensively used for the detection of chemical and biological analytes in a solution [65].

SPR nanosensors consist of an optical system and a transducing medium that correlates the chemical (or biological) properties with optical outputs along with an electronic database. In this sensor, the optoelectronic components allow the processing of data through the transducing medium. These two components contribute in determining the capabilities of this sensor, that is, sensor stability, sensitivity, and resolution, and completely depend on the properties of the optical and transducing systems [66]. This sensor produces a surface wave that moves parallel to the dielectric axis and the surface of the metal in between them while producing a field strength that demonstrates evanescent characteristics. The strength reduces rapidly as a function of distance from the surface, usually up to several hundred nanometers. When the target analyte is captured by any biorecognition molecule like antibodies, enzymes, or receptors, a change occurs in the reflection intensity, which is then converted into the refractive index. Usually, SPR is treated as a label-free detection system, as it does not require any distinguishing fluorophore or reagents at the sample preparation phase [67]. However, for the specificity of the target analyte, biorecognition molecules or antibodies are occasionally predeposited onto the SPR surface.

This sensor has been used to detect damaged and live cells of *L. monocytogenes* at 10^6 cell mL^{-1} using rabbit antibodies or phage displayed single-chain fragment variable antibodies on a sensor platform that is covalently restrained [68].

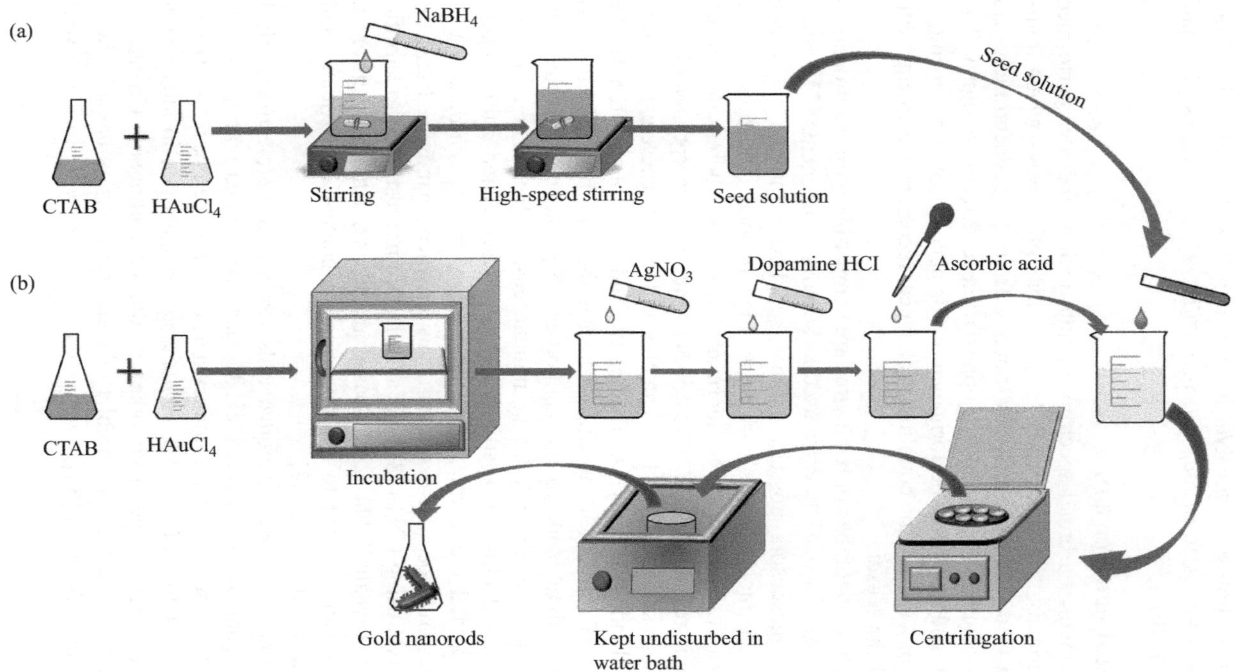

Figure 10.2 *Seed mediated synthesis of gold nanorods: (a) Process denotes the preparation of seed solution, (b) process denotes the preparation of growth solution. After the growth solution is prepared, seed solution is mixed in it with high speed and then kept undisturbed for accumulation of the precipitate. Finally the gold nanorods are observed through localized surface plasmon resonance technique*

Detection of *Salmonella enterica* serovar Typhimurium, at a concentration of 10^2 CFU mL^{-1} using protein G molecules immobilized with anti-*Salmonella* antibodies on the gold surface, was studied [69]. SPR biosensors can also be used effectively for the detection of *Salmonella* in milk at 1.25×10^5 CFU mL^{-1} concentration [70], enterotoxins from *Staphylococcus* [71,72], mycotoxins [73], and *E. coli* O157:H7 [74–76].

10.2.1.4 Chemical nanosensors

These nanosensors involve a broad spectrum of transducing processes performed optically, calorimetrically, gravimetrically, or in various other ways which transform data acquired from chemical analytes (ranging from the concentration of a single specific component to a mixed component) into a detectable signal. The chemical selectivity and analytical performance of this nanosensors are dependent on the transduction process and the coupling of chemical recognition elements with the physical part of the sensor.

Nanomaterials like nanotubes or nanowires are compatible with the properties of chemical nanosensors; hence, they are used as sensing components [77]. Electrochemical-based nanosensors possess another standard technique by which nanomaterial-based chemical sensors with application in food industries can be extensively compared to colorimetric or fluorimetric methods, whereas the former approach would be more useful for food matrices as the light absorption and scattering from the various surface of the food can be avoided [78]. These nanosensors mainly function by binding selective antibodies to conductive nanomaterials like carbon nanotubes (CNTs) and then applying changes to the material conductivity as soon as the target analyte binds to the antibodies [79]. Other electrochemical systems based on nanomaterials includes an immunosensor, which is composed of a CeO_2 nanoparticle and chitosan (partially or fully deacetylated chitin) nanocomposite to detect a vital foodborne fungal contaminant that is, *ochratoxin*-A. It could also detect *staphylococcal* enterotoxin B using silicon nanowire transistor [80] and *Cholera*-toxin using CNTs [81]. As analytes are not restricted to harmful substances, a study revealed that CNTs-based electrochemical nanosensor in microfluidic devices can be used to quantify vitamin content, antioxidant, and flavor in vanilla beans and apples [82].

Apart from chemical analytes, nanomaterials also aid in the electrochemical detection of microorganisms. A conductive TiO_2 nanowire coated with antibodies has been developed for screening *L. monocytogenes*. This method was able to detect about 4.7×10^2 CFU mL^{-1} of *L. monocytogenes* pathogen in almost 1 h without any unusual interference from other foodborne pathogens. Change in resistance or conductance across the circuits of these nanosensors has also been done for the detection of *Bacillus* [83], *Salmonella* [84,85], *E. coli* [86,87] as well as viruses [85].

10.2.2 Biosensors and biological nanosensors

The invention of biosensors has proved to be a promising tool in food safety, medicine, environmental protection, and many other fields. The safety and quality of food have been the main concern for food technologists and environmental scientists. Biosensors provide a rapid and efficient way to assess safety and quality of food and related products [88]. Currently, industries need quick and economical methods that enable development of products of best quality. High selectivity, specificity, quick response, and cost affordability make biosensors as one of the most efficient tools available to the food industry. A biosensor is defined as a device integrated with a biological sensing element of receptor along with one or more transducers capable of relaying selective or semiquantitative information to an output device [88]. The integration of a diverse group of cross-disciplinary expertise, such as biochemistry, physical science, molecular engineering, biotechnology, and material sciences, enables the possibility of detecting changes at nanolevel, makes biosensors the most versatile analytical tools ever used. In food industries, a major area of biosensor use is in the detection of pathogens, pesticides, and toxins present in various food samples and packaging materials. Traditional methods of detection of contaminants include physiochemical and biochemical tests that take days or weeks to yield results. However, the results often lack sufficient selectivity and sensitivity with a significant investment of money and time for preparation of samples. Biosensors with all its advantages of selectivity, sensitivity, simplicity, low cost, and rapid processing time are found to be a great nondestructive alternative approach. The method implemented in its construction also allows its usage in determining the composition of raw and processed food samples and on-line control of fermentation process. Biosensors are expected to improve the quality of life in upcoming years.

Regarding the conceptual and fundamental mode of operation, there are three components of a typical biosensor: bioreceptor/biorecognition element, transducer, and detector. The main sensing elements of biosensors are enzymes, microorganisms, antibodies, biocatalysts, antigens, nucleic acids, whole cells, etc. [89]. The organic element acts together selectively with the target analytes (pathogens, microorganism, pesticides, or toxins), providing the discriminatory power of the sensors. Speed and selectivity of biorecognition elements are the most important and crucial aspect of biosensors. Enzymes and antibodies are most widely used biological recognition elements. The second key component is a transducer, which converts the response resulting from the chemical interaction between bioanalyte and bio receptor into an electrical signal. The name itself defines the ability to convert one form of energy into another as the word "trans" means change and "ducer" means energy. The transducers may convert biochemical energy into electrical energy. Some of the interactions are temperature-dependent with heat output or electrochemical reactions with the electronic output. Some other mode of signal transduction may include, light output or light absorbance between reactant and product which are based on the mass of the reactant, absorption or emission of electromagnetic radiation, mass and micro-viscosity alteration of wave propagation, etc. [90]. The third component is the detector, which receives electrical energy and processes it. The detector mainly consists of a microprocessor or other electrical or optical processing tools. In many cases, the signal produced by the transducer is so weak that it cannot be detected by the processor. In that scenario, another component, an amplifier, can also be installed in the detector to amplify the signal received for further improvements of processing of the signal [91]. The specificity of the sensor for a specific target analyte is assured by biorecognition elements; whereas, the accuracy and sensitivity is influenced by the type of transducer employed. In addition to these separate components, a critical structural feature of biosensors is governed by the conjugation chemistry to immobilize the bio receptor on the sensor platform. Proper immobilization is important to ensure the accuracy and speed of the biosensor. However, the process is affected by temperature, pH, types of contaminants, and other physiochemical changes [92].

10.2.3 Nanotechnology-based biosensors

In past decades, with the progress and successes in nanotechnology, most of the nanomaterials and their properties have been studied extensively paving the way to their applications for the development of biosensors [93]. The use of diverse groups of nanomaterials, such as nanoparticles, nanorods, nanotubes, and nanowires, has enabled faster detection with enhanced reproducibility of the target analytes. The main advantage of nanomaterials is their dimension. Nanomaterials with their high surface area allow more efficient and better use of the surface-to-volume ratio as compared to other materials on a sensor platform. Nano biosensors are biosensors that use nanomaterials as an integral part of their design.

10.2.3.1 Nanomaterials-based biosensors

Carbon nanotube

Nanotubes are long and narrow tubes that can be assayed onto a strong, stable, and malleable honeycomb-like structure. Nanotube-based biosensors are popular for their large surface area and ability to immobilize large numbers of biomolecules. They usually have a tubular or porous membrane with very active mechanical, electrical, chemical, and optical properties. Nanotubes can be reinforced into composite structures with high fracture and thermal stability. Among nanotubes made out of different materials, CNTs are the most popular and have a wide range of applications [94,95]. They have applicability in many fields due to their graphite structure, highly effective surface chemistry, electronic potential, and typical nanostructure. The CNTs have various applications in biomedicine, food processing, chemistry, bioanalysis, and biosensing. For example, polymeric CNTs have good mechanical properties and high electrical conductivities, which makes them ultrasensitive and effective electrochemical sensor [96]. The interesting properties of CNTs in electrochemical measurement are their abilities to provide large active surface area at electrodes of relatively smaller magnitudes. The high rate of electron transmission and sorption capacity, and with their potential to be used as effective electrode material for immobilization of substrate on the surface, makes CNT as one of the most popular materials used in biosensors for detection of pesticide and food analysis [97]. CNTs can be classified as single CNTs or multi CNTs. Single CNTs have a single layer of graphite sheet enclosed in a cylindrical tube, whereas, multi CNTs consist of an array of single nanotubes nested in a concentric manner [82,98]. They can be coupled with microchips as a novel electrochemical material for analysis and detection of food additives, sugars, flavors, vitamins, and isoflavones. These biosensors have the possibility to supplement and even replace the old method of pesticide detection through a simple, fast, and cost-effective method of enzymatic biosensors. Enzymatic biosensors used to screen for carbamate pesticides through the enzymatic inhibition of acetylcholinesterase and their substrate, acetylcholine. The carbamate inhibits the catalytic activity through binding to the active stearic site of the enzyme and blocking the serine residue in the catalytic triad of AChE by phosphorylation or carbamylation. For detection of carbamate pesticide in food and vegetables, the core–shell structure of CNTs has a compact layer of polyaniline, and detection principle is based on immobilization of acetylcholinesterase on it is surface so that it can be used to quantify methomyl and carbamyl in fruit and vegetable samples through chronoamperometry [97,99].

Quantum dots

Food contaminants such as, added pesticides, and bacterial toxins (e.g., botulinum toxin, enterotoxin produced by *S. aureus*, and *E. coli*) can be detected by the use of water-soluble bi-conjugated QDs. The advantage of these aqueous synthesized QDs includes photo stability for longer periods of time, broad absorption, stability, and a very specific emission spectrum with very high compatibility. They can be combined with various biomolecules to develop an integrated and hybrid form of biosensors with the combination of unique optical and magnetic properties having specific and sensitive detection capability. These characteristics make QDs an attractive fluorescent probe for quantitative and qualitative analysis [100]. QDs can be arranged in an assembly for detection of toxin compound (paraoxon) produced by the organophosphorus insecticide parathion by covering it with layers of chitosan, organophosphorus hydrolase, and thioglycolic acid. This detection regimen is based on the principle of inhibition of acetylcholinesterase, resulting in the accumulation of acetylcholine in cholinergic synapses. The unique ability of QDs makes it an essential tool for bioanalysis and food safety [101].

Nanofibers

Nanofibers are electrospunfibers with diameters of around 100 nm. These nanofibers with large surface-to-volume ratios offer a great prospecting detection scenario in which high porosity is required. A nanofiber converted into a porous structure can be a dynamic source for detection in which pore size and shape can be changed depending upon requirement [102]. Biosensors made from nanofibers similar to nanofiber mats and arrays have the advantage of an extremely large surface area, which increases the availability of immobilization sites for biorecognition elements in biosensors [103]. In addition to the physical properties, the chemical characteristics of the sensor can be modified by blending with different polymers, nanocomposites made of organic or inorganic materials, or some bioactive ingredients. Its novel properties can be utilized for sample preparation and analyte detection.

10.2.3.2 Other sensors utilizing nanoscale materials

Electronic nose

An electronic nose is a device that consists of groups of chemical sensors with a high degree of electivity, sensitivity, and the ability to recognize and analyze simple and complex odors arising from test samples. Ideally, this sensor can replace human olfactory system [104]. The versatility of the electronic nose to recognize and differentiate between different types of odor with accurate results makes it a very popular tool in food, clinical, and biomedical industries. Electronic noses can be defined as "an intelligent system of chemical-array sensors that mimic the human olfactory system" [105]. The concept of the electronic nose has been introduced in food for quality inspection, shelf life investigation, freshness evaluation, and process monitoring [106]. The smaller aspect ratio structures of nanoparticles, nano spheres, and nanocubes are typically used as a film for sensing element, whereas high aspect ratio structures can be utilized as a single or in multiple films for fabrication of an electronic nose [105]. Considerable work has been done to develop these sensors and tested on food samples including fish, vegetables, meats, cheese, mushroom, wine, etc. [104,105].

The fabrication of electronic nose devices for detection and discrimination between food samples and odor consists of an array of broadly tuned sensors which are treated with odor-sensitive biological or chemical materials [106]. The chemical sensors that are the key components of the system comprise of inorganic crystalline materials like semiconductors, organic materials, polymers, and biologically derived materials. The sensing element depends on the type of analyte to be detected. The odor stimulus generates a signal or fingerprint from the sensor array (Figure 10.3). The signals are then captured by data acquisition system for processing through a processor or computer software [107]. The control and data acquisition system can be integrated into a single device called microcontroller. Specific signals from known analyte are used to develop a database and a pattern recognition system is created to enable classification, identification, and quantification of the odor based on the data stored in the database [105]. The system is produced to have long-term usage with repeatability and reproducibility [104]. The implementation of nanotechnology in its construction offers the advantage of simple operation, ease of fabrication, low cost, compatibility with microelectronics processing, and low power consumption.

Electronic noses have several applications in food processing and quality assurance of raw and manufactured food products. The determination of the quality of harvested fruits and vegetables is an example in which electronic nose system is used. One of the main concerns of the food industry currently is to monitor and control the quality variation of fresh harvest fruits and vegetables [108]. Electronic noses have the specificity and precision to differentiate between numerous types of fruits and to calculate the extent of damage (such as spoilage) and the duration of storage [109]. The measurement is done by aroma transformation, changes in defect caused by skin cuts or over-ripening, changes in starch or sugar composition, etc. [108]. Application of the electronic nose greatly depends on upon the type of sensor array used and gives an idea

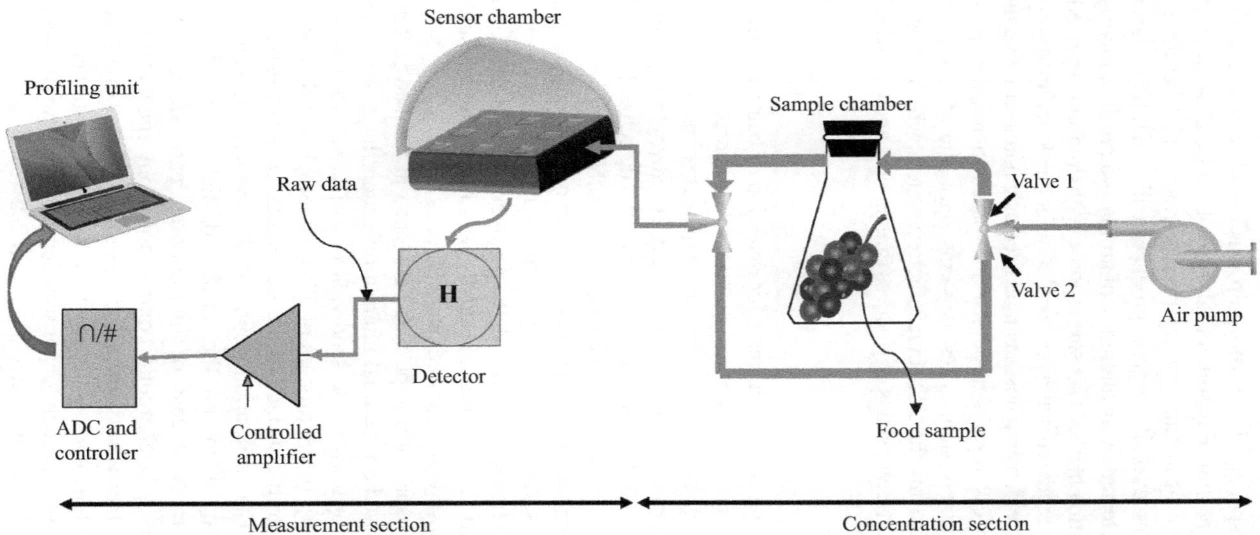

Figure 10.3 *Schematic of an electronic nose. The system comprises two sections. In the concentration section, the food sample is kept in a closed chamber and left for certain time so that aroma containing volatile compounds get accumulated in the headspace. Meanwhile, synthetic gas from the pump is used to carry the concentrated flavor to the sensor chamber. In the measurement section, the sensor creates a specified fingerprint for aroma that is detected by the detector and forwarded for amplification. At the end, the analog signal after amplification is converted into digital and used for further analysis*

about the contribution of the sensor array in the analysis of food. Electronic noses composed of quartz crystal microbalance coated by modified metal porphyrins and related compounds can be used to understand the level of spoilage in red wine and pulped tomato [110]. The spoilage in tomato pulp was detected by estimating the amount of acetic acid present in it. Experiments show that the LOD of an electronic nose is much lower than that of human sensory perceptions [111]. Therefore, these devices may yield very sensitive results for fresh produce quality control. In the case of red wine, spoilage can be detected by an alteration of aroma as a result of mixing with the air. In this case, a correlation between sensor response with time and expected wine taste is reported [112] which rendered it as a powerful tool for quick quality assessment of food products. Moreover, electronic noses can be used as a nondestructive method for altering the ripening process in fruits through a SnO-based chemical array sensor and neural pattern recognition system. These systems sense the aromatic volatile compounds emitted by fruits using the electronic olfactory system, which is capable of classifying fruits into three different states of ripeness (green, ripe, and over-ripe) with great accuracy [113,114].

Array sensors

Array-based sensors consisting of nanoparticles for detection of bacterial pathogens have been developed by different groups to improve microbial diagnostics [115–117]. In such sensing strategies, array of polymer constructs (such as nanoparticle-fluorescent polymer) or (poly(*para*-phenyleneethynylene)) with conjugated gold-nanoparticles (AuNP) were utilized for bacterial detection [116,117]. In this sensor, arrays of analyte receptors conjugated with AuNPs are assembled as "chemical noses" to discriminate and detect analytes based on their unique response diagrams. The hydrophobic surfaces of bacteria are recognized by arrays consisting of AuNPs (diameter 1.6 nm) functionalized with the polymer (poly-lysine), following a complementary electrostatic interactions of AuNPs with live bacteria resulting in the self-assembly of the arrays with captured bacterial cells [117].

Another "Array Biosensor" prototype was developed at the Naval Research Laboratory (NRL) and is based on the total internal reflection fluorescence (TIRF) (Figure 10.4). This biosensor utilizes the sandwich immunoassay format for analysis and detection [118,119]. In the process of TIRF, fluorophores that are either attached to or in proximity to the surface of the waveguide are selectively excited via an evanescent wave. The infiltration depth of the evanescent wave allows measurement of the binding event on the surface of the waveguide without significant interference with an excess of fluorophores, particulate, and other compounds in the bulk fluid. Background contribution from endogenous fluorophores is minimized with the more widespread use of fluorescent dyes, such as Cy5 (far red fluorescent cyanine dye having excitation wavelength in between 633 and 647 nm), which are excited at much longer wavelength than naturally occurring fluorophores [120]. Array biosensors have been utilized to detect and analyze toxins, large pathogens, and bacteria from food samples [121]. Sandwich and fluoroimmunoassay have been developed to identify low or high molecular weight toxins in complex food samples [120]. NRL array biosensor was employed to

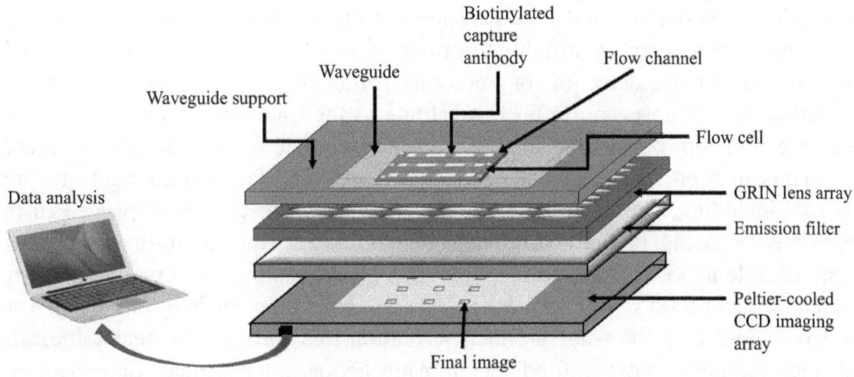

Figure 10.4 *Schematic of an array sensor. Physically isolated patterning technique is used for printing the antibody sample on the waveguide surface (made up of borosilicate glass slides). Flow cell (with polydimethylsiloxane in its flow channels capable of molding and forming 3D structures) is placed perpendicular to the columns of the recognition antibodies so that assays could be analyzed independently. A diode laser is directed toward the waveguide and its excitation light gets coupled in the flow cell (by reflective cladding). The fluorescence emission from the cell automatically focuses toward the graded index (GRIN) lens which then passes through the optical filter onto a Peltier-cooled charge-coupled device (CCD) imaging array. Data are extracted from the image using appropriate software*

interrogate up to 12 different samples for the presence of multiple contaminants [120]. Among other major applications, the array sensors were utilized for detection of *Shigella* and *Campylobacter* at a very low concentration [118,121]. The biosensor was also successfully employed in the detection of deoxynivalenol (DON) (a mycotoxin produced by various *Fusarium* species), which is a common contaminant of cereal grains, such as barley, corns, oats, and wheat. The traditional method uses chromatography for its detection, but the method is time-consuming and not appropriate for large-scale detection. Hence, the application of the array biosensor increases the speed and eliminates highly complex method for quantification and identification of DON. The analysis does not require any sample clean-up and sample preconcentration; therefore, it minimizes the cost and time [119].

10.3 Concluding remarks

Nanotechnology has transformed the food safety diagnostics field. It has created new possibilities to improve the resolution of sample preparation and downstream by presenting innovative classes of materials with novel properties. The application of nanoscience and nanosensors goes beyond microbial food safety and provide various efficient applications to the broader food sector, including detection of crop pesticides, food quality, and food packaging. Conventional techniques can be very inexpensive or deliver quantifiable information; however, they require time to yield final results for the detection of microbial pathogens or allergens or other food contaminants. Nanosensors offer a substitute to the traditional methods, allowing multiple and rapid real-time analysis that are crucial for the detection of contaminants in food, especially in perishable items. The key requirements for any sensors including, sensitivity, speed, and specificity are greatly improved in the nanosensors owing to the size advantages. Moreover, the flexibility of creating customizable nanostructures for specific sensing applications is very conducive for agriculture and food safety applications, as the diagnostic needs in this sector vary widely depending on the products. Nanosensors provide a technologically advanced solution to detect food contaminants addressing a critical component of a complex public health issue of food safety.

References

[1] Buzby J.C., Roberts T., Lin C.T.J., MacDonald J.M. *Bacterial Foodborne Disease: Medical Costs and Productivity Losses*. United States Department of Agriculture, Economic Research Service; Washington, DC, Report, 741; 1996.

[2] Wilcock A., Ball B. 'Food safety'. *Practical Food Safety: Contemporary Issues and Future Directions*. John Wiley & Sons, Ltd; 2014. pp. 11–29.

[3] Ivnitski D., Abdel-Hamid I., Atanasov P., Wilkins E. 'Biosensors for detection of pathogenic bacteria'. *Biosensors and Bioelectronics*. 1999;**14**(7):599–624.

[4] Knura S., Gymnich S., Rembialkowska E., Petersen B. 'Agri-food production chain'. In: Luning, P, Devlieghere, F, Verhé, R, editors. *Safety in the Agri-Food Chain*. Wageningen Academic Publishers: 2006. p. 19–65.

[5] Grunert K.G. 'Food quality and safety: consumer perception and demand'. *European Review of Agricultural Economics*. 2005;**32**(3):369–91.

[6] Wilcock A., Pun M., Khanona J., Aung M. 'Consumer attitudes, knowledge and behaviour: a review of food safety issues'. *Trends in Food Science and Technology*. 2004;**15**(2):56–66.

[7] Baruah S., Dutta J. 'Nanotechnology applications in pollution sensing and degradation in agriculture: a review'. *Environmental Chemistry Letters*. 2009;**7**(3):191–204.

[8] Shrivastava S., Dash D. 'Nanotechnology in food sector and agriculture'. *Proceedings of the National Academy of Sciences, India Section B: Biological Sciences*. 2012;**82**(1):29–35.

[9] Scott N., Chen H. 'Nanoscale science and engineering for agriculture and food systems'. *Industrial Biotechnology*. 2013;**b**(1):17–18.

[10] Moraru C.I., Panchapakesan C.P., Huang Q., Takhistov P., Liu S., Kokini J.L. 'Nanotechnology: a new frontier in food science'. *Food Technology*. 2003;**57**(12):24–9.

[11] Masciangioli T., Zhang W.-X. 'Peer reviewed: environmental technologies at the nanoscale'. *Environmental Science and Technology*. 2003;**37**(5):102A–8A.

[12] Donsì F., Annunziata M., Sessa M., Ferrari G. 'Nanoencapsulation of essential oils to enhance their antimicrobial activity in foods'. *Journal of Biotechnology*. 2011;**44**(9):1908–14.
[13] Duncan T.V. 'Applications of nanotechnology in food packaging and food safety: barrier materials, antimicrobials and sensors'. *Journal of Colloid and Interface Science*. 2011;**363**(1):1–24.
[14] Chen H., Yada R. 'Nanotechnologies in agriculture: new tools for sustainable development'. *Trends in Food Science & Technology*. 2011;**22**(11): 585–94.
[15] Ditta A. 'How helpful is nanotechnology in agriculture?' *Advances in Natural Sciences: Nanoscience and Nanotechnology*. 2012;**3**(3):033002.
[16] Mousavi S.R., Rezaei M. 'Nanotechnology in agriculture and food production'. *Journal of Applied Environmental and Biological Sciences*. 2011;**1**(10): 414–19.
[17] Scrinis G., Lyons K. 'The emerging nano-corporate paradigm: nanotechnology and the transformation of nature, food and agri-food systems'. *International Journal of Sociology of Food and Agriculture*. 2007;**15**(2): 22–44.
[18] Prasad R., Kumar V., Prasad K.S. 'Nanotechnology in sustainable agriculture: present concerns and future aspects'. *African Journal of Biotechnology*. 2014;**13**(6):705–13.
[19] Rhim J.W., Park H.M., Ha C.S. 'Bio-nanocomposites for food packaging applications'. *Progress in Polymer Science*. 2013;**38**(10):1629–52.
[20] Coles D., Frewer L. 'Nanotechnology applied to European food production—a review of ethical and regulatory issues'. *Trends in Food Science & Technology*. 2013;**34**(1):32–43.
[21] Di Natale C., Paolesse R., Macagnano A., *et al.* 'Electronic nose and electronic tongue integration for improved classification of clinical and food samples'. *Sensors and Actuators B: Chemical*. 2000;**64**(1):15–21.
[22] Mendoza-Madrigal A.G., Chanona-Pérez J., Guadarrama-Fernández L., *et al.* 'Nanobiosensors in food science and technology'. *Food Nanoscience and Nanotechnology*. Berlin: Springer; 2015. pp. 213–30.
[23] Sharma H., Mutharasan R. 'Review of biosensors for foodborne pathogens and toxins'. *Sensors and Actuators B: Chemical*. 2013;**183**:535–49.
[24] Banerjee P., Bhunia A.K. 'Mammalian cell-based biosensors for pathogens and toxins'. *Trends in Biotechnology*. 2009;**27**(3):179–88.
[25] Banerjee P., Bhunia A.K. 'Cell-based biosensor for rapid screening of pathogens and toxins'. *Biosensors and Bioelectronics*. 2010;**26**(1):99–106.
[26] Banerjee P., Franz B., Bhunia A. 'Mammalian cell-based sensor system'. In: Belkin S, Gu MB, editors. *Advances in Biochemical Engineering/ Biotechnology: Whole Cell Sensing Systems I*. 117. Berlin: Springer; 2010. pp. 21–55.
[27] Arora P., Sindhu A., Dilbaghi N., Chaudhury A. 'Biosensors as innovative tools for the detection of food borne pathogens'. *Biosensors and Bioelectronics*. 2011;**28**(1):1–12.

[28] Iqbal S.S., Mayo M.W., Bruno J.G., Bronk B.V., Batt C.A., Chambers J.P. 'A review of molecular recognition technologies for detection of biological threat agents'. *Biosensors and Bioelectronics*. 2000;**15**(11):549–78.

[29] Banerjee P., Kintzios S., Prabhakarpandian B. 'Biotoxin detection using cell-based sensors'. *Toxins (Basel)*. 2013;**5**(12):2366–83.

[30] Sze S.M. *Semiconductor Sensors*. New York: Wiley; 1994.

[31] Lim T.C., Ramakrishna S. 'A conceptual review of nanosensors'. *Zeitschrift für Naturforschung A*. 2006;**61**(7–8):402–12.

[32] Sasaki K., Shi Z.Y., Kopelman R., Masuhara H. 'Three-dimensional pH microprobing with an optically-manipulated fluorescent particle'. *Chemistry Letters*. 1996;**2**:141–2.

[33] Aylott J.W. 'Optical nanosensors—an enabling technology for intracellular measurements'. *Analyst*. 2003;**128**(4):309–12.

[34] Lobnik A., Turel M., Urek Š.K., Košak A. 'Nanostructured materials use in sensors: their benefits and drawbacks'. *Carbon and Oxide Nanostructures*. Berlin: Springer; 2011. pp. 307–54.

[35] Vo-Dinh T., Kasili P. 'Fiber-optic nanosensors for single-cell monitoring'. *Analytical and Bioanalytical Chemistry*. 2005;**382**(4):918–25.

[36] Miyajima K., Koshida T., Arakawa T., *et al.* 'Fiber-optic fluoroimmunoassay system with a flow-through cell for rapid on-site determination of *Escherichia coli* O157:H7 by monitoring fluorescence dynamics'. *Biosensors*. 2013;**3**(1):120–31.

[37] DeMarco D.R., Saaski E.W., McCrae D.A., Lim D.V. 'Rapid detection of *Escherichia coli* O157:H7 in ground beef using a fiber-optic biosensor'. *Journal of Food Protection*. 1999;**62**(7):711–16.

[38] Geng T., Uknalis J., Tu S.I., Bhunia A.K. 'Fiber-optic biosensor employing Alexa-Fluor conjugated antibody for detection of *Escherichia coli* O157:H7 from ground beef in four hours'. *Sensors*. 2006;**6**(8): 796–807.

[39] Nanduri V., Kim G., Morgan M.T., *et al.* 'Antibody immobilization on waveguides using a flow—through system shows improved *Listeria monocytogenes* detection in an automated fiber optic biosensor: RAPTORTM'. *Sensors*. 2006;**6**(8):808–22.

[40] Valadez A.M., Lana C.A., Tu S.I., Morgan M.T., Bhunia A.K. 'Evanescent wave fiber optic biosensor for *Salmonella* detection in food'. *Sensors*. 2009;**9**(7):5810–24.

[41] Tempelman L.A., King K.D., Anderson G.P., Ligler F.S. 'Quantitating staphylococcal enterotoxin B in diverse media using a portable fiber-optic biosensor'. *Analytical Biochemistry*. 1996;**233**(1):50–7.

[42] Ishihara N. 'Syntheses and properties of syndiotactic polystyrene'. *Macromolecular Symposia*. 1995;**89**(1):553–62.

[43] von Bultzingslowen C., McEvoy A.K., McDonagh C., *et al.* 'Sol–gel based optical carbon dioxide sensor employing dual luminophore referencing for application in food packaging technology'. *Analyst*. 2002;**127**(11): 1478–83.

[44] Sinha R., Kulldorff M., Gunter M.J., Strickland P., Rothman N. 'Dietary benzo[*a*]pyrene intake and risk of colorectal adenoma'. *Cancer Epidemiology Biomarkers and Prevention*. 2005;**14**(8):2030–4.

[45] Buck S.M., Koo Y.E., Park E., *et al*. 'Optochemical nanosensor PEBBLEs: photonic explorers for bioanalysis with biologically localized embedding'. *Current Opinion in Chemical Biology*. 2004;**8**(5):540–6.

[46] Martin-Orozco N., Touret N., Zaharik M.L., *et al*. 'Visualization of vacuolar acidification-induced transcription of genes of pathogens inside macrophages'. *Molecular Biology of the Cell*. 2006;**17**(1):498–510.

[47] Söderholm S., Roos Y.H., Meinander N., Hotokka M. 'Raman spectra of fructose and glucose in the amorphous and crystalline states'. *Journal of Raman Spectroscopy*. 1999;**30**(11):1009–18.

[48] Fleischmann M., Hendra P.J., McQuillan A.J. 'Raman spectra of pyridine adsorbed at a silver electrode'. *Chemical Physics Letters*. 1974;**26**(2):163–6.

[49] Li Y.S., Church J.S. 'Raman spectroscopy in the analysis of food and pharmaceutical nanomaterials'. *Journal of Food and Drug Analysis*. 2014; **22**(1):29–48.

[50] Zhang X., Young M.A., Lyandres O., Van Duyne R.P. 'Rapid detection of an anthrax biomarker by surface-enhanced Raman spectroscopy'. *Journal of the American Chemical Society*. 2005;**127**(12):4484–9.

[51] Cînta Pînzaru S., Pavel I., Leopold N., Kiefer W. 'Identification and characterization of pharmaceuticals using Raman and surface-enhanced Raman scattering'. *Journal of Raman Spectroscopy*. 2004;**35**(5):338–46.

[52] Yonzon C.R., Haynes C.L., Zhang X., Walsh J.T., Van Duyne R.P. 'A glucose biosensor based on surface-enhanced Raman scattering: improved partition layer, temporal stability, reversibility, and resistance to serum protein interference'. *Analytical Chemistry*. 2004;**76**(1):78–85.

[53] Talley C.E., Jusinski L., Hollars C.W., Lane S.M., Huser T. 'Intracellular pH sensors based on surface-enhanced Raman scattering'. *Analytical Chemistry*. 2004;**76**(23):7064–8.

[54] Goodacre R., Shann B., Gilbert R.J., *et al*. 'Detection of the dipicolinic acid biomarker in *Bacillus* spores using Curie-point pyrolysis mass spectrometry and Fourier transform infrared spectroscopy'. *Analytical Chemistry*. 2000; **72**(1):119–27.

[55] Chin J., American Public Health Association. *Control of Communicable Diseases Manual (Vol 17)*. Washington, DC: American Public Health Association; 2000.

[56] He L., Liu Y., Mustapha A., Lin M. 'Antifungal activity of zinc oxide nanoparticles against *Botrytis cinerea* and *Penicillium expansum*'. *Microbiological Research*. 2011;**166**(3):207–15.

[57] Müller C., David L., Chis V., Pînzaru S.C. 'Detection of thiabendazole applied on citrus fruits and bananas using surface enhanced Raman scattering'. *Food Chemistry*. 2014;**145**:814–20.

[58] Saute B., Narayanan R. 'Solution-based direct readout surface enhanced Raman spectroscopic (SERS) detection of ultra-low levels of thiram with dogbone shaped gold nanoparticles'. *Analyst*. 2011;**136**(3):527–32.

[59] Gu B., Ruan C., Wang W. 'Perchlorate detection at nanomolar concentrations by surface-enhanced Raman scattering'. *Applied Spectroscopy*. 2009;**63**(1):98–102.

[60] Liu Y., Chen Y.R., Nou X., Kim M.S., Chao K. 'Label-free SERS for rapid species identification of *Escherichia coli, Listeria monocytogenes*, and *Salmonella typhimurium* bacteria'. *Spectroscopy*. 2008;**23**(2):48.

[61] Fan C., Hu Z., Riley L.K., Purdy G.A., Mustapha A., Lin M. 'Detecting food- and waterborne viruses by surface-enhanced Raman spectroscopy'. *Journal of Food Science*. 2010;**75**(5):M302–7.

[62] Najafi R., Mukherjee S., Hudson J., Sharma A., Banerjee P. 'Development of a rapid capture-cum-detection method for *Escherichia coli* O157 from apple juice comprising nano-immunomagnetic separation in tandem with surface enhanced Raman scattering'. *International Journal of Food Microbiology*. 2014;**189**:89–97.

[63] Liedberg B., Nylander C., Lunström I. 'Surface plasmon resonance for gas detection and biosensing'. *Sensors and Actuators*. 1983;**4**:299–304.

[64] Cusano A., Arregui F.J., Giordano M., Cutolo A. *Optochemical Nanosensors*. Boca Raton, FL: CRC Press; 2012.

[65] Shankaran D.R., Gobi K.V., Miura N. 'Recent advancements in surface plasmon resonance immunosensors for detection of small molecules of biomedical, food and environmental interest'. *Sensors and Actuators B: Chemical*. 2007;**121**(1):158–77.

[66] Homola J., Yee S.S., Gauglitz G. 'Surface plasmon resonance sensors: review'. *Sensors and Actuators B: Chemical*. 1999;**54**(1):3–15.

[67] Vollmer F., Arnold S. 'Whispering-gallery-mode biosensing: label-free detection down to single molecules'. *Nature Methods*. 2008;**5**(7):591–6.

[68] Leonard P., Hearty S., Quinn J., O'Kennedy R. 'A generic approach for the detection of whole *Listeria monocytogenes* cells in contaminated samples using surface plasmon resonance'. *Biosensors and Bioelectronics*. 2004;**19**(10):1331–5.

[69] Oh B.K., Kim Y.K., Park K.W., Lee W.H., Choi J.W. 'Surface plasmon resonance immunosensor for the detection of *Salmonella typhimurium*'. *Biosensors and Bioelectronics*. 2004;**19**(11):1497–504.

[70] Mazumdar S.D., Hartmann M., Kämpfer P., Keusgen M. 'Rapid method for detection of *Salmonella* in milk by surface plasmon resonance (SPR)'. *Biosensors and Bioelectronics*. 2007;**22**(9):2040–6.

[71] Homola J., Dostalek J., Chen S., Rasooly A., Jiang S., Yee S.S. 'Spectral surface plasmon resonance biosensor for detection of staphylococcal enterotoxin B in milk'. *International Journal of Food Microbiology*. 2002;**75**(1):61–9.

[72] Rasooly A. 'Surface plasmon resonance analysis of staphylococcal enterotoxin B in food'. *Journal of Food Protection*. 2001;**64**(1):37–43.

[73] Schnerr H., Vogel R., Niessen L. 'A biosensor-based immunoassay for rapid screening of deoxynivalenol contamination in wheat'. *Food and Agricultural Immunology*. 2002;**14**(4):313–21.

[74] Fratamico P., Strobaugh T., Medina M., Gehring A. 'Detection of *Escherichia coli* 0157:H7 using a surface plasmon resonance biosensor'. *Biotechnology Techniques*. 1998;**12**(7):571–6.

[75] Oh B.K., Kim Y.K., Bae Y., Lee W., Choi J.W. 'Detection of *Escherichia coli* O157:H7 using immunosensor based on surface plasmon resonance'. *Journal of Microbiology and Biotechnology*. 2002;**12**(5):780–6.

[76] Subramanian A., Irudayaraj J., Ryan T. 'A mixed self-assembled monolayer-based surface plasmon immunosensor for detection of *E. coli* O157:H7'. *Biosensors and Bioelectronics*. 2006;**21**(7):998–1006.

[77] Khanna V.K. Nanosensors: *Physical, Chemical, and Biological*. Boca Raton, FL: CRC Press; 2011.

[78] Das R., Sharma M.K., Rao V.K., *et al*. 'An electrochemical genosensor for *Salmonella typhi* on gold nanoparticles-mercaptosilane modified screen printed electrode'. *Journal of Biotechnology*. 2014;**188**:9–16.

[79] Wang L., Chen W., Xu D., *et al*. 'Simple, rapid, sensitive, and versatile SWNT—paper sensor for environmental toxin detection competitive with ELISA'. *Nano Letters*. 2009;**9**(12):4147–52.

[80] Mishra N.N., Maki W.C., Cameron E., *et al*. 'Ultra-sensitive detection of bacterial toxin with silicon nanowire transistor'. *Lab on a Chip*. 2008;**8**(6):868–71.

[81] Viswanathan S., Wu L.C., Huang M.R., Ho J.A. 'Electrochemical immunosensor for cholera toxin using liposomes and poly (3,4-ethylenedioxythiophene)-coated carbon nanotubes'. *Analytical Chemistry*. 2006;**78** (4):1115–21.

[82] González Crevillén A., Avila M., Pumera M., González M.C., Escarpa A. 'Food analysis on microfluidic devices using ultrasensitive carbon nanotubes detectors'. *Analytical Chemistry*. 2007;**79**(19):7408–15.

[83] Pal S., Alocilja E.C., Downes F.P. 'Nanowire labeled direct-charge transfer biosensor for detecting *Bacillus* species'. *Biosensors and Bioelectronics*. 2007;**22**(9):2329–36.

[84] Villamizar R.A., Maroto A., Rius F.X., Inza I., Figueras M.J. 'Fast detection of *Salmonella* Infantis with carbon nanotube field effect transistors'. *Biosensors and Bioelectronics*. 2008;**24**(2):279–83.

[85] De La Rica R., Mendoza E., Lechuga L.M., Matsui H. 'Label-free pathogen detection with sensor chips assembled from peptide nanotubes'. *Angewandte Chemie International Edition*. 2008;**47**(50):9752–5.

[86] Lin Y.H., Chen S.H., Chuang Y.C., *et al*. 'Disposable amperometric immunosensing strips fabricated by Au nanoparticles-modified screen-printed carbon electrodes for the detection of foodborne pathogen *Escherichia coli* O157:H7'. *Biosensors and Bioelectronics*. 2008;**23**(12):1832–7.

[87] So H.M., Park D.W., Jeon E.K., *et al*. 'Detection and titer estimation of *Escherichia coli* using aptamer-functionalized single-walled carbon-nanotube field-effect transistors'. *Small*. 2008;**4**(2):197–201.

[88] Terry L.A., White S.F., Tigwell L.J. 'The application of biosensors to fresh produce and the wider food industry'. *Journal of Agricultural and Food Chemistry*. 2005;**53**(5):1309–16.

[89] Castillo J., Gáspár S., Leth S., *et al.* 'Biosensors for life quality: design, development and applications'. *Sensors and Actuators B: Chemical*. 2004;**102**(2):179–94.

[90] Thakur M., Ragavan K. 'Biosensors in food processing'. *Journal of Food Science and Technology*. 2013;**50**(4):625–41.

[91] Bala C. 'New challenges in the design of bio (sensors) for biological warfare agents'. *Portable Chemical Sensors*. Berlin: Springer; 2012. pp. 15–41.

[92] Malik P., Katyal V., Malik V., Asatkar A., Inwati G., Mukherjee T.K. 'Nanobiosensors: concepts and variations'. *ISRN Nanomaterials*. 2013; **2013**. Available from http://dx.doi.org/10.1155/2013/ 327435 [Accessed on 12 March 2016].

[93] Li Y., Cheng P., Gong J., *et al.* 'Amperometric immunosensor for the detection of *Escherichia coli* O157:H7 in food specimens'. *Analytical Biochemistry*. 2012;**421**(1):227–33.

[94] Jianrong C., Yuqing M., Nongyue H., Xiaohua W., Sijiao L. 'Nanotechnology and biosensors'. *Biotechnology Advances*. 2004;**22**(7):505–18.

[95] Rizvi S.S., Moraru C.I., Bouwmeester H., Kampers F.W. 'Nanotechnology and food safety'. *Ensuring Global Food Safety: Exploring Global Harmonization*. Academic press; 2009. p. 263.

[96] Zhang X., Guo Q., Cui D. 'Recent advances in nanotechnology applied to biosensors'. *Sensors*. 2009;**9**(2):1033–53.

[97] Cesarino I., Moraes F.C., Lanza M.R., Machado S.A. 'Electrochemical detection of carbamate pesticides in fruit and vegetables with a biosensor based on acetylcholinesterase immobilized on a composite of polyaniline–carbon nanotubes'. *Food Chemistry*. 2012;**135**(3):873–9.

[98] Crevillén A.G., Ávila M., Pumera M., González M.C, Escarpa A. 'Food analysis on microfluidic devices using ultrasensitive carbon nanotubes detectors'. *Analytical Chemistry*. 2007;**79**(19):7408–15.

[99] Kirsch J., Siltanen C., Zhou Q., Revzin A., Simonian A. 'Biosensor technology: recent advances in threat agent detection and medicine'. *Chemical Society Reviews*. 2013;**42**(22):8733–68.

[100] Vinayaka A., Thakur M. 'Focus on quantum dots as potential fluorescent probes for monitoring food toxicants and foodborne pathogens'. *Analytical and Bioanalytical Chemistry*. 2010;**397**(4):1445–55.

[101] Valdés M.G., González A.C.V., Calzón J.A.G., Díaz-García M.E. 'Analytical nanotechnology for food analysis'. *Microchimica Acta*. 2009;**166**(1–2):1–19.

[102] Ramakrishna S., Fujihara K., Teo W.E., Yong T., Ma Z., Ramaseshan R. 'Electrospun nanofibers: solving global issues'. *Materials Today*. 2006;**9**(3): 40–50.

[103] Matlock-Colangelo L., Baeumner A.J. 'Recent progress in the design of nanofiber-based biosensing devices'. *Lab on a Chip*. 2012;**12**(15):2612–20.

[104] Schaller E., Bosset J.O., Escher F. '"Electronic noses" and their application to food'. *LWT—Food Science and Technology*. 1998;**31**(4):305–16.

[105] Ramgir N.S. 'Electronic nose based on nanomaterials: issues, challenges, and prospects'. *ISRN Nanomaterials*. 2013;**2013**. Available from http://dx.doi.org/10.1155/2013/941581 [Accessed on 12 March 2016].

[106] Peris M., Escuder-Gilabert L. 'A 21st century technique for food control: Electronic noses'. *Analytica Chimica Acta*. 2009;**638**(1):1–15.

[107] Rock F., Barsan N., Weimar U. 'Electronic nose: current status and future trends'. *Chemical Reviews*. 2008;**108**(2):705–25.

[108] Brezmes J., Llobet E. 'Electronic noses for monitoring the quality of fruit'. *Electronic Noses and Tongues in Food Science*. Amsterdam: Elsevier; 2016. pp. 49–58.

[109] Di Natale C., Macagnano A., Martinelli E., Paolesse R., Proietti E., D'Amico A. 'The evaluation of quality of post-harvest oranges and apples by means of an electronic nose'. *Sensors and Actuators B: Chemical*. 2001;**78**(1):26–31.

[110] Baietto M., Wilson A.D. 'Electronic-nose applications for fruit identification, ripeness and quality grading'. *Sensors (Basel)*. 2015;**15**(1):899–931.

[111] Santos J.P., Lozano J., Aleixandre M., *et al*. 'Threshold detection of aromatic compounds in wine with an electronic nose and a human sensory panel'. *Talanta*. 2010;**80**(5):1899–906.

[112] García M., Aleixandre M., Gutiérrez J., Horrillo M. 'Electronic nose for wine discrimination'. *Sensors and Actuators B: Chemical*. 2006;**113**(2):911–16.

[113] Brezmes J., Llobet E., Vilanova X., Saiz G., Correig X. 'Fruit ripeness monitoring using an electronic nose'. *Sensors and Actuators B: Chemical*. 2000;**69**(3):223–9.

[114] Wang P., Liu Q., Wu C., Hsia K.J. *Bioinspired Smell and Taste Sensors*. Berlin: Springer; 2015.

[115] Miranda O.R., Creran B., Rotello V.M. 'Array-based sensing with nanoparticles: "chemical noses" for sensing biomolecules and cell surfaces'. *Current Opinion in Chemical Biology*. 2010;**14**(6):728–36.

[116] You C.C., Miranda O.R., Gider B., *et al*. 'Detection and identification of proteins using nanoparticle-fluorescent polymer "chemical nose" sensors'. *Nature Nanotechnology*. 2007;**2**(5):318–23.

[117] Phillips R.L., Miranda O.R., You C.C., Rotello V.M., Bunz U.H. 'Rapid and efficient identification of bacteria using gold-nanoparticle-poly (para-phenyleneethynylene) constructs'. *Angewandte Chemie International Edition*. 2008;**47**(14):2590–4.

[118] Sapsford K.E., Rasooly A., Taitt C.R., Ligler F.S. 'Detection of *Campylobacter* and *Shigella* species in food samples using an array biosensor'. *Analytical Chemistry*. 2004;**76**(2):433–40.

[119] Rowe-Taitt C.A., Golden J.P., Feldstein M.J., Cras J.J., Hoffman K.E, Ligler F.S. 'Array biosensor for detection of biohazards'. *Biosensors and Bioelectronics*. 2000;**14**(10–11):785–94.

[120] Ligler F.S., Taitt C.R., Shriver-Lake L.C., Sapsford K.E., Shubin Y., Golden J.P. 'Array biosensor for detection of toxins'. *Analytical and Bioanalytical Chemistry*. 2003;**377**(3):469–77.

[121] Sapsford K.E., Ngundi M.M., Moore M.H., *et al*. 'Rapid detection of foodborne contaminants using an array biosensor'. *Sensors and Actuators B: Chemical*. 2006;**113**(2):599–607.

Chapter 11
POC in biowarfare detection and defence applications: an update

Petr Skládal[1]

11.1 Introduction

Detection and identification of dangerous microbial and other biowarfare agents is a challenging task. This includes the prevention and treatment of infectious diseases due to possible bioterrorist attacks or other pandemic outbreaks that are current threats for the society. The most dangerous pathogens were compiled by the Centers for Disease Control (http://emergency.cdc.gov/agent/agentlist.asp). The list includes bacteria as *Bacillus anthracis* (anthrax), *Yersinia pestis* (plague), *Francisella tularensis, Brucella* spp., *Burkholderia pseudomallei*, the parasitic organism *Coxiella burnetii* and most dangerous viruses such as *Variola major* (smallpox), *Venezuelan equine encephalitis* virus, filoviruses (Marburg, Ebola) and haemorrhagic fever. Selected toxins are botulotoxin, ricin and *Staphylococcal* enterotoxin B. The 'less dangerous' microbial pathogens which might be present at common situations or seasonally are *Mycobacterium tuberculosis, Streptococcus pneumoniae, Neisseria meningitidis* and *Legionella*, influenza virus A (H1N1), bird influenza virus (H5N1), *Rubella*, SARS, enteroviruses and rhinoviruses [1].

The on-site and near real-time bioanalytical procedures, applicable for the point-of-care (POC), or perhaps more precisely, point-of-incidence, use will be addressed here. Suitable procedures (polymerase chain reaction, PCR; enzyme-linked immunosorbent assays, ELISA) and relevant devices become available; however, slow operation, complicated portability and high running costs limit their widespread use. The detection of bioagents is further complicated by only minor differences between dangerous and commonly present microbial species. Numerous types of biosensors are considered as well suited for this purpose [2–6]. The previous demand on detection of bioagents originated mainly from military, however, civil rescue and security services, protection of public buildings as well as homeland security represent the main current fields of interest. The increasing activity of terrorist organisations transforms the potential danger of using biological warfare agents (BWAs) into a real threat and portable, rapid and simple

[1]Department of Biochemistry and CEITEC RG Nanobio, Masaryk University, Brno, Czech Republic

instrumentation based on immunoanalytical and nucleic acid assay principles is widely supported [7–9].

For bacteria and viruses as bioagents, various types of immunochemical devices are preferred for the early response and good sensitivity; the capability of continuous monitoring is also a quite attractive advantage. The detection occurs on the phenotype level, and no extraction of the genetic material from the bioagent is required, which is the necessary case for methods based on PCR and general detection of specific nucleic acid sequences. ELISA as a classic immunoanalytical format inspired various types of immunosensors. Here, the optical [10] and electrochemical [11] immunosensors will be addressed as the approach combining high sensitivity, simple construction and portability of the sensing part with the excellent specificity of antibodies and other affinity-based recognition systems.

11.2 Biodetection technologies

11.2.1 Paper and lateral-flow-based assays

Low-cost assays have broad applications in human health, diagnostics and food safety inspection, environmental analysis and biodetection. These assays are attractive for rural areas and developing countries, where financial resources and high-level laboratories are limited. The assays function on the principle of spontaneous capillary flow of the applied sample drop through different zones of a porous strip (Figure 11.1). The sample first dissolves the tracer – a conjugate of secondary detection antibody with a label suitable for direct visual evaluation (gold or carbon nanoparticle, chromophore, fluorophore and enzyme); immunoreaction starts and the analyte forms immunocomplex with the tracer. Next, it becomes trapped in the zone of capture antibody – this is the signal generation zone, the evaluated intensity corresponds to the concentration of analyte. Furthermore, the remaining part of tracer is bound in the control zone confirming correct function. For user convenience, the strip can be embedded in a suitable plastic folder with sample application opening and the exposed evaluation zone. The multiplex LF version can be also based on multicolour silver nanoparticles, and this expands the usefulness; it was possible to distinguish dengue, yellow fever and Ebola viruses [12].

Paper-based microfluidic devices emerged as fully paper-based and hybrid platforms [13]. Paper systems for viable pathogenic bacteria detection with the naked eye employed isothermal amplification of the hlyA mRNA gene marker from *Listeria monocytogenes* [14]. The amplicons were applied to the paper-based platform to perform a visual evaluation using sandwich hybridisation assay. The products migrated along the platform by capillary action, resulting in the gold nanoparticles accumulating at the designated area. As little as 0.5 ng mL^{-1} genomic RNA was detected, corresponding to 20 CFU mL^{-1} in food samples. However, assay requires few hours to complete all steps.

11.2.2 Microfluidic and lab-on-chip concepts

Typically, microfluidic chips integrate miniaturised electronics and optical sensing elements, fluid-handling components and, together with acquisition software,

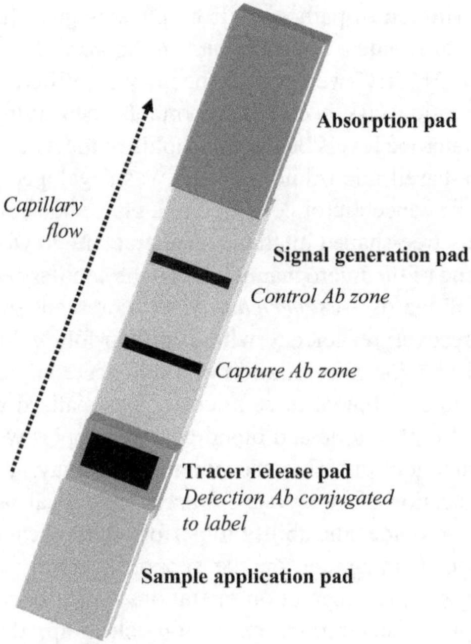

Figure 11.1 Immunostrip (lateral flow immunoassay). The strip consists of sample application zone, in the following tracer zone the immunoreaction starts – the analyte forms immunocomplex with the secondary detection antibody (Ab), the labelled complex is bound in the capture zone by the immobilised primary antibody. The remaining part of the tracer becomes bound in the control zone verifying correct function of the strip

provide a portable self-contained device. For example, pre-concentration was enabled by photo polymerising a thin, nanoporous membrane with a molecular weight (MW) cut-off of 10 kDa in the sample loading zone of the chip [15]; polymeric gels with larger pores were adjacent to the size exclusion membrane to perform electrophoretic separation of the antibody–analyte complex and the unbound excess antibody. Quantification of the bound and unbound forms employed laser-induced fluorescence detection. With off-chip mixing and no sample pre-concentration, the limits of detection (LOD) were 300 pM for Staphylococcal enterotoxin B (SEB), 500 pM for Shiga toxin I, and 20 nM for ricin. With a 10-min on-chip pre-concentration, the LOD for SEB was under 10 pM. Progress in the lab-on-a-chip (LOC) systems for viral detection was recently reviewed [16].

The integrated microfluidic device for complex physiological matrices such as blood was fabricated using SlipChip technologies; it integrated four channels processing independent samples and identifying up to 20 pathogens. Briefly, diluted blood was directly injected, pathogens were extracted by dielectrophoresis, retained in an array of grooves and identified by multiplex array PCR in nanolitre volumes with end-point fluorescence detection [17]. The universality of the dielectrophoretic separation of pathogens from physiological fluids was evaluated with a panel of clinical isolates; *Pseudomonas aeruginosa, Staphylococcus aureus* and *Escherichia coli* O157:H7 were simultaneously identified within 3 h.

The pre-concentration task becomes important when trying to analyse liquid samples containing microbe levels below thresholds of the assay procedure. A rapid and portable agarose-based microfluidic device was developed to concentrate biological fluids [18]. The concentrator consisted of a glass slide covered by an agarose layer forming a binary tree-shaped microchannel structure, in which pathogens were concentrated at the end of the microchannel due to the capillary effect and the strong water permeability of agarose. *Escherichia coli* (fluorescent strain OP50) demonstrated 90 per cent recovery efficiency with a million-fold volume reduction from 0.4 mL down to 0.4 nL. For concentration of 1,000 bacterial cells mL^{-1}, approximately 10 million-fold enrichment in cell density was realised with volume reduction from 0.1 mL to 1.6 pL. Urine and blood plasma samples were used to validate the method. In conjugation with fluorescence immunoassay, it was applied to the concentration and detection of infectious *S. aureus* in clinical samples.

LOC technology provides the ability to perform biological assays in resource-poor areas. The concept integrates on the supporting chip various miniaturised elements performing sample preparation operations typically carried out in analytical laboratories (pre-concentration, extraction, clean-up, dilution, addition of reagents and mixing); more sophisticated operations may include cell lysis, extraction of nucleic acids, even PCR. Eventually, the (bio)sensing elements providing the response signal might be embedded, too.

The ELISA-LOC concept integrated fluid delivery and a miniature 96-well plate, a simplified non-enzymatic reporter – a gold nanoparticle–antibody conjugate [19]. For higher visual signal, silver enhancement and carbon nanotubes were used. The device was tested for detection of SEB using three detection modes, visual, camera and document scanner; the resulting LODs were 0.5, 0.1 and 0.5 ng mL^{-1}.

The Lab-on-Disk system belongs to rotating CD-ROM like platforms [20]. It was employed for detection of *Bacillus atrophaeus* subsp. globigii spores. The complete assay comprised cellular lysis, PCR amplification, amplicon digestion and microarray hybridisation on a plastic support. The fluidic robustness and operating efficiency resulted from optimisation of microfluidic tools enabling beneficial implementation of capillary valves and accurate control of timing procedures. Thermoplastic elastomer (TPE) was selected as the material for CD fabrication and assembly, allowing both high quality hot-embossing and injection moulding manufacturing processes; in this way, the gap between proof-of-concept and final device can be overcome within a reasonably short time. Furthermore, the low-temperature operations and pressure-free assembly and bonding properties of TPE material ensured simple and efficient loading and storage of reagents and other on-board components.

Lithographically fabricated micron-scale forms of cubic retroreflectors served as reporter labels for use in sensitive immunoassays [21]. Retroreflector cubes as labels are first mixed with the sample, then the cubes are allowed to settle onto an immuno-capture surface, followed by inversion for gravity-driven removal of non-specifically bound cubes. Cubes bridged to the capture surface by the microbe are detected using inexpensive low-numerical aperture optics. For model bacterial and viral pathogens, sensitivity in 10 times diluted human serum was 10^4 bacterial cells mL^{-1} and 10^4 virus particles mL^{-1}.

Conveniently, widespread and cheap personal glucometers are transformed for other unique purposes, including detection of microbial pathogens. The glucoamylase-quarternised magnetic nanoparticle conjugates were disrupted by the competitive multivalent interactions with pathogenic bacteria, resulting in the release of glucoamylase, magnetic separation and catalysed hydrolysis of amylose into glucose [22]. LOD down to 20 cells mL^{-1} was achieved.

11.2.3 Electrochemical biosensors

Electrochemical immunosensors combine high sensitivity of electrochemical methods, simple and miniature construction of the required instrumentation with excellent specificity of antibodies as recognition elements for biosensing. Electrochemical technology is also suitable for measurement of nucleic acids extracted from the target microbes. The principles of electrochemical immunosensors are well known [23] and their potential for detection of pathogens was realised as well [24,25]. The comprehensive review summarising commercially available biosensors focused on pathogens and biothreat agents [26]. Recent advances in this field were summarised and discussed in reference [11].

Generally, the electrochemical measuring system is highly sensitive and quite cheap. Progress in electronics allows miniaturising the whole detector to a single-chip format; the embedded digitally controlled potentiostat LMP91000 (Texas Instruments) is programmed through serial interface (I2C) and consumes minimum power. Potentiometric techniques can be realised with most digital multimeters. Even the advanced pulsed, voltammetric and galvanostatic techniques are available as hand-held instruments from several companies: PalmSens and EmStat from PalmInstruments, µStat from DropSens, PG581 from Uniscan Instruments, 910 PSTAT mini from Metrohm, as well as other prototypes designed directly in laboratories. As the measuring element, the screen-printed electrodes (SPE) are widely applied due to easy and reproducible fabrication at both laboratory and mass production scales. The low production costs allow single use of the resulting immunosensors; thus, no complicated regeneration procedures are required. The measuring formats are either indirect with suitable enzyme labels (horse radish peroxidase, alkaline phosphatase) generating electrochemically measurable products (Figure 11.2(a)). Simplified direct formats (Figure 11.2(b)) usually employ electrochemical impedance spectroscopy in which formation of immunocomplex blocks the sensing surface towards access of a redox probe (e.g. ferricyanide, ferrocene and ruthenium complexes).

11.2.4 DNA analysis on chips

The need for diagnostic tests focused on nucleic acid was stimulated by the available information on hereditary genetic diseases following completion of the Human

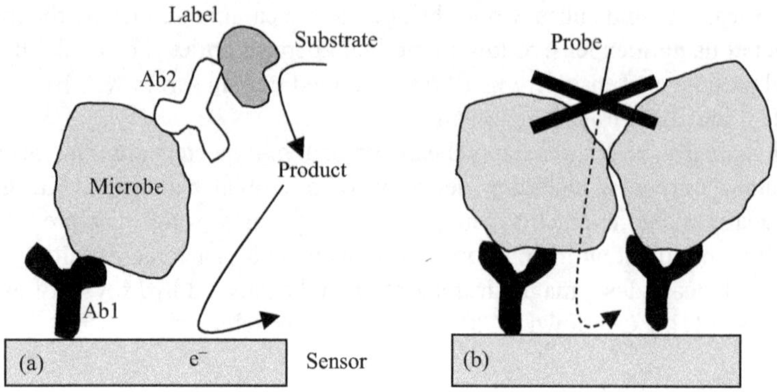

Figure 11.2 *Concepts of label-based ((a) enzyme label generates product measured by the transducer) and direct ((b) signal is due to the capture of the microbe; a blocked access of probe to the surface helps to indicate formation of the immunocomplex) immunosensor assays for microbes*

Genome Project, presence of new rapidly spreading viral threats (SARS and Ebola outbreaks) or microorganisms considered for biological warfare [27]. Rapid diagnoses of specific nucleic acid sequence are required, new strategies have to be developed, including compact bioanalytical systems as biosensors.

Significant effort was devoted to the miniaturisation of PCR into a portable or even on-chip systems. An integrated biochip device consisted of microfluidic mixers, valves, pumps, channels, chambers, heaters and DNA microarray sensors [28]. Sample preparation (magnetic bead-based cell capture, cell pre-concentration, purification and lysis), PCR, hybridisation of amplicons with probes and eventual electrochemical detection were performed in a fully automated mode. Cavitations based micro-streaming enhanced capture and accelerated the DNA hybridisation step. Thermally actuated paraffin microvalves regulated flows. Electrochemical and thermo-pneumatic pumps on the chip moved liquid solutions. The device was completely self-contained, thus eliminating sample contamination and simplifying device operation. Pathogenic bacteria detection was successfully demonstrated.

The multiplex asymmetric PCR was able to detect several target organisms simultaneously; however, optimisations of primer concentrations and staggered additions of primers are required for equal amplification of multiplex genes. To overcome this shortcoming, a novel method based on multiplex asymmetric PCR and paper-based nucleic acid diagnostics was proposed [29]. A universal primer was introduced to break the bottlenecks of poor sensitivity and self-inhibition among different sets of primers. Amplification using the multiplex asymmetric PCR boosted the quantity of single-stranded amplicons containing the same sequence at the 5′ end. In this way, only one gold nanoparticle-based signal probe was needed for simultaneous detection of three genes with the naked eye, providing LOD of 1 ng mL^{-1} of genomic DNA.

A multi-pathogen analysis array was fabricated through immobilizing three specific polyT-capture probes for rfbE gene (*E. coli* O157:H7), invA gene (*Salmonella enterica*), inlA gene (*L. monocytogenes*) on the plastic substrates [30]. After PCR amplification, biotin labelling and capture of the target, the biotinylated target DNA was stained with avidin-peroxidase conjugate and biotinylated anti-HRP antibody for amplification of the signal through multiple binding cycles; the pathogens from food samples were detected as colour change with naked eyes. LODs were around five copies of the target DNA.

Loop-mediated isothermal amplification (LAMP) is alternative to the common PCR. DNA-paramagnetic silica bead aggregation in a rotating magnetic field facilitates the quantification of DNA with femtogram sensitivity; an original aggregation inhibition for the detection of specific sequences was developed [31]. The fragments generated via LAMP are able to passivate the bead surface and effectively inhibit bead aggregation by longer 'trigger' DNA. Aggregation inhibition served for the detection of bacterial and viral pathogens with sensitivity close to single copies of the target, successful tests were done for food-borne *E. coli* O157:H7 and *S. enterica*, as well as Rift Valley fever virus, a bioweapon considered virus of national security concern.

LAMP coupled to the electrochemical measurement (methylene blue as label) was integrated on a laser etched indium tin oxide multiplex microfluidic chip [32]. Three bacteria – *M. tuberculosis, Haemophilus influenza* and *Klebsiella pneumonia* were detected at LODs equal to 28, 17 and 16 copies mL^{-1}, respectively. The assay was finished in 45 min.

Integrated electrochemical sensors offer a particularly promising solution to genetic detection because they do not require optical instrumentation and are compatible with both integrated circuit and microfluidics. The development of generally applicable microfluidic electrochemical platforms integrating sample treatment and amplification (both PCR and LAMP were tested) as well as quantitative and multiplexed detection remains a challenging and not yet completely solved technical issue. The E-DNA platform for specific nucleic acid sequences detection employs label-free single-step electrochemical sensors [33]. The electrode-bound redox-reporter-modified DNA 'probe' generates a change of current after undergoing a hybridisation-induced conformational change.

LAMP was also coupled to roll-to-roll ribbon fluid-handling device for electrochemical (osmium complex) real-time bacteria detection, LODs were 30 CFU mL^{-1} of *E. coli* and 200 CFU mL^{-1} of *S. aureus* [34]. Strand exchange nucleic acid circuitry was used to transduce LAMP products into signals readable on an off-the-shelf glucometer. Combining LAMP method with a thermostable invertase, direct transduction of Middle-East respiratory syndrome coronavirus and Zaire Ebola virus (cell lysates and templates) into glucose signals was achieved [35]; sensitivity was 20–100 copies μL^{-1}; an OR gate that coordinated triggering on viral amplicons further guaranteed fail-safe virus detection.

11.2.5 Smartphones for analysis

POC devices need to combine adequate sensitivity with low cost of production and operational simplicity and speed. In this context, smartphones can provide substantial simplification of design and construction by providing the compact control microcomputer with suitable communication tools (Bluetooth for local links, wireless Wi-Fi and digital cellular networks for internet links) and also transduction option – the camera module. In addition, the internal battery can provide power for the sensing module, if the requirements are reasonable, and the display serves as graphic user interface for the control and presentation of results. Recently, also the options for sending data (e.g. images) to the central server for storage, proper chemometric treatment and returning back as the 'simplified' results, becomes attractive. A hand-held phone-based colorimetric micro-plate reader used a 3D-printed opto-mechanical attachment to hold and illuminate a 96-well plate using a light emitting diode (LED) array. This light is transmitted through each well, collected via 96 optical fibres and images of this fibre-bundle are transmitted to servers for processing using a machine learning algorithm; results are delivered to the user within 1 min and visualised on the display [36].

For detection of viruses, a cradle unit with disposable sensing cartridge, a tiny magnetic stirrer and a few passive optical components was designed [37]. The detection principle was the 'Reflective Phantom', measurement of the intensity of light reflected by the amorphous fluoropolymer substrate with various antibody spots. The reflectivity of spots was monitored by the camera using the embedded flash LED for illumination. Immunoglobulins and antigens used as markers for hepatitis B and HIV were detected at few ng mL^{-1}, using the rate of increase of the signal after the addition of the sample in comparison with the subsequent addition of standard. The D3 (digital diffraction & diagnosis) system used microbeads to generate unique diffraction patterns, this was acquired by smartphone and processed by a remote server. Detection of human papillomavirus (HPV) DNA was completed in 45 min [38].

The smartphone camera served also as a spectrometer for a label-free photonic crystal biosensor. A custom cradle provided fixed alignment with optical components required for accurate and repeatable measurements of resonant wavelength shifts [39]. External broadband light passed through entrance pinhole, it was collimated and linearly polarised before passing through the biosensor, which reflects only a narrow band of wavelengths at resonance. A diffraction grating spreads the remaining wavelengths over a camera resulting in a high resolution transmission spectrum. The exchangeable biosensing element was made on a plastic substrate attached to glass slide. The resonant wavelength was measured with 0.009 nm accuracy.

Even the acoustic part of smartphones might be useful; lysis of *Mycobacterium marinum* and *Staphylococcus epidermidis* bacteria utilised a portable audio device coupled with a simple electromagnetic coil [40]. The resulting alternating magnetic field rotated a magnet in a tube with the sample and glass beads, lysing the bacterial cells before further processing.

11.2.6 Surface plasmon resonance based approaches

The surface plasmon resonance (SPR) system as desktop devices exist for nearly three decades, Biacore variants being most popular among life scientists. Miniaturised SPR system was once represented by Spreeta. The palm-sized, battery-operated biodetection based on localised SPR modified the classic construction [41]. The spectrum analyser was replaced by four pulsed light-emitting diodes with different emission spectra. The reflected light beams from all LEDs were detected by a single photodiode, in which the composite output was demultiplexed by a four-channel lock-in amplifier.

Printable multianalyte biochips should enable simultaneous quantitative detection of multiple biomarkers in POC and resource-limited settings. However, preserving the functionality of biomolecules during conventional printing remains challenging. As alternative, plasmonic calligraphy involved a regular ballpoint pen filled with functionalised gold nanorods as plasmonic ink for creating individual test domains (sized 5 mm^2) on paper. The achieved sensitivity was significantly better than in the case of bioplasmonic paper fabricated using complete immersion procedures [42].

Plasmonic microarray coupled with a lens-free computational imaging allowed multiplexed and high-throughput screening in a compact format (60 g, 7.5 cm tall); the diffraction patterns of plasmonic nanostructures were uniformly illuminated by a single LED tuned to the plasmonic mode of the nanoapertures [43]. An iterative phase retrieval-based image reconstruction method offered digital imaging of a highly multiplexed array suitable for high-throughput in field settings. A wide working range from 10^3 to 10^9 viral particles mL^{-1} was achieved with optimised fluidic system [44].

11.2.7 Bioaerosols

So far, it was expected that microbes are detected in typical clinical samples, which are based on body liquids. However, when detecting microbes in the form of bioaerosol, the microorganisms from litres of air need to be concentrated into a small volume of water for subsequent 'common' analysis. This is a typical situation for misused pathogens as bioweapons; on the other hand, such monitoring becomes very important for hospital settings and even large public spaces (airports, underground networks, transportation transfer stations). Microbes are spread as individual cells and spores, in the form of microdroplets, or adsorbed on dust or other particles [45]. The levels of pathogens which should be reliably detected are derived from the infectious dose of *B. anthracis* (around 100 spores) and the air volume of 1 m^3 per one person in public space; at the same time, the natural background of non-pathogenic microbes can achieve up to 17,000 spores m^{-3} (general schools) [46]. At such situation, the purely physical detection should be replaced (or supplemented) by approaches relaying on biospecific recognition.

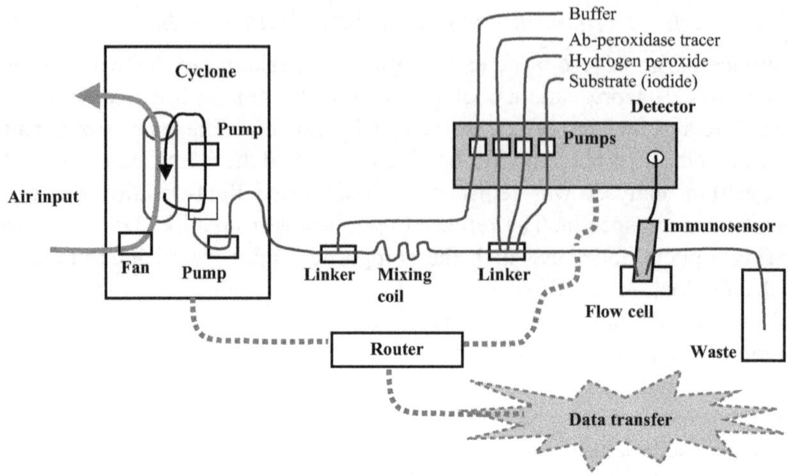

Figure 11.3 Schema of combination of the cyclone air sampler SASS 2300 (left, consisting of air-pumping fan, cyclone unit, liquid sample container and two peristaltic pumps) with the electrochemical detector ImmunoSMART (four peristaltic pumps and digitally controlled potentiostat) for measurements with immunosensors. The fluidic system consists of two linker elements; the first one is merging the sample from the cyclone with the buffer path from the immunodetector; the sample is delayed in the holding coil and passes through the second linker (mixing with other reagents) to the flow-through cell containing the immunosensor. The serial interfaces of both instruments are converted into the TCP-IP network and routed to the external computer located outside the chamber. The thick arrow indicates flow of air in the cyclone, thinner lines corresponds to the fluidics and dashed lines are communication paths

For capture of microbes, filtration of the analysed air is the simplest option; the captured microbes are evaluated afterwards. More complex samplers suitable for near-online monitoring include cyclones, impactors and electrostatic precipitators; a comprehensive review on such combined devices consisting of the air sampling and sensing components is available [47]. In our laboratory, the cyclone-based sampler SASS-2300 (Research International) was successfully used together with either electrochemical [48] or piezoelectric [49] immunosensors. The integrated air sampling/electrochemical immunosensing system allows stand-alone remotely controlled operation (Figure 11.3). This is critical for the testing phase, when experiments are carried out in aerosol chambers and field trials. In both situations, the operators are away from the actual measuring place. Similar conditions will exist also in real application situations.

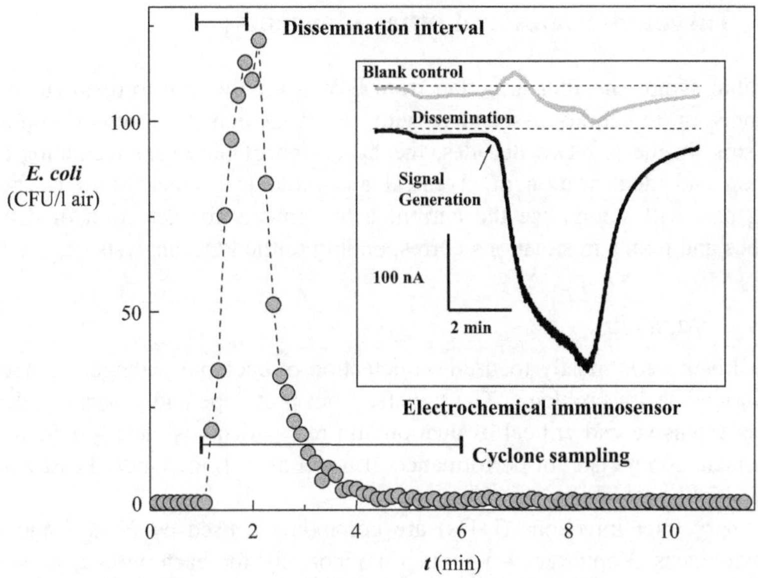

Figure 11.4 Testing of the biosensor system for detection of microbes in air. The main graph shows independent monitoring of viable cells in the air inside the aerosol chamber (data from slit samplers with rotating agar plates). The dissemination of Escherichia coli cells and cyclone sampling intervals are marked. The inserted signal traces from the immunodetector are the current in time plots corresponding to the enzyme reaction of peroxidase used as a label. The upper trace represents control blank signal in the absence of E. coli, the lower trace indicates positive detection of microbes

Typical result from model air analysis using such system – detection of *E. coli* in the bioaerosol generated inside experimental chamber – is shown in Figure 11.4. The microbe was disseminated in air from water suspension using nebuliser, the concentration profile (in colony forming units per litre of air) was generated using an array of sequentially triggered slit samplers with rotating agar plates. The cyclone was conducting repeated 5-min sampling, the captured microbes were measured with the sandwich immunosensor based on an SPE. The amperometric responses for disseminated microbe and control indicated that less than 100 CFU L^{-1} air are detectable in around 15 min.

Even such complex system can become miniaturised; a microfluidic lab-on-chip was also coupled to an electrostatic precipitator, in which corona charging and electrophoretic transport served to capture aerosol droplets onto the chip [50]. The fabrication of an electrostatic precipitator prototype exhibited collection efficiencies comparable to Biosampler impinger, with the significant advantage of providing samples that are at least 10 times more concentrated. The device has also a high potential for POC breath-based diagnostics.

11.3 Target microbes and other bioagents

The initial symptoms after infection from BWAs are difficult to distinguish from symptoms of infections associated with rather benign or common biological organisms. In the last two decades, the expansion of biosensor technologies for detection and identification of chemical and biological agents started. The following part will summarise the current achievements for detection of different microbes and toxins in situations corresponding to the POC analysis.

11.3.1 Bacteria

Recent literature is mostly focused on detection of bacterial pathogens associated with various health problems. The transfer from prototype and laboratory devices towards extensive and critical evaluations on real patients is clearly evident. The independent comparison of performance of novel as well as standard assay procedures is highly valuable.

Urinary tract infections (UTIs) are commonly caused by *E. coli* and other G-uropathogens. RapidBac – a rapid immunoassay for bacteriuria (Silver Lake Research Corp.) was compared with standard culture using 966 clean-catch urine specimens [51]. RapidBac provided a positive or negative result within 20 min, and it identified as positive 245/285 (sensitivity 86 per cent, specificity 94 per cent) samples with significant bacteriuria (presence of G-uropathogen or *Staphylococcus saprophyticus*) above 1,000 CFU mL^{-1}. The sensitivities for G-microbes at 10^4 and 10^5 CFU mL^{-1} were 96 per cent and 99 per cent, respectively. The RapidBac test may be effective as an aid in POC diagnosis of UTIs in emergency and primary care settings.

The mariPOC system for rapid detection of respiratory tract infections was compared with bacterial culture for detecting group A *Streptococcus* (GAS) in 219 pharyngitis patients and 109 healthy controls [52]. In addition, 42 patient samples were analysed by qPCR. Of the 219 patient samples, 32 were positive in a GAS bacterial culture (prevalence, 15 per cent) and 65 (30 per cent) in the mariPOC test. The amount of GAS in samples, reported positive by the mariPOC test and negative by culture was, on average, 10-fold less than that of those positive in both methods. This indicated that the negative results in cultures were due to lower sensitivity. The qPCR results were positive and in line with the mariPOC results in 43 per cent of the discordant samples. Two GAS culture-positive samples were negative by the mariPOC test. In conclusion, the mariPOC antigen test was more sensitive than the conventional culture for the detection of GAS among symptomatic patients.

Centrifugal microfluidic LabDisk with pre-stored reagents featured fully automated and integrated DNA extraction, consensus multiplex PCR preamplification and geometrically multiplexed species-specific real-time PCR [53]. Processing requires loading of the sample and DNA extraction reagents with minimal hands-on time of approx. 5 min. Detection sensitivity was very good, LODs were 15 CFU mL^{-1} of *Staphylococcus warneri*, 1,000 CFU mL^{-1} of *Streptococcus agalactiae*, 25 CFU mL^{-1} of *E. coli* and 10 CFU mL^{-1} of *Haemophilus influenzae* in serum.

The time of the complete analysis from was below 4 h. The advantage is easy-to-use diagnostic platform for relatively rapid (compared to classic culture) and highly sensitive detection of bacterial pathogens, without long hands-on time and complex instrumentation.

Microarray-based LOC was able to detect above 1,000 CFU mL^{-1} of *S. pneumoniae* in blood samples [54]. All individual working steps as well as the overall procedure were properly analysed and optimised and a control strategy was proposed according to recommendations of the US Food and Drug Administration agency.

Immunochromatographic nitrocellulose test strips with smartphone fluorescence readout were used in the detection of foodborne pathogens *Salmonella* spp. and *E. coli* O157 [55]. Silica nanoparticles doped with fluoresceine isothiocyanate (FITC) and Ru(bpy) were conjugated to antibodies and used in a lateral flow immunoassay. Fluorescence was read in a smartphone-based fluorimeter light weight (40 g) optical module, containing LED light source, filters and lens. The images were processed in the phone and provided results within few minutes, and LOD was 10^5 CFU mL^{-1} without pre-enrichment. The reported LOD appeared 10-times better compared to alternative labels – gold nanoparticles.

Clostridium difficile is the primary cause of antibiotic associated diarrhoea in humans and is a significant cause of morbidity and mortality, its rapid and accurate identification in clinical samples such as faeces is a key step in reducing the devastating impact. A rapid assay based on microwave-accelerated metal-enhanced fluorescence was capable of detecting the presence of 10 bacteria in unprocessed human faeces within 40 s [56].

11.3.2 Viruses

Each year, outbreaks of viral infections cause illness, disability, death and economic loss. As learned from past incidents, the detrimental impact grows exponentially without effective quarantine. Therefore, rapid on-site detection and analysis are highly desired. Furthermore, for high-risk areas of viral contamination, close monitoring should be provided during the potential disease incubation period [16]. The critical factors in ensuring success of viral diagnostic at POC were reviewed [57] with a focus on infrastructure and workflow limitations in clinical settings in both the developed and developing parts of world. Requirements include being low cost, easy-to-use, accurate and adapted for the intended laboratory and healthcare environment; furthermore, information that appropriately directs clinical treatment decisions should be provided. The challenges and implications of linking diagnostics to clinical decision-making at POC were demonstrated on three examples: respiratory viruses in the developed world, differential fever diagnosis in the developing world and HPV detection in resource-limited settings.

The technologies used to produce POC viral diagnostics range from highly sophisticated to commonly available resources. The laser cutter served to prepare microchip support using transparency sheet and template for electrodes using masking paper [58]. After deposition of silver ink and baking, the fluidics was completed by a piece of double-sided adhesive. The assay employed streptavidin-modified magnetic beads for attachment of biotinylated specific antibodies, the incubation with samples (plasma, saliva) continued for 30 min. After separation and lysis, the lysate was introduced to the microchip and change of capacitance was obtained using LCR meter. The working range was from 100 to 10^5 copies mL^{-1} and the system was used for detection of several HIV subtypes, Epstein-Barr virus and sarcoma-associated herpes virus.

Human enterovirus 71 (EV71) causes hand, foot and mouth disease, which generally leads to neurological diseases and fatal complications among children. As the early clinical symptoms are similar to Coxsackie-virus B3 (CVB3) infection, a robust and sensitive detection method that can be used to distinguish EV71 and CVB3 is urgently needed for prompting medical treatment. Immunomagnetic nanobeads and fluorescent semiconductor CdSe quantum dots (QDs) were combined for simultaneous POC detection of EV71 and CVB3 [59]. The detection of EV71 and CVB3 virions was completed within 45 min with LODs around 1,800 copies mL^{-1}. Validation on 20 human throat swabs obtained from EV71 or CVB3 positive cases resulted in 93.3 per cent consistency with those by the real-time PCR method, demonstrating the potential of this method for clinical quantification of EV71 and CVB3. The method may also facilitated the prevention and treatment of the diseases.

The progressive spread of the dengue virus and its rising incidence require rapid diagnosis suitable for developing countries in temperate climates. Recent advances in bioelectronics, micro- and nanofabrication technologies focused on POC devices and analytical platforms suited for rapid detection of infections were reviewed [60]. Starting from the available tests for dengue diagnosis, emerging rapid micro/nanotechnologies-based tools including label-free, biosensor methods, microarray and microfluidic platforms were examined.

The R-Biopharm RIDA (R) QUICK immunochromatography assay for norovirus detection was examined using faecal material from Australian gastroenteritis incidents. The study involved the analysis of three groups of specimens, including norovirus open reading frame (ORF) 1 RT-PCR positive specimens, ORF 1 norovirus negative specimens and specimens containing common gastroenteritis viruses other than norovirus. The RIDA QUICK (N1402) assay detected norovirus and had an overall sensitivity of 87 per cent [61].

11.3.2.1 Influenza viruses

Respiratory viruses are responsible for a large proportion of acute, respiratory illness in adults as well as children and are associated with a huge socio-economic burden worldwide. Accurate molecular platforms are now able to test for a comprehensive range of viruses, can be operated by non-laboratory staff and generate a result in approximately 1 h. The potential clinical benefits include a reduction in unnecessary antibiotic use, improved antiviral prescribing for influenza and rationalisation of isolation facilities [62]. The mariPOC assay allowed automated detection of eight respiratory viruses:, influenza A and B viruses, syncytial virus, adenovirus, human metapneumovirus and parainfluenza viruses 1–3 [63]. Positive results from samples with high viral load were available in 20 min. Sensitivity, specificity, positive predictive value and negative predictive values were 85.4 per cent, 99.2 per cent, 95.9 per cent and 97 per cent, respectively, and 84.6 per cent of positive results were obtained in 20 min.

11.3.2.2 Ebola

For Ebola virus disease (EVD), reverse transcription PCR technology is currently the standard; mobile rapid diagnostics were, and still are, not readily available for immediate and definitive diagnosis, a stunning strategic flaw that needs correcting worldwide [64]. The lack of laboratory facilities resulted in diagnostic complications during the West African EV outbreak in 2013 to 2015, thus compromising outbreak control; 27,748 confirmed, probable and suspected cases were reported by 29 July 2015. Owing to limited laboratory capacity and local transport infrastructure, the delays from sample collection to test results have often been two days or more [65]. The review summarised Ebola rapid diagnostic tests approved by the World Health Organisation and those currently in development. Such rapid diagnostic tests could allow early triaging of patients, thereby reducing the transmission. Despite the lower test accuracy, rapid diagnosis may be beneficial because of the reduced time spent by uninfected individuals in health-care settings in which they may be at increased risk of infection; this also frees up hospital beds. Based on modelling, it was estimated that if such tests had been available throughout the epidemic, their use in combination with confirmatory PCR might have reduced the scale of the epidemic by over a third for Sierra Leone.

The diagnostic accuracy of the EV bedside rapid diagnostic antigen test based on lateral flow assay (RDT, developed in UK Defence Science and Technology Laboratory) was compared with PCR for diagnosis of suspected cases admitted to Ebola holding units [66]. One hundred thirty-eight participants were enrolled, EVD prevalence was 11.5 per cent; all EVD cases were identified by a positive RDT result giving a sensitivity of 100 per cent; the corresponding specificity was 96.6 per cent. This suggests that RDT could be used as a 'rule-out' screening test for rapid EVD case identification.

Surface acoustic wave (SAW) immunosensors rapidly detected Ebola antigens from fragmented Ebola viral particles with LOD corresponding to 1.9×10^4 PFU mL^{-1}, which was below the average level of viraemia detected on the first day of symptoms by PCR [67]. The SAW system was easily portable and operated on common AA batteries.

11.3.2.3 Hepatitis viruses

Hepatitis B virus infection is one of the major causes of hepatitis, liver cirrhosis and liver cancer. Luminescent QD beads in the sandwich immunochromatographic assay served for detection of hepatitis B virus surface antigen (HBsAg) in serum [68]. The sensor achieved LOD of 75 pg mL^{-1}, which is much better than that of the routinely used gold nanoparticle based assay. In addition, the quantitative method developed showed no false positive results in an analysis of 49 real HBsAg-negative serum samples and exhibited excellent agreement with a commercial chemiluminescence immunoassay kit in identifying 47 HBsAg-positive serum samples.

Hepatitis C virus (HCV) infection has emerged as one of the most significant causes of chronic liver disease worldwide with an estimated prevalence ranging from 2.2 to 3.0 per cent. Considering that acute HCV infections are usually asymptomatic, early diagnosis is rare. Thus, new initiatives are needed to identify patients with chronic viral hepatitis and to propose controls and antiviral treatments to avoid the progression. The San Raffaele Scientific Institute in Milan realised a prevention program using the oral OraQuick HCV rapid antibody test (OraSure technologies); from January 2011 to April 2014, 29,600 subjects were approached and 4,507 volunteers agreed to perform this HCV tests [69]. Twenty-seven subjects (0.6 per cent of the total) turned HCV oral test reactive, confirmed by a conventional test. All 27 patients were asymptomatic and without a history of HCV symptoms. The results from this analysis suggest that the promotion of alternative HCV test screening has not yet been fully developed as a strategy. The recent introduction of rapid oral HCV antibody test could completely change the HCV diagnosis approach by facilitating the possibility of testing millions of people worldwide.

11.3.2.4 Human immunodeficiency virus

The need to screen sub-populations for several sexually transmitted and blood-borne infections including HIV simultaneously with multiplex POC platforms was considered [70]. This seems to be encouraging and promising approach; many factors need to be considered before implementation, integration and uptake of these technologies across global settings. Careful planning and investments in training health care professionals, improving test and treat algorithms, rapid protocols on communicating results to providers and timely action will bring about the desired impact in patient's lives and public health and social impact.

11.3.3 Toxins

Botulinum neurotoxin type E (BoNT/E) was detected using its protease activity [71]. It cleaves a peptide bond between two specific amino acid residues. The method is based on a two-step proteolytic cleavage using a target BoNT/E, light chain (BoNT/E-LC) and supplemented exopeptidase L-leucine-aminopeptidase (LAP). BoNT/E-LC cleaves a peptide bond between arginine and isoleucine in the IDTQNRQIDRI-4-amino-1-naphthol conjugate (oligopeptide-AN) to generate isoleucine-AN. Subsequently, LAP cleaves a bond between isoleucine and AN to free electroactive AN species. The AN participates in electrochemical–chemical–chemical (ECC) redox cycling with $[Ru(NH_3)_6]^{3+}$, which allows a high signal amplification on indium tin oxide electrodes. BoNT/E-LC could be detected at 2.0 pg mL^{-1}, 0.2 and 3 ng mL^{-1} after 4 h, 2 h and 15 min pre-incubation steps, respectively.

11.3.4 Antimicrobial antibodies

Instead of detecting very low levels of the primary microbes responsible for the disease, the levels of the corresponding antimicrobe antibodies might be assessed with sufficient sensitivity prior to any manifestation of clinical symptoms [72]. Lateral flow immunochromatographic assay (LFIA) relied on the gold magnetic nanoparticles conjugated with anti-human IgM antibody for the detection of immunoglobulin M antibodies related to toxoplasmosis, Rubella virus (German Measles), Cytomegalovirus and Herpes simplex virus infections [73]. The corresponding antigens were immobilised on the LFIA surface. Capture was based on gold magnetic nanoparticles. Following modification with poly(methacrylic acid), the gold magnetic nanoparticles conjugated with an anti-human IgM antibody (μ-chain specific) to construct a probe. The LFIA strips were used to assess 41 seropositive and 121 seronegative serum samples, providing 100 per cent sensitivity and specificity as well.

11.4 Conclusion

The progress from recent few years of the immunochemical and nucleic acid sensing devices used or planned for detection of microbial agents in POC situations was briefly summarised. At present, two types of assays are in focus.

The heterogeneous sandwich immunoassay is robust and reliable, sensitivity is provided by enzyme or other types of labels generating measurable or directly visible products. Complications include several incubations and the required washing steps, though microfluidic and lab-on-chip concepts help to realise such steps without manual workload. As straightforward alternative, the direct assay formats employ different strategies how to evaluate capture of the target (microbes for immunoassays, DNA for nucleic acid sensing) without any label. This area seems very promising, as novel surface preparation techniques, use of nanotechnologies (nano-particles/wires/tubes) resulted in very sophisticated devices and assay formats potentially exhibiting excellent analytical parameters.

However, the proof of reliable performance in real life should be addressed more intensively. The transfer of the detector system from laboratory to the real world usually demonstrates several more or less significant problems which associated together make function of the detector rather unreliable. However, the experience gained during this phase of testing from purely research approaches to practical evaluations under unpredictable conditions is invaluable and helps to correct weak parts of the biosensor.

Finally, the conclusion whether the target bioagent was detected or not, is presently carried out by the user looking on the measured trace of signals. Such evaluation and decision-making should be implemented in the control software, and this might be quite challenging, too. The combination of the analytical devices with chemometric procedures should provide unambiguous results. It seems that computing capability of smartphones will play a major role; also transfer of measured data (signals, images) to central servers for database storage, near-real-time evaluation and return of result already appeared. All these existing individual approaches integrated together should soon provide really smart POC devices suitable for automated measurements and early detection of potentially danger microorganisms. Recent outbreak of the Ebola virus clearly indicated that such technology is urgently needed and it will be immediately beneficial to patients and other users.

References

[1] Fronczek C.F., Yoon J.Y. 'Biosensors for monitoring airborne pathogens'. *Journal of Laboratory Automation*. 2015;**20**:390–410.

[2] Ivnitski D, Abdel-Hamid I, Atanasov P., Wilkins E. 'Biosensors for detection of pathogenic bacteria'. *Biosensors and Bioelectronics*. 1999;**14**:599–624.

[3] Laczka O., Del Campo F.J., Munoz F.X. 'Pathogen detection: a perspective of traditional methods and biosensors'. *Biosensors and Bioelectronics*. 2007;**22**:1205–17.

[4] Ligler F.S., Sapsford K.E., Golden J.P., *et al*. 'The array biosensor: portable, automated systems'. *Analytical Sciences*. 2007;**23**:5–10.

[5] Caygill R.L., Blair G.E., Millner P.A. 'A review on viral biosensors to detect human pathogens'. *Analytica Chimica Acta*. 2010;**681**:8–15.

[6] Yadav R., Dwivedi S., Kumar S., Chaudhury A. 'Trends and perspectives of biosensors for food and environmental virology'. *Food and Environmental Virology*. 2010;**2**:53–63.

[7] Lim D.V., Simpson J.M., Kearns E.A., Kramer M.F. 'Current and developing technologies for monitoring agents of bioterrorism and biowarfare'. *Clinical Microbiology Reviews*. 2005;**18**:583–607.

[8] Gooding J.J. 'Biosensor technology for detecting biological warfare agents: recent progress and future trends'. *Analytica Chimica Acta*. 2006;**559**:137–51.

[9] Heo J., Hua S.Z. 'An overview of recent strategies in pathogen sensing'. *Sensors*. 2009;**9**:4483–502.

[10] Ligler F.S. 'Biosensors for identification of biological warfare agents'. *Biosensors and Bioelectronics*. 2000;**14**:749.

[11] Skládal P. 'Electrochemical detection for biological identification'. In Schaudies P.R. (Ed.). *Biological Identification, DNA Amplification and Sequencing, Optical Sensing, Lab-On-chip and Portable Systems*. Amsterdam: Elsevier; 2014. pp. 131–52.

[12] Yen C.W., de Puig H., Tam J.O., *et al*. 'Multicolored silver nanoparticles for multiplexed disease diagnostics: distinguishing dengue, yellow fever, and Ebola viruses'. *Lab on a Chip*. 2015;**15**:1638–41.

[13] Dou M., Sanjay S.T., Benhabib M., Xu F., Li, XJ. 'Low-cost bioanalysis on paper-based and its hybrid microfluidic, platforms'. *Talanta*. 2015;**145**:43–54.

[14] Liu H., Zhan F., Liu F., Zhu M., Zhou X., Xing Da. 'Visual and sensitive detection of viable pathogenic bacteria by sensing, of RNA markers in gold nanoparticles based paper platform'. *Biosensors and Bioelectronics*. 2014;**62**:38–46.

[15] Meagher R.J., Hatch A.V., Renzi R.F., Singh A.K. 'An integrated microfluidic platform for sensitive and rapid detection of, biological toxins'. *Lab on a Chip*. 2008;**8**:2046–53.

[16] Yeh Y.T., Nisic M., Yu X., Xia Y., Zheng S.Y. 'Point-of-care microdevices for blood plasma analysis in viral infectious diseases'. *Annals of Biomedical Engineering*. 2014;**42**:2333–43.

[17] Cai D., Xiao M., Xu P., Xu Y.C., Du W. 'An integrated microfluidic device utilizing dielectrophoresis and, multiplex array PCR for point-of-care detection of pathogens'. *Lab on a Chip*. 2014;**14**:3917–24.

[18] Li Y., Yan X., Feng X., *et al.* 'Agarose-based microfluidic device for point-of-care concentration and detection of pathogen'. *Analytical Chemistry*. 2014;**86**:10653–9.

[19] Yang M., Sun S., Kostov Y., Rasooly A. 'A simple 96-well microfluidic chip combined with visual and densitometry, detection for resource-poor point of care testing'. *Sensors and Actuators B: Chemical*. 2011;**153**: 176–81.

[20] Roy E., Stewart G., Mounier M., *et al.* 'From cellular lysis to microarray detection, an integrated thermoplastic, elastomer (TPE) point of care Lab on a Disc'. *Lab on a Chip*. 2015;**15**:406–16.

[21] Garvey G., Shakarisaz D., Ruiz-Ruiz F., *et al.* 'Microretroreflector-sedimentation immunoassays for pathogen detection'. *Analytical Chemistry*. 2014;**86**:9029–35.

[22] Wang Z., Chen Z., Gao N., Ren J., Qu X. 'Transmutation of personal glucose meters into portable and highly, sensitive microbial pathogen detection platform'. *Small*. 2015;**11**:4970–5.

[23] Skládal P. 'Advances in electrochemical immunosensors: review'. *Electroanalysis*. 1997;**9**(10):737–45.

[24] Sadik O.A., Land W.H., Wang J. 'Targeting chemical and biological warfare agents at the molecular level'. *Electroanalysis*. 2003;**15**:1149–59.

[25] Shah J., Wilkins E. 'Electrochemical biosensors for detection of biological warfare agents'. *Electroanalysis*. 2003;**15**:157–67.

[26] Bahadir E.B., Sezginturk M.K. 'Applications of commercial biosensors in clinical, food, environmental, and biothreat/biowarfare analyses'. *Analytical Biochemistry*. 2015;**478**:107–20.

[27] Hahn S., Mergenthaler S., Zimmermann B., Holzgreve W. 'Nucleic acid based biosensors: the desires of the use'. *Bioelectrochem*. 2005;**67**:151–4.

[28] Liu R.H., Yang J.N., Lenigk R., Bonanno J., Grodzinski P. 'Self-contained, fully integrated biochip for sample preparation, polymerase chain reaction amplification, and DNA microarray detection'. *Analytical Chemistry*. 2004;**76**:1824–31.

[29] Liu F., Liu H., Liao Y., Wei J., Zhou X., Xing D. 'Multiplex detection and genotyping of pathogenic bacteria on paper-based, biosensor with a novel universal primer mediated asymmetric PCR'. *Biosensors and Bioelectronics*. 2015;**74**:778–85.

[30] Zhang H., Zhang Y., Lin Y., *et al.* 'Ultrasensitive detection and rapid identification of multiple foodborne, pathogens with the naked eyes'. *Biosensors and Bioelectronics*. 2015;**71**:186–93.

[31] DuVall. J.A., Borba J.C., Shafagati N., *et al.* 'Optical imaging of paramagnetic bead-DNA aggregation inhibition allows, for low copy number detection of infectious pathogens'. *PLoS ONE*. 2015;**10**:UNSP e0129830.

[32] Luo J., Fang X., Ye D., *et al.* 'A real-time microfluidic multiplex electrochemical loop-mediated, isothermal amplification chip for differentiating bacteria'. *Biosensors and Bioelectronics*. 2014;**60**:84–91.

[33] Hsieh K., Ferguson B.S., Eisenstein M., Plaxco K.W., Soh H.T. 'Integrated electrochemical microsystems for genetic detection of pathogens at the point of care'. *Accounts of Chemical Research*. 2015; **48**:911–20.

[34] Safavieh M., Ahmed M.U., Ng A., Zourob M. 'High-throughput real-time electrochemical monitoring of LAMP for pathogenic bacteria detection'. *Biosensors and Bioelectronics*. 2014; **58**:101–6.

[35] Du Y., Hughes R.A., Bhadra S., Jiang Y.S., Ellington A.D., Li B. 'A sweet spot for molecular diagnostics: coupling isothermal, amplification and strand exchange circuits to glucometers'. *Scientific Reports*. 2015;**5**:AR 11039.

[36] Berg B., Cortazar B., Tseng D., *et al.* 'Cellphone-based hand-held microplate reader for point-of-care testing of enzyme-linked immunosorbent assays'. *ACS Nano*. 2015;**9**:7857–66.

[37] Giavazzi F., Salina M., Ceccarello E., *et al.* 'A fast and simple label-free immunoassay based on a smartphone'. *Biosensors and Bioelectronics*. 2014;**58**:395–402.

[38] Im H., Castro C.M., Shao H., *et al.* 'Digital diffraction analysis enables low-cost molecular diagnostics on a smartphone'. *Proceedings of the National Academy of Sciences of the United States of America*. 2015;**112**:5613–18.

[39] Gallegos D., Long K.D., Yu H., *et al.* 'Label-free biodetection using a smartphone'. *Lab on a Chip*. 2013;**13**:2124–32.

[40] Buser J.R., Wollen A., Heiniger E.K., *et al.* 'Electromechanical cell lysis using a portable audio device: enabling, challenging sample preparation at the point-of-care'. *Lab on a Chip*. 2015;**15**:1994–7.

[41] Neuzil P., Reboud J. 'Palm-sized biodetection system based on localized surface plasmon, resonance'. *Analytical Chemistry*. 2008;**80**:6100–3.

[42] Tian L., Tadepalli S., Park S.H., *et al.* 'Bioplasmonic calligraphy for multiplexed label-free biodetection'. *Biosensors and Bioelectronics*. 2014;**59**:208–15.

[43] Cetin A.E., Coskun A.F., Galarreta B.C., *et al.* 'Handheld high-throughput plasmonic biosensor using computational on-chip, imaging'. *Light: Science and Applications*. 2014;**3**:AR e122.

[44] Huang M., Galarreta B.C., Cetin A.E., Altug H. 'Actively transporting virus like analytes with optofluidics for rapid and ultrasensitive biodetection'. *Lab Chip*. 2013;**13**:4841–7.

[45] Fung A.O., Mykhaylova N. 'Analysis of airborne biomarkers for point-of-care diagnostics'. *Journal of Laboratory Automation*. 2014;**19**:225–31.

[46] Brauner E.V., Frederiksen M., Kolarik B., Gunnarsen L. 'Typical benign indoor aerosol concentrations in public spaces and designing biosensors for pathogen detection: a review'. *Building and Environment*. 2014;**82**:190–202.

[47] Svabenska E. 'Systems for detection and identification of biological aerosols'. *Defence Science Journal*. 2012;**62**:404–11.

[48] Skládal P., Svabenska E., Zeravik J., Přibyl J., Tjarnhage T., Gustafson I. 'Electrochemical immunosensor coupled to cyclone air sampler for detection of *Escherichia coli* DH5α in bioaerosols'. *Electroanalysis*. 2012;**24**:539–45.

[49] Kovář D., Farka Z., Skládal P. 'Detection of aerosolized biological agents using the piezoelectric immunosensor'. *Analytical Chemistry*. 2014;**86**:8680–5.

[50] Pardon G., Ladhani L., Sandstrom N., Ettori M., Lobov G., van der Wijngaart W. 'Aerosol sampling using an electrostatic precipitator integrated with a microfluidic interface'. *Sensors and Actuators B: Chemical*. 2015;**212**:344–52.

[51] Stapleton A.E., Cox M.E., DiNello R.K., *et al.* 'Performance of a new rapid immunoassay test kit for point-of-care, diagnosis of significant bacteriuria'. *Journal of Clinical Microbiology*. 2015;**53**:2805–9.

[52] Vakkila J., Koskinen J.O., Brandt A., *et al.* 'Detection of group A *Streptococcus* from pharyngeal swab samples by bacterial culture is challenged by a novel mariPOC point-of-care test'. *Journal of Clinical Microbiology*. 2015;**53**:2079–83.

[53] Czilwik G., Messinger T., Strohmeier O., *et al.* 'Rapid and fully automated bacterial pathogen detection on a, centrifugal-microfluidic LabDisk using highly sensitive nested PCR with integrated sample preparation'. *Lab on a Chip*. 2015;**15**:3749–59.

[54] Van Dorst B., Cremers A., Jans K., *et al.* 'Integration of clinical point-of-care requirements in a DNA microarray, genotyping test'. *Biosensors and Bioelectronics*. 2014;**61**:605–11.

[55] Rajendran V.K., Bakthavathsalam P., Ali B.M.J. 'Smartphone based bacterial detection using biofunctionalized fluorescent, nanoparticles'. *Microchimica Acta*. 2014;**181**:1815–21.

[56] Joshi L.T., Mali B.L., Geddes C.D., Baillie L. 'Extraction and sensitive detection of toxins A and B from the human, pathogen clostridium difficile in 40 seconds using microwave-accelerated, metal-enhanced fluorescence'. *PLoS ONE*. 2014;**9**:AR e104334.

[57] Weidemaier K., Carrino J., Curry A., Connor J.H., Liebmann-Vinson A. 'Advancing rapid point-of-care viral diagnostics to a clinical setting'. *Future Virology*. 2015;**10**:313–28.

[58] Shafiee H., Kanakasabapathy M.K., Juillard F., *et al.* 'Printed flexible plastic microchip for viral load measurement through, quantitative detection of viruses in plasma and saliva'. *Scientific Reports*. 2015;**5**:AR 9919.

[59] Wang J.J., Jiang Y.Z., Lin Y., *et al.* 'Simultaneous point-of-care detection of enterovirus 71 and coxsackievirus B3'. *Analytical Chemistry*. 2015;**87**:11105–12.

[60] Zhang B., Salieb-Beugelaar G.B., Nigo M.M., Weidmann M., Hunziker P. 'Diagnosing dengue virus infection: rapid tests and the role of, micro/nanotechnologies'. *Nanomedicine: Nanotechnology, Biology and Medicine*. 2015;**11**:1745–61.

[61] Bruggink L.D., Dunbar N.L., Marshall J.A. 'Evaluation of the updated RIDPOQUICK (Version N1402), immunochromatographic assay for the

detection of norovirus in clinical, specimens'. *Journal of Virological Methods*. 2015;**223**:82–7.

[62] Brendish N.J., Schiff H.F., Clark T.W. 'Point-of-care testing for respiratory viruses in adults: the current, landscape and future potential'. *Journal of Infection*. 2015;**71**:501–10.

[63] Sanbonmatsu-Gamez S., Perez-Ruiz M., Lara-Oya A., Pedrosa-Corral I., Riazzo-Damas C.M., Navarro-Mari J. 'Analytical performance of the automated multianalyte point-of-care, mariPOC (R) for the detection of respiratory viruses'. *Diagnostic Microbiology and Infectious Disease*. 2015;**83**:252–6.

[64] Kost G.J., Ferguson W., Truong A.T., *et al.* 'Molecular detection and point-of-care testing in Ebola virus disease and other threats: a new global public health framework to stop outbreaks'. *Expert Review of Molecular Diagnostics*. 2015;**15**:1245–59.

[65] Nouvellet P., Garske T., Mills H.L., *et al.* 'The role of rapid diagnostics in managing Ebola epidemics'. *Nature*. 2015;**528**(7580):S109–16.

[66] Walker N.F., Brown C.S., Youkee D., *et al.* 'Evaluation of a point-of-care blood test for identification of Ebola virus disease at Ebola holding units, Western Area, Sierra Leone, January to February 2015'. *Eurosurveillance*. 2015;**20**(12):AR 21073.

[67] Baca J.T., Severns V., Lovato D., Branch D.W., Larson R.S. 'Rapid detection of Ebola virus with a reagent-free, point-of-care, biosensor'. *Sensors*. 2015;**15**:8605–14.

[68] Shen J., Zhou Y., Fu F., *et al.* 'Immunochromatographic assay for quantitative and sensitive detection of, hepatitis B virus surface antigen using highly luminescent quantum dot-beads'. *Talanta*. 2015;**142**:145–9.

[69] Parisi M.R., Soldini L., Vidoni G., *et al.* 'Point-of-care testing for HCV infection: recent advances and implications for alternative screening'. *New Microbiologica*. 2014;**37**:449–57.

[70] Pai N.P., Daher J. 'Multiplexed testing for HIV and related bacterial and viral, co-infections at the point-of-care: quo vadis?'. *Expert Review of Molecular Diagnostics*. 2015;**15**:463–9.

[71] Park S., Shin Y.M., Song J.J., Yang H. 'Facile electrochemical detection of botulinum neurotoxin type E using a two-step proteolytic cleavage'. *Biosensors and Bioelectronics*. 2015;**72**:11–17.

[72] Pohanka M., Pavlis O., Skládal P. 'Diagnosis of tularemia using piezoelectric biosensor technology'. *Talanta*. 2007;**71**:981–5.

[73] Li X., Zhang Q., Hou P., *et al.* 'Gold magnetic nanoparticle conjugate-based lateral flow assay for the, detection of IgM class antibodies related to TORCH infections'. *International Journal of Molecular Medicine*. 2015;**36**:1319–26.

Chapter 12

POC in travel, marine, and airport security monitoring

M. Drancourt[1]

12.1 Introduction

In developed countries, the main situation is that any individual is close enough to a laboratory platform to get laboratory test results within one hour in order to assist doctor in providing an accurate diagnosis of the majority of medical syndromes and situations [1]. However, in both developed and developing countries, part of the populations may not be in such a favorable situation in the fact that laboratory test results may not be provided within one hour of consultation. These situations could be characterized as medical deserts and may result from several situations, including the geographical distance from a healthcare structure and laboratory; or even the time distance in some large urban areas in the case of concentrated laboratory facilities. Plane and boat travels are among these situations in which individuals may be far from any readily reachable laboratory, leaving them as in a medical desert. This holds true in particular for febrile syndromes which are presumed to be because of infection, and this review will focus on infectious diseases. A large series reviewing in-flight emergencies indicated that infectious diseases accounted for 2.8 percent of emergencies and no death [2]. In this situation, physicians present aboard provided almost half of the initial in-flight medical care [2].

Infectious diseases in travelers not only pose questions regarding the immediate management of the febrile patient, but also for the collectivity in the case of contagious infection. Therefore, it may be desirable to make a rapid diagnosis of infectious disease during travel not only to optimize the medical management and prognosis of patients but also to help preventing the extension of any epidemic by the timely and appropriate confinement of any patient diagnosed with a contagious infection.

In order to anticipate how modern technologies for medical laboratory and internet transmission of data may help reducing the medical deserts during travel, it is useful to briefly review the medical syndromes and situations most frequently encountered during travel, and then review how point-of-care (POC) laboratories may help resolving this issue.

[1]Aix Marseille University, INSERM, CNRS, IRD, URMITE, Marseille, France

Indeed, POCs have been invented to provide patients and doctors with near-to-patient, rapid diagnosis of some urgent diseases requiring some rapid medical decisions regarding the necessity for hospitalization, the necessity for isolating a contagious patient and the necessity to start a specific medical treatment such as an appropriate antimicrobial in the case of diagnosed infectious disease [3].

12.2 Remote detection of fever in travelers

Fever is a major, albeit non-specific, clinical sign of infection. Remote detection fever is therefore of interest to select individuals presenting an increased probability of infectious disease, who may benefit from rapid diagnosis testing. Individual remote devices are commercially available for the monitoring of body temperature and detection of fever, yet not validated. Nevertheless, these devices may be connected to a medical platform in order to advice the person to further consult. Also, several airports and other buildings have been equipped with thermal camera for an automated, noninvasive detection of fever in order to propose some travelers a specific medical check-up [4,5]. Fever detection could be achieved by questionnaire in travelers as well as using thermal camera [4]. Both have been shown to be correlated with the tympanic temperature and thus useful tools for the screening of travelers at risk of infection [4,5]. In airports and other buildings with public, persons detected with fever may be proposed to have a free medical consultation and appropriate POC tests to assist in both rapid medical management and isolation when appropriate [1,3].

12.3 Documented infections during mass plane traveling

Aboard planes, communicable respiratory tract pathogens and digestive tract pathogens may challenge the plane personals as well as the airport medical staff. Although mass transportation has been shown to promote at-destination spread of influenza and coronavirus epidemics by air-flights but not by cruise ships [6], Influenza A, including the past H1N1 influenza epidemics, is a potential albeit low risk for flight travelers [7–9]. These reports highlighted that the potential of in-flight transmission of H1N1 Influenza virus was low but not null, with one secondary case in one investigation [7], a measured risk of about 3.5 percent [8] and an attack rate of 2.4 percent and 5.2 percent [9]. Significantly, secondary cases were mainly detected seating within two rows from index cases [7,8] and direct contacts by speaking with coughing index case were identified as the major situation for transmission [7]. These first observations led WHO to propose limiting contact tracing to people seated two rows around the index case. This WHO two-row rule has been further challenged [10,11]. In-flight patients with pulmonary tuberculosis have been shown to pose few risks to other travelers [12,13]. In fact, risks for in-flight transmission include close contacts with highly spreader during long-flight [14]. There is no documented report of transmission of pulmonary tuberculosis in an airport.

It has been observed that many travelers are boarding aircraft when presenting with traveler's diarrhea related to rotavirus, toxinogen *Escherichia coli*, and *Salmonella* spp. [15]. Noroviruses diarrhea is a definite risk for in-flight transmission [16,17]. It is because it is causing public vomiting and is resistant to many biocides used for decontamination of aircrafts. Moreover, inappropriate decontamination of an aircraft has been shown to provoke recurrent Norovirus infection [18,19]. *Salmonella* spp. enteritis has been reported in connection with in-flight catering [20]. Such infections have been linked to consumption of fresh foodstuffs, such as egg products and milk tart [21]. Historical deadly cholera has also been reported [22].

At last, when no case of in-flight malaria has ever been reported, "airport malaria" has been reported as limited outbreaks of malaria in connection with airport in non-endemic countries, due to inadvertent transportation of Plasmodium-infected mosquitoes [23].

12.4 Documented infections during cruise ship traveling

Contagious infections are a well-recognized health issue during mass cruise and on-board commercial ships [24]. Indeed, surveying more than 1.5 million person-days aboard cargo ships found that 21 percent of visits to the infirmary were due to probable infection, including 68 outbreaks essentially of respiratory tract infections [25]. During cruises, major outbreaks have been related to diarrhea due to enterotoxigenic *E. coli* [26] and *Cyclospora* [27], to public vomiting due to Norovirus [28–30], and to respiratory tract infection due to influenza [31,32] and *Legionella* spp. [33]. Accordingly, Influenza outbreaks aboard cruise ships have been reported to affect 2–7 percent of passengers [6]. Among other limited but health-threatening infections are *Neisseria meningitidis* meningitis [34,35], HEV hepatitis [36], and varicella [37].

12.5 Proposed management of febrile patients in airport and cruise facilities

A traveler may self-declare sick and febrile because of his own feeling or indication of any remote device measuring body temperature including airport thermal camera (Figure 12.1). In flight, the person may contact in-flight personal, and the captain may directly inform the airport medical team in order to manage rapid care including POC diagnosis as detailed below. Most companies are using on-ground telemedical assistance [38]. Alternatively, detection of fever is made into the airport by remote thermal camera and the patient will be proposed to wear protective clothes, mask, and gloves and to benefit a medical advice including POC diagnosis, following a pathway dedicated to people suspected of contagious infection, including dedicated room and disposable instruments in the medical facility (Figure 12.1). This proposed pattern is enforced by the observation that investigation of potential

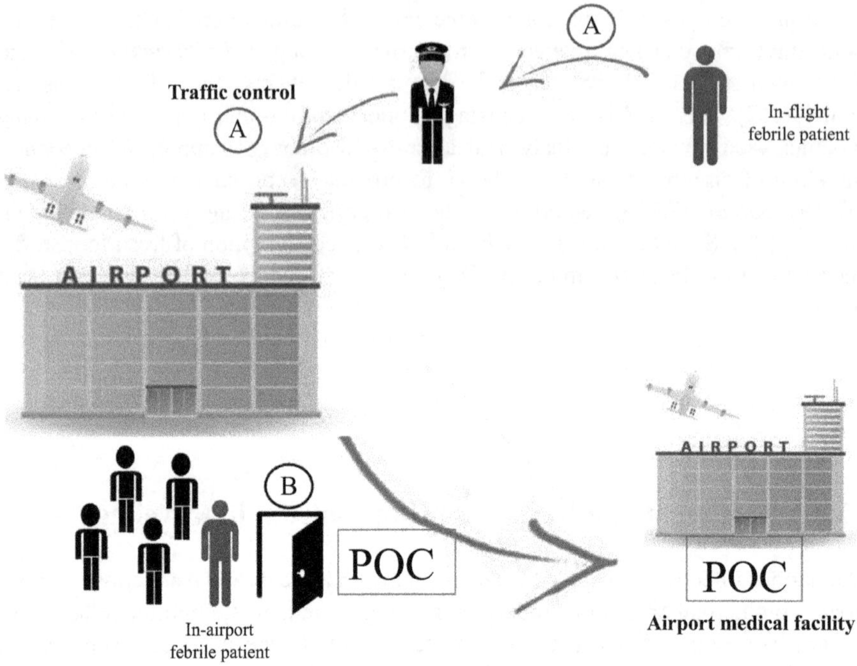

Figure 12.1 Proposed organization of medical management of febrile patients in the airport

in-flight transmission of contagious pathogen is laborious, once individuals have left the airport [8].

12.6 Setting-up POC diagnosis in airports and travel facilities

POC laboratories are now commercially available for the rapid diagnosis of major pathogens that may challenge the medical management of travelers detected with fever in the airports and other mass travel facilities. These POC laboratories are small, smart, and connected laboratories allowing any trained personal to rapidly perform and interpret tests. The POC laboratory occupies about one square meter and requires only one electric plug so that it could be easily located in the medical facility of the airport and even closer to the thermal camera location. It features confined work station and instrumentation to perform the rapid diagnosis by immunochromatographic tests and real-time polymerase chain reaction (PCR) tests [3]. A POC laboratory can be run by a doctor, a nurse, or a laboratory technician. Alternatively, we assessed the effectiveness of four-hour training for non-health care personals [39]. Efficiently, POC tests could be grouped into syndromic POC kits featuring all the recipients and tests to rapidly assess the microbes most frequently encountered as responsible for the syndrome, as well as contagious pathogens and highly dangerous pathogens. As reported above, the most frequently

Table 12.1 Proposed syndromic kits to be delivered at airport POC

Pharyngitis	*Streptococcus* A Epstein–Barr virus
Enteritis	Norovirus *Salmonella* spp.
Respiratory	Influenza viruses
Tropical fever	*Plasmodium* spp.
Febrile rash	Zika virus Dengue Chikungunya virus

encountered syndromes in airports and mass travel facilities include tropical fever, febrile rash, diarrhea, respiratory tract infection, and pharyngitis. We propose that these four syndromes may be included in the minimal list of clinical syndromes to be POC-investigated in airports and related facilities. Efficiently, the patient detected with fever may be proposed a menu of signs and symptoms to tick in order to determine the syndrome he is suffering and the appropriate POC kit to be delivered. Samples including urines, stools, throat swab, and blood drop could be obtained by auto-sampling or collected by the POC personal. POC personal then manipulates clinical samples into the confined POC laboratory according to one POC syndromic kit instructions in order to get results within 15–60 min (Table 12.1). A doctor can then reach medical decisions regarding continuing isolation of a patient with contagious infection, immediate anti-infective treatment, and hospitalization in the case of a health-threatening infection requiring specific health care such as intensive cares.

12.7 Perspectives

New technologies will offer opportunities for the remote detection of fever and additional clinical signs and symptoms in travelers, indicative of potential organ infection. These medical information may be useful for in-flight and aboard cruise ship care delivery, or appropriate management of the patient as soon as he is arriving the airport and cruise facility. There, POC laboratory is a useful piece of the sanitary network to rapidly confirm the diagnosis, isolate contagious patients, and to assist in proper anti-infectious treatment prescription. Further, diagnosis data provided by remote POC laboratory onboard cruise and cargo ships [39] and other remote locations could be assembled into a surveillance network of infections worldwide.

12.8 Conflicts of interest

MD is a co-inventor of a patented model of POC laboratory and stock holder of POCRAMé, a French start-up devoted to the development and commercialization of POC laboratories and POC solutions.

References

[1] Fournier P.E., Drancourt M., Colson P., Rolain J.M., La Scola B., Raoult D. 'Modern clinical microbiology: new challenges and solutions'. *Nature Reviews Microbiology*. 2013;**11**:574–85.

[2] Peterson D.C., Martin-Gill C., Guyette F.X., *et al.* 'Outcomes of medical emergencies on commercial airline flights'. *New England Journal of Medicine*. 2013;**368**:2075–83.

[3] Drancourt M., Michel-Lepage A., Boyer S., Raoult D. 'The point-of-care laboratory in clinical microbiology'. *Clinical Microbiology Reviews*. 2016;**29**:429–47.

[4] Cho K.S., Yoon J. 'Fever screening and detection of febrile arrivals at an international airport in Korea: association among self-reported fever, infrared thermal camera scanning, and tympanic temperature'. *Epidemiology and Health*. 2014;**36**:e2014004.

[5] Nguyen A.V., Cohen N.J., Lipman H., *et al.* 'Comparison of 3 infrared thermal detection systems and self-report for mass fever screening'. *Emerging Infectious Diseases*. 2010;**16**:1710–17.

[6] Browne A., Ahmad S.S., Beck C.R., Nguyen-Van-Tam J.S. 'The roles of transportation and transportation hubs in the propagation of influenza and coronaviruses: a systematic review'. *Journal of Travel Medicine*. 2016;**23**: pii: tav002.

[7] Han K., Zhu X., He F., *et al.* 'Lack of airborne transmission during outbreak of pandemic (H1N1) 2009 among tour group members, China, June 2009'. *Emerging Infectious Diseases*. 2009;**15**:1578–81.

[8] Baker M.G., Thornley C.N., Mills C., *et al.* 'Transmission of pandemic A/H1N1 2009 influenza on passenger aircraft: retrospective cohort study'. *British Medical Journal*. 2010;**340**:c2424.

[9] Neatherlin J., Cramer E.H., Dubray C., *et al.* 'Influenza A(H1N1)pdm09 during air travel'. *Travel Medicine and Infectious Disease*. 2013;**11**:110–8.

[10] Young N., Pebody R., Smith G., *et al.* 'International flight-related transmission of pandemic influenza A(H1N1)pdm09: an historical cohort study of the first identified cases in the United Kingdom'. *Influenza and Other Respiratory Viruses*. 2014;**8**:66–73.

[11] Shankar A.G., Janmohamed K., Olowokure B., *et al.* 'Contact tracing for influenza A(H1N1)pdm09 virus-infected passenger on international flight'. *Emerging Infectious Diseases*. 2014;**20**:118–20.

[12] Gerard E. 'Infectious diseases in air travelers arriving in the UK'. *Journal of the Royal* Society *for the Promotion* of *Health*. 2002;**122**:86–8.

[13] Kotila S.M., Payne Hallström L., Jansen N., Helbling P., Abubakar I. 'Systematic review on tuberculosis transmission on aircraft and update of the European Centre for Disease Prevention and Control risk assessment guidelines for tuberculosis transmitted on aircraft (RAGIDA-TB)'. *Eurosurveillance*. 2016;**21**:pii:30114.

[14] Kenyon T.A., Valway S.E., Ihle W.W., Onorato I.M., Castro K.G. 'Transmission of multidrug-resistant *Mycobacterium tuberculosis* during a long airplane flight'. *New England Journal of Medicine*. 1996;**334**:933–8.

[15] Steffen R., Collard F., Tornieporth N., *et al.* 'Epidemiology, etiology, and impact of traveler's diarrhea in Jamaica'. *Journal of the American Medical Association*. 1999;**281**:811–17.

[16] Kirking H.L., Cortes J., Burrer S., *et al.* 'Likely transmission of norovirus on an airplane, October 2008'. *Clinical Infectious Diseases*. 2010;**50**:1216–21.

[17] Widdowson M.A., Glass R., Monroe S., *et al.* 'Probable transmission of norovirus on an airplane'. *Journal of the American Medical Association*. 2005;**293**:1859–60.

[18] Thornley C.N., Emslie N.A., Sprott T.W., Greening G.E., Rapana J.P. 'Recurring norovirus transmission on an airplane'. *Clinical Infectious Diseases*. 2011;**53**:515–20.

[19] Lopman B. 'Air sickness: vomiting and environmental transmission of norovirus on aircraft'. *Clinical Infectious Diseases*. 2011;**53**:521–2.

[20] Osimani A., Aquilanti L., Clementi F. 'Salmonellosis associated with mass catering: a survey of European Union cases over a 15-year period'. *Epidemiology and Infection*. 2016:1–13.

[21] Rebolledo J., Garvey P., Ryan A., *et al.* 'International outbreak investigation of *Salmonella* Heidelberg associated with in-flight catering'. *Epidemiology and Infection*. 2014;**142**:833–42.

[22] Sutton R.G. 'An outbreak of cholera in Australia due to food served in flight on an international aircraft'. *The Journal of Hygiene*. 1974;**72**:441–51.

[23] Guillet P., Germain M.C., Giacomini T., *et al.* 'Origin and prevention of airport malaria in France'. *Tropical Medicine* and *International Health*. 1998;**3**:700–5.

[24] Minooee A., Rickman L.S. 'Infectious diseases on cruise ships'. *Clinical Infectious Diseases*. 1999;**29**:737–43.

[25] Schlaich C.C., Oldenburg M., Lamshöft M.M. 'Estimating the risk of communicable diseases aboard cargo ships'. *Journal of Travel Medicine*. 2009;**16**:402–6.

[26] Daniels N.A., Neimann J., Karpati A., *et al.* 'Traveler's diarrhea at sea: three outbreaks of waterborne enterotoxigenic *Escherichia coli* on cruise ships'. *Journal of Infectious Diseases*. 2000;**181**:1491–5.

[27] Gibbs R.A., Nanyonjo R., Pingault N.M., *et al.* 'An outbreak of *Cyclospora* infection on a cruise ship'. *Epidemiology and Infection*. 2013;**141**:508–16.

[28] Wikswo M.E., Cortes J., Hall A.J., *et al.* 'Disease transmission and passenger behaviors during a high morbidity Norovirus outbreak on a cruise ship, January 2009'. *Clinical Infectious Diseases*. 2011;**52**:1116–22.

[29] Koopmans M., Harris J., Verhoef L., Depoortere E., Takkinen J., Coulombier D. International Outbreak Investigation Team. 'European investigation into recent norovirus outbreaks on cruise ships: update'. *Eurosurveillance*. 2006;**11**:E060706.5.

[30] Vivancos R., Keenan A., Sopwith W., *et al.* 'Norovirus outbreak in a cruise ship sailing around the British Isles: investigation and multi-agency management of an international outbreak'. *Journal of Infection*. 2010;**60**:478–85.
[31] Tarabbo M., Lapa D., Castilletti C., *et al.* 'Retrospective investigation of an influenza A/H1N1pdm outbreak in an Italian military ship cruising in the Mediterranean Sea, May–September 2009'. *PLoS ONE*. 2011;**6**:e15933.
[32] Ward K.A., Armstrong P., McAnulty J.M., Iwasenko J.M., Dwyer D.E. 'Outbreaks of pandemic (H1N1) 2009 and seasonal influenza A (H3N2) on cruise ship'. *Emerging Infectious Diseases*. 2010;**16**:1731–7.
[33] Joseph C.A., Ricketts K.D., Yadav R., Patel S. European Working Group for *Legionella* Infections. 'Travel-associated Legionnaires' disease in Europe in 2009'. *Eurosurveillance*. 2010;**15**:19683.
[34] Neri A., Fazio C., Ciammaruconi A., *et al.* 'Draft genome sequence of C: P1.5-1,10-8:F3-6:ST-11 meningococcal clinical isolate associated with a cluster on a cruise ship'. *Genome Announcements*. 2014;**2**: pii: e01263-14.
[35] Stefanelli P., Fazio C., Neri A., *et al.* 'Cluster of invasive *Neisseria meningitidis* infections on a cruise ship, Italy, October 2012'. *Eurosurveillance*. 2012;**17**: pii:20336.
[36] Said B., Ijaz S., Kafatos G., *et al.* Hepatitis E Incident Investigation Team. 'Hepatitis E outbreak on cruise ship'. *Emerging Infectious Diseases*. 2009;**15**:1738–44.
[37] Acevedo F., Diskin A.L., Dahl E. 'Varicella at sea: a two-year study on cruise ships'. *International Maritime Health*. 2011;**62**:254–61.
[38] Silverman D., Gendreau M. 'Medical issues associated with commercial flights'. *Lancet*. 2009;**373**:2067–77.
[39] Bouricha M., Samad M.A., Levy P.Y., Raoult D., Drancourt M. 'Point-of-care syndrome-based, rapid diagnosis of infections on commercial ships'. *Journal of Travel Medicine*. 2014;**21**:12–16.

Chapter 13

Biosensor applications in veterinary science

Georgia Moschopoulou[1]

13.1 Introduction

The adoption of biosensor systems in the management of animal care, either at the farm or home is not new. Sensor approaches have been developed since the early 1990s, in parallel with the emergence of automated processes, such as automated milking systems in dairy farming aimed to reduce labor requirements and associated costs [1,2]. On a different level, customized portable blood glucose meters (PBGM) for monitoring diabetic pets have been commercially available for more than 15 years [3,4]. Nowadays, a rapid expansion in the number of point-of-care (POC) systems for veterinary science is observed, the scope of which exceeds by far the respective applications for human medicine. In the present chapter, the progress in this continuously evolving field during the last 5 years is briefly reviewed.

13.2 Biosensors in animal husbandry

13.2.1 Disease surveillance

Particular effort has been invested in the development of either working principles or even operational systems for the on-site detection of pathogens and/or disease biomarkers in farm animals. The most prominent case is mastitis, followed by influenza viruses, bovine herpes virus, various porcine pathogenic viruses, and noninfectious diseases like ketosis.

13.2.1.1 Mastitis

Mastitis is the infection of the mammary gland, due to a number of bacteria species including *Staphylococcus, Streptococcus*, and *Corynebacterium* species [5–7] and can be associated with >90 percent of milk production losses [8]. Although the prevalence of clinical mastitis is fairly low (in the order of 0.04 percent), a particular challenge is the detection of subclinical mastitis (the incidence of which can be as high as 50 percent), since alterations in milk are not visible at this stage [9,10].

[1]Laboratory of Cell Technology, Department of Biotechnology, Agricultural University of Athens, Athens, Greece

Biosensors for the detection of mastitis are discriminated according to their application in-line (i.e., monitoring milk in its continuous flow from the cow) vs on-line (i.e., monitoring a sample) [2]. The majority of biosensors for mastitis detection are based on the measurement of milk electrical conductivity (EC), the value of which is dependent on the extrusion of sodium ions by infected mammary gland cells [11]. EC mastitis biosensors outperform, by far, other in-line sensors such as the assay of l-lactate dehydrogenase [12], *N*-acetyl-*b*-d-glucosaminidase [13] or on-line sensors such as color detectors [14]. On the other hand, a popular method to detect mastitis is somatic cell count (SCC) in milk, which is increased in pathological conditions, due to the release of polymorphonuclear leukocytes from infected mammary glands [15]. SCC is facilitated by means of DNA gel-staining [16]. As an on-line sensor, SCC complies very well with the standard California Mastitis Test [17]. EC measurements can be successfully combined with SCC techniques to increase the efficiency of mastitis detection, even though the size of the assay time window has been found to be critical and should not be longer than 48 h, although shorter time windows (e.g., 24-h long) may be associated with false negative results [2,711]. Alternative, promising biosensor approaches for mastitis detection, especially at subclinical level, have also been reported. For example, Tan *et al.* [18] reported an electrochemical immunosensor for the detection of haptoglobin (Hp), a mastitis biomarker, at a concentration range of 15–100 mg L^{-1} and a limit of detection of 0.63 mg L^{-1}. A fair correlation with conventional enzyme-linked immunosorbent assay (ELISA) tests was established, though not at the upper concentration range, possibly due to interferences involved with the conventional immunoassays. This approach is apparently closer to POC Hp detection than previously reported optical biosensor methods, including an surface plasmon resonance (SPR) sensor [19]. More recently, Koop *et al.* [20] reported a stand-alone, user-friendly, protease-activity based, fluorescence mastitis test for used with frozen milk samples. Finally, biosensors may be applied for evaluating the antibiotic resistance of mastitis causative agents, as reported, for example, by Guntupalli *et al.* [21] for the assessment of the resistance of *Staphylococcus aureus* strains against methicillin using a quartz crystal microbalance with dissipation tracking (QCM-D) covered with bacteriophage 12,600 as the biorecognition probe for *S. aureus* strains, in combination with the resistance-conferring penicillin-binding protein 2a membrane protein.

13.2.1.2 Porcine circovirus and classical swine fever virus

Porcine circovirus type 2 (PCV2) is a major swine pathogen associated with animal losses due to post weaning multisystemic wasting syndrome disease [22–24]. Various ELISA-like tests have been developed for the fast and sensitive detection of the viral antibodies, including colloidal gold immunochromatographic test strips [25]. More recently, an improvement in portable immunoassays for PCV2 has been reported on the basis of the fusion of PCV2-specific single domain antibodies with alkaline phosphatase [26], which led to a 5-fold increase of affinity, as documented both with SPR and an immunocytochemistry assay. In turn, this led to a faster, simpler, and more sensitive (with an limit of detection of 0.05 μg mL^{-1}) PCV2 detection compared to both ELISA and Western blot. The same researchers were able to increase the sensitivity of the assay by immobilizing the capturing antibodies on magnetic nanobeads and using quantum dots for the visualization of the captured virus particles at concentrations as low as 10^3 copies mL^{-1} [27]. In similar fashion, an SPR immunoassay for the direct detection of PCV2 moieties was reported by Hu *et al.* [28], with a theoretical limit of detection of 0.04 μg mL^{-1}.

Classical swine fever virus (CSFV) is another swine pathogen associated with both chronic and acute disease, the symptomatology of the latter being prominently characterized by hemorrhagic fever [29]. Among the structural proteins of the virus envelope, the major immunodominant E2 glycoprotein has attracted attention due to its ability to induce neutralizing antibodies in infected CSFV pigs. In this context, a novel immunoreactive, multi-epitope recombinant protein (GST-BT21) has been constructed as a means to improve the initiation of protective host-immune responses. The affinity and specific interaction of GST-BT21 with CSFV-positive serum has been evaluated with SPR techniques [30]. Surface plasmon resonance approaches have been previously reported for the detection of CSFV antibodies using the recombinant gp55 protein as the target antigen [31], with a limit of detection of 0.01 μg mL^{-1}. Quite recently, Guo *et al.* [32] reported a portable impedance/magnetoelastic biosensor for CSFV antibody detection with a limit of detection of 0.6 μg mL^{-1} (an in-depth presentation and analysis of magnetoelastic biosensor technology is provided in Chapter 5).

13.2.1.3 Influenza viruses

Influenza viruses are the causative agents of potentially devastating respiratory diseases, often characterized by cross-species infection, including humans, pigs, and birds [33]. An example is avian influenza virus (AIV) H5N1, with a high mortality in humans, as observed in mini-epidemics in recent years [34]. Impedimetric biosensors for the direct detection of H5N1 seem to be the current approach of choice for the development of POC tests for AIV. Pioneering work in this direction has been done by Wang *et al.* [35]. A further sophisticated impedimetric immunosensor based on polyclonal antibodies was developed by Lum *et al.* [36] for the relatively rapid (~1 h), direct detection of H5N1. The sensitivity of the assay, associated with the increase of the impedance due to antibody–virus binding, was further improved by the co-binding of chicken red blood cells, which acted as resistors in the system and amplified considerably the impedance signal. At the same time, however, this approach led to nonquantitative responses, which presented a limitation of the biosensor, together with observed cross-reactivity with other influenza virus subtypes. Still, this approach is promising for the development of a POC test for H5N1. More recently, the same group reported a significantly improved version of the impedimetric biosensor (assay time <30 min, zero cross-reactivity with other virus subtypes) by using aptamers as the biorecognition elements [37]. Equally satisfactory results, using a monoclonal antibody were reported for the low-cost impedimetric biosensor developed by Lin *et al.* [38]. Rapid (assay time ~20–30 min) optical biosensor techniques, including SPR and lateral flow tests based on monoclonal antibodies have been used for the detection of another influenza subtype, swine-origin influenza A (H1N1) virus [39,40]. Quite recently, Krishna *et al.* [41] demonstrated the applicability of a multi-array (64×) giant magnetoresistance-based immunoassay, based on magnetic nanoparticles coupled with monoclonal antibodies, for the detection of H1N1.

13.2.1.4 Other infectious disease agents

Nurul Najian *et al.* [42] have recently reported a lateral flow dipstick biosensor for the detection of pathogenic *Leptospira* species based on multiplex loop mediated isothermal amplification (m-LAMP) and using colloidal gold nanoparticles functionalized with anti-FITC antibodies. The biosensor targeted the *Leptospira LipL32* gene. Even though the dipstick assay was rapid, the m-LAMP assay was considerably longer (>70 min). A lateral flow assay using gold nanoparticles was also developed by Toubanaki *et al.* [43] for nervous necrosis virus, a nodavirus responsible for a highly lethal marine fish disease known as viral nervous necrosis or vacuolating encephalopathy and retinopathy. Following RNA extraction and reverse transcription-polymerase chain reaction amplification from fish samples, the lateral flow assay was completed within 20 min. The biosensor produced fairly specific and reproducible results (4 percent coefficient of variation), with an estimated limit of detection of 135 pg of initial total RNA, corresponding to 125 fmol of target DNA. The use of lateral-flow biosensors in combination with loop-mediated isothermal amplification is becoming a rather common practice for the detection of pathogenic viruses in marine organisms [44], not at least due to the sensor's suitability for use in quality control screening by aquatic industries irrespective of size and capacity.

The Vantix^{TM} system is a nice example of a rather simple immunopotentiometric biosensor that can be used in different veterinary detection applications, based on electrical charge generation following target antigen–(electrode-immobilized) antibody binding and reaction with a 3,3′,5,5′-tetramethylbenzidine substrate. As demonstrated in the case of detection of the bovine herpesvirus-1 (BHV-1), the etiological agent for bovine respiratory disease, the biosensor assay was completed in ~15 min, much shorter than ELISA (3–19 h) [45]. Quite recently, Tarasov *et al.* [46] reported a biosensor for BHV-1 assay based on the covalent immobilization of viral glycoprotein gE on the gold surface of an extended gate field-effect transistor via amine coupling to the carboxyl groups of a self-assembled monolayer of thiol linker molecules. In this way, the glycoprotein served as a capture antigen for anti-gE specific antibodies circulating in animals in response to BHV-1 infection. Using the miniaturized biosensor, the authors were able to complete virus detection in less than 10 min. Biosensor-assisted screening for BHV-1 can greatly support global "test and cull" disease eradication programs [47].

Bacterial and protozoan diseases are equally important to viruses for veterinary medicine, with the additional risk of infecting humans. Therefore, the need for rapid and portable screening methods is an indisputable priority. Various biosensor approaches have been reported for protozoa and bacteria detection in animals. A recent example is presented by Gokce *et al.* [48], who applied differential pulsed voltammetry to monitor guanine oxidation as a result of the hybridization of the target *Toxoplasma gondii* DNA with the respective capture oligonucleotide probe. The assay lasted approx. 1 h and a limit of detection of 1.78 mg mL^{-1} was observed. Wei *et al.* [49] developed a simple enzyme-amplified fluorescence immunoassay method using magnetic particles for the detection of antibodies against virulent exotoxins of *Actinobacillus pleuropneumoniae*, the causative agent of severe swine respiratory disease. The limit of detection (1:10,240 dilution) was considerably lower than that of ELISA (1:320), whereas the assay was successfully tested against several (n = 78) clinical porcine sera samples. A wide series of different biosensor approaches as well as actual commercial devices and kits are available for the POC screening of bacteria responsible for urinary tract infections (UTIs), mainly gram negative *Enterobacteriaceae* species and a few gram positive species, such as *Enterococcus faecalis, Staphylococcus saprophyticus*, and *Actinobaculum schaali* [50]. Depending also on the target UTI pathogen (in particular its endotoxin), biosensor methods are based on a variety of detection principles, such as magnetoelastic sensing [51], DNA-aptamer impedance sensing [52], and antibody-based microcantilever sensing [53]. An excellent review of the biosensor approaches for UTI pathogen detection is given by Kumar *et al.* [54], whereas a detailed study on the commercially available biosensor kits is presented by Das *et al.* [55] of the same group.

Finally, the work of Bayn *et al.* [56] for monitoring the infection of bisons by *Brucella* species is worth mentioning as an example of noninvasive approach. The authors used a nanoscale sensor array comprising chemically diverse chemiresistors that were based either on gold nanoparticles or single-walled carbon nanotubes. The array was used for the rapid (15 min) analysis of animal breath samples and the identification of volatile organic compounds (VOCs) possibly associated with seropositive individuals. For five VOCs (2-ethyl-1-hexanol, acetophenone, octanal, heptanal, and benzaldehyde), a significantly lower concentration in seropositive animals was confirmed.

13.2.1.5 Noninfectious disease monitoring

Ketosis, especially subclinical ketosis (SCK), is a common cow disease occurring at early lactation and characterized by high concentration of ketone bodies in blood [57]. In its extreme version, ketoacidosis can lead to severe metabolic imbalance and pathological reduction of blood pH value (below 7.3). Although there are several semiquantitative urine test strips (e.g., Ketolac BHB, KetoStix) with very good sensitivity and specificity [58] for the indirect measurement of ketosis (more precisely ketonuria, which is dependent on the concentration of acetoacetate ketone bodies in urine), these are not sensitive or specific enough to detect SCK on a routine basis. This is particularly important for cattle, in which the incidence of SCK (40–60 percent) is much higher than its clinical version (on average 7–8 percent) [59]. Biosensor-assisted determination of representative ketone bodies, such as β-hydroxybutyrate (βHBA), is a more reliable approach; commercial kits are available, including the electrochemical Precision Xtra [60] and OptiumXceed [61] Glucose and Ketone Monitoring Systems (both by Abbott). Recent developments in the field of SCK-detecting biosensors include a rapid, low-cost microfluidic optical biosensor with an assay time of just 1 min and a detection limit of 50 μM βHBA [62]. The working principle of the biosensor was based on the enzymatic conversion of βHBA to acetoacetate by βHBA dehydrogenase and the parallel reduction of the enzyme cofactor nicotineamide adenine dinucleotide (NAD+) to reduced nicotinamide adenine dinucleotide. The latter reacts with the colorimetric detector water-soluble tetrazolium-1 to produce a formazan dye, which is the actual assayed compound. The same group has presented an alternative version based on the same working principle but having NAD+ pre-conjugated with fluorescent quantum dots [63], thus achieving an improved detection limit of 35 μM *β*HBA.

Another area of biosensor application in cattle health assessment is the monitoring of circulating nonesterified fatty acid (NEFA) levels, a reliable marker of negative energy balance. This condition may lead, among other symptoms, to severe ketosis and reduced fertility [64]. Major NEFA components in the serum, such as oleic, palmitic, palmitoleic, linoleic, and linolenic acid, can be determined with the aid of colorimetric assays, for example, the HR Series NEFA-HR(2) by Wako Diagnostics [65]. The latter product is based on the acylation of coenzyme A (CoA) by NEFAs in the presence of added acyl-CoA synthetase. The resulting acyl-CoA is oxidized by added acyl-CoA oxidase with concomitant synthesis of hydrogen peroxide. The latter is used by peroxidase in order to achieve the oxidative condensation of 3-methy-*N*-ethyl-*N*(*β*-hydroxyethyl)-aniline with 4-aminoantipyrine to form a colored product for colorimetric assay. Quite recently, Veerapandian *et al.* [66] reported an electrochemical biosensor for NEFA detection, based on ruthenium bipyridyl complex-modified grapheme oxide nanosheets ($[Ru(bpy)_3]^{2+}$-GO) and using lipoxygenase as the specific biorecognition element. Following this approach, NEFA was determined in linear fashion in the range of 0.1–1 mM (equivalent oleic acid concentration).

13.2.2 Estrus and fertility monitoring

Automated estrus monitoring is an essential element of automated farm management [2]. Estrus is indirectly assessed via monitoring cow activity with the help of attached sensors, such as pedometers and 3D-accelerometers or even cow-to-cow contact sensors (e.g., HeatWatch) [67]. On the other hand, progesterone (P4) determination in milk is a far more accurate and specific predictor of estrus and is thus considered the gold standard. Although readily available commercial progesterone biosensors for cattle have not yet been realized, a number of recent developments in environmental hormone testing are pointing towards this direction. This is due to the vivid interest by the global community for assessing the endocrine-disrupting potential of environmental samples. For example, an aptamer-based impedimetric biosensor was recently reported able to detect progesterone in water with a detection limit of 0.90 ng mL^{-1} [68]. An electrochemical microfluidic immunosensor has also been reported for the detection of P4 in bovine serum samples with a limit of detection of 0.2 ng mL^{-1} [69]. Generally, electrochemical biosensing seems to be the method of choice for hormone detection [70].

13.2.3 Other applications

Gumus *et al.* [71,72] applied a miniaturized implantable electrochemical sensor to monitor the levels of uric acid in chickens. The biosensor comprised a Pt/Ir wire and Ag/AgCl paste, whereas the biorecognition element was uricase immobilized on a Nafion/cellulose inner membrane. The system was able to assay, in pH-dependent linear fashion, uric acid concentrations in the range of 0.05–0.6 mM.

13.3 Biosensors in pet care

13.3.1 Glucose and lactate monitoring

Similarly to humans, diabetes mellitus is a common disease among the two main pet groups, cats, and dogs, as a result of indoor living, reduced exercise and dietary excesses. Cats, in particular, demonstrate higher natural blood glucose levels as documented by the established minimum concentrations in the species-specific glucose curves (cats: 100–150 mg dL^{-1}; dogs: 80–120 mg dL^{-1}) and are frequently diabetic [4,73]. In any case, the use of PBGM has become a common practice in veterinary medicine, with a large number of commercial products being available during the last 20 years. The majority of these meters are directly adapted from existing devices for human diabetic patients (e.g., Accu-Chek, MediSense Precision). However, PBGM for pet glucose monitoring are not without their own challenges: First, measured glucose levels in blood may be erroneously over- (with concurrent anemia) or underestimated (e.g., in ketotic and/or dehydrated animals). Second, and more important, is the efficiency of blood sampling from pets: This should be preferably done by the pet owner at home, thus reducing stressing the animal by frequently exposing it to the unfamiliar environment of the veterinary clinic (which can also lead to glucose overestimation – the so-called Somogyi rebound effect [74]). On the other hand, reliable home-based, routine glucose monitoring depends on the skill and consistency of the pet owner. Fortunately, commercial systems (e.g., MicroletVaculance, BayerDiagnostics) have been developed for obtaining capillary blood by means of lancing devices. These have been originally manufactured for human use but are very popular in veterinary practice as they allow easy and relatively painless or at least tolerable blood sampling through the skin [75].

A similar philosophy has been used for developing portable blood lactate meters: Elevated blood lactate levels are prognostic in several canine pathological conditions, including tissue hypoxia [76]. A number of electrochemical (amperometric) lactate sensors have been available since the last 20 years (e.g., Lactate Pro by Arkray, Japan and Lactate Scout by SensLab, Germany). The reliability of these products has been verified by comparison of their performance with that of established lactate determination methods [77].

13.3.2 Screening for infectious disease

Contrary to pathological conditions due to indoor, sedentary lifestyle, pets are usually at risk of contagious disease from contact with wild animals. In several cases, like rabies, the disease is not only lethal but can also spread among other animals and humans; hence, the necessity of POC, rapid screening tests for common and/or high risk profile infectious disease agents in pets.

There is a number of recent developments in this field. Waner *et al.* [78] reported a novel in-clinic point-of-care polymerase chain reaction test for the detection of *Ehrlichiacanis*, the causative agent of canine monocyticehrlichiosis in dogs. Visceral leishmaniasis (VL, kala-azar) is another canine disease caused by protozoan parasites of the *Leishmania* genus. The development of commercial POC tests for the diagnosis of VL has been greatly facilitated by the construction of the recombinant antigen rK39, a kinesin-related protein, as the biorecognition element for disease-specific antibodies in seropositive individuals. Consequently, a number of user-friendly, rapid (~15 min) dipstick immunoassays based on rK39 have emerged and are practically used worldwide (e.g., Corixa Corp®, InBiosInc®, and DiaMed IT®) [79].

Canine parvovirus (CPV) is another dog pathogen, causing highly contagious, highly lethal enteric disease, the causes of which cannot be easily inferred through the assessment of symptoms [80]. Recently, a biosensor was developed for the rapid, POC detection of CPV [81]. The sensor was based on a gold-coated QCM, on the surface of which anti-CPV monoclonal antibodies were immobilized. The specificity and sensitivity of the sensor tested with clinical dog fecal samples were very high (>95 percent).

Finally, and in spite of successful disease eradication projects implemented in several countries, rabies remains an important disease in a large part of the world, associated with more than 40,000 new infections per year and a very high mortality rate (close to 100 percent) if not the affected person is treated immediately [82]. Although a standard rapid test for rabies based on direct fluorescent antibody is in routine use for more than 40 years [83], research towards the development of more sensitive and specific systems is going on. For example, Hnaien *et al.* [84] reported an impedimetric immunosensor based on the immobilization of specific anti-rabies polyclonal antibodies onto a functionalized gold microelectrode. The detection limit of this sensor was 0.5 $\mu g\ mL^{-1}$. More recently, Jeon *et al.* [85] reported a colorimetric ELISA-on-a-chip immunosensor able to complete a test for rabies virus in canine serum within 20–25 min from sample application.

13.4 Conclusion

The remarkable progress in developing biosensors and related POC systems for use in veterinary medicine is expected to increase in the future. This forecast is based on (i) the economic value associated with animal products, (ii) the ever-expanding pet market, (iii) the infection risk for humans by animal diseases, (iv) the easy transferability and adaptation of existing POC technologies for humans in the veterinary field, and (v) the identification of novel disease biomarkers, which could serve as novel targets for advanced portable assay platforms.

References

[1] Svennersten-Sjaunja K.M., Pettersson G. 'Pros and cons of automatic milking in Europe'. *Journal of Animal Science*. 2008;**86**:37–46.

[2] Rutten C.J., Velthuis A.G.J., Steeneveld W., Hogeveen H. 'Sensors to support health management on dairy farms'. *Journal of Dairy Science*. 2013;**96**:1928–52.

[3] Wess G, Reusch C. 'Evaluation of five portable blood glucose meters for use in dogs'. *Journal of the American Veterinary Medical Association*. 2000;**216**:203–9.

[4] Stein J.E., Greco D.S. 'Portable blood glucose meters as a means of monitoring blood glucose concentrations in dogs and cats with diabetes mellitus'. *Clinical Techniques in Small Animal Practice*. 2002;**17**:70–2.

[5] Pitkälä A., Haveri M., Pyörälä S., Myllys V., Honkanen-Buzalski T. 'Bovine mastitis in Finland 2001—prevalence, distribution of bacteria, and antimicrobial resistance'. *Journal of Dairy Science*. 2004;**87**:2433–41.

[6] Tenhagen B.A., Köster G., Wallmann J., Heuwieser W. 'Prevalence of mastitis pathogens and their resistance against antimicrobial agents in dairy cows in Brandenburg, Germany'. *Journal of Dairy Science*. 2006;**89**:2542–51.

[7] Deb R., Kumar A., Chakraborty S., *et al.* 'Trends in diagnosis and control of bovine mastitis: a review'. *Pakistan Journal of Biological Sciences*. 2013;**16**:1653–61.

[8] Hogeveen H., Huijps K., Lam T.J. 'Economic aspects of mastitis: new developments'. *New Zealand Veterinary Journal*. 2011;**59**:16–23.

[9] Sears P.M., Smith B.S., English P.B., Herer P.S., Gonzalez R.N. 'Shedding pattern of *Staphylococcus aureus* from bovine intramammary infections'. *Journal of Dairy Science*. 1990;**73**:2785–9.

[10] Hogeveen H., Ouweltjes W. 'Mastitis therapy and control—automatic online detection of abnormal milk' in Roginski H., Fuquay J.W., Fox P.F. (eds.). *Encyclopedia of Dairy Sciences*. London: Academic Press; 2002. pp. 1735–40.

[11] Hogeveen H., Kamphuis C., Steeneveld W., Mollenhorst H. 'Sensors and clinical mastitis—the quest for the perfect alert'. *Sensors*. 2010; **10**:7991–8009.

[12] Chagunda M.G.G., Friggens N.C., Rasmussen M.D., Larsen T. 'A model for detection of individual cow mastitis based on an indicator measured in milk'. *Journal of Dairy Sciences*. 2006;**89**:2980–98.

[13] Viguier C., Arora S., Gilmartin N., Welbeck K., O'Kennedy R. 'Mastitis detection: current trends and future perspectives'. *Trends in Biotechnology*. 2009;**27**:486–93.

[14] Song X., Zhuang S., Van der Tol P.P.J. 'New model to detect clinical mastitis in Astronaut A3 next milking robot' in Hillerton J.E. (ed.). *Mastitis Research into Practice*. Wellington: VetLearn; 2010. pp. 2689–93.

[15] Paape M.J., Wergin W.P., Guidry A.J., Pearson R.E. 'Leukocytes 2nd line of defense against invading mastitis pathogens'. *Journal of Dairy Sciences*. 1979;**62**:135–53.

[16] Hillerton J.E., Berry E.A., Gravenor M.B., Middleton N. 'Errors associated with milk cell counting'. *Veterinary Record*. 2004;**155**:445–8.

[17] Neitzel A.C., Stamer E., Junge W., Thaller G. 'Calibration of an automated California mastitis test with focus on the device-dependent variation'. *Springerplus*. 2014;**3**:760.

[18] Tan X., Ding S.Q., Hu Y.X., Li J.J., Zhou J.Y. 'Development of an immunosensor assay for detection of haptoglobin in mastitic milk'. *Veterinary Clinical Pathology*. 2012;**41**:575–81.

[19] Akerstedt M, Björck L., Persson Waller K., Sternesjö A. 'Biosensor assay for determination of haptoglobin in bovine milk'. *Journal of Dairy Research*. 2006;**73**:299–305.

[20] Koop G., van Werven T., Roffel S., Hogeveen H., Nazmi K., Bikker F.J. 'Protease activity measurement in milk as a diagnostic test for clinical mastitis in dairy cows'. *Journal of Dairy Sciences*. 2015;**98**:4613–18.

[21] Guntupalli R., Sorokulova I., Olsen E., Globa L., Pustovyy O., Vodyanoy V. 'Biosensor for detection of antibiotic resistant *Staphylococcus bacteria*'. *Journal of Visualized Experiments*. 2013;75:e50474.

[22] Allan G., Meehan B., Todd D., *et al.* 'Novel porcine circoviruses from pigs with wasting disease syndromes'. *Veterinary Record*. 1998;**142**:467–8.

[23] Ge X.N., Wang F., Guo X., Yang H.C. 'Porcine circovirus type 2 and its associated diseases in China'. *Virus Research*. 2012;**164**:100–6.

[24] Segales J. 'Porcine circovirus type 2 (PCV2) infections: clinical signs, pathology and laboratory diagnosis'. *Virus Research*. 2012;**164**:10–19.

[25] Zhang W.T., Wei F., Wang J.L., Xiao Y.Q., Sheng Z.Q. 'Establishment of immunochromatographic method for rapid detection of antibody of porcine circovirus type 2'. *Chinese Journal of Preventive Veterinary Medicine*. 2012;**34**:728–31.

[26] Yang S., Shang Y., Yin S., *et al.* 'A phage-displayed single domain antibody fused to alkaline phosphatase for detection of porcine circovirus type 2'. *Journal of Virological Methods*. 2015;**213**:84–92.

[27] Yang S., Shang Y., Wang D., Yin S., Cai J., Liu X. 'Diagnosis of porcine circovirus type 2 infection with a combination of immunomagnetic beads, single-domain antibody, and fluorescent quantum dot probes'. *Archives of Virology*. 2015;**160**:2325–34.

[28] Hu J., Wang T., Wang S., *et al.* 'Development of a surface plasmon resonance biosensing approach for the rapid detection of porcine circovirus type 2 in sample solutions'. *PLoS ONE*. 2014;**9**:e111292.

[29] Moennig V. 'Introduction to classical swine fever: virus, disease and control policy'. *Veterinary Microbiology*. 2000;**73**:93–102.

[30] Hou X., Tian H., Wu J., *et al.* 'Real-time analysis of the interaction of a multiple-epitope peptide with antibodies against classical swine fever virus using surface plasmon resonance'. *Journal of Biotechnology*. 2012;**161**:221–7.

[31] Cho H.S., Park N.Y. 'Serodiagnostic comparison between two methods, ELISA and surface plasmon resonance for the detection of antibodies of classical swine fever'. *Journal of Veterinary Medical Science*. 2006;**68**:1327–9.

[32] Guo X., Gao S., Sang S., *et al.* 'Detection system based on magnetoelastic sensor for classical swine fever virus'. *Biosensors and Bioelectronics*. 2016;**82**:127–31.

[33] Trombetta C., Piccirella S., Perini D., Kistner O., Montomoli E. 'Emerging influenza strains in the last two decades: a threat of a new pandemic?'. *Vaccines*. 2015;**3**:172–85.

[34] Mertz D., Kim T.H., Johnstone J., *et al.* 'Populations at risk for severe or complicated Avian Influenza H5N1: a systematic review and meta-analysis'. *PLoS ONE*. 2014;**9**:e89697.

[35] Wang R., Lin J., Lassiter K., *et al.* 'Evaluation study of a portable impedance biosensor for detection of avian influenza virus'. *Journal of Virological Methods*. 2011;**178**:52–8.

[36] Lum J., Wang R., Lassiter K., *et al.* 'Rapid detection of avian influenza H5N1 virus using impedance measurement of immuno-reaction coupled with RBC amplification'. *Biosensors and Bioelectronics*. 2012;**38**:67–73.

[37] Lum J., Wang R., Hargis B., *et al.* 'An impedance aptasensor with microfluidic chips for specific detection of H5N1 avian influenza virus'. *Sensors*. 2015;**15**:18565–78.

[38] Lin J., Wang R., Jiao P., *et al.* 'An impedance immunosensor based on low-cost microelectrodes and specific monoclonal antibodies for rapid detection of avian influenza virus H5N1 in chicken swabs'. *Biosensors and Bioelectronics*. 2015;**67**:546–52.

[39] Apiwat C., Wiriyachaiporn N., Maneeprakorn W., *et al.* 'Simultaneous discrimination and detection of influenza A(H1N1)pdm09 and seasonal influenza A viruses using a rapid immunogold biosensor'. *Archives of Virology*. 2014;**159**:1603–11.

[40] Shi L., Sun Q., He J., *et al.* 'Development of SPR biosensor for simultaneous detection of multiplex respiratory viruses'. *Biomedical Material Engineering*. 2015;**26**(Suppl 1):S2207–16.

[41] Krishna V.D., Wu K., Perez A.M., Wang J.P. 'Giant magnetoresistance-based biosensor for detection of influenza A virus'. *Frontiers in Microbiology*. 2016;**7**:400.

[42] Nurul Najian A.B., Engku NurSyafirah E.A., Ismail N., Mohamed M., Yean C.Y. 'Development of multiplex loop mediated isothermal amplification (m-LAMP) label-based gold nanoparticles lateral flow dipstick biosensor for detection of pathogenic *Leptospira*'. *Analytica Chimica Acta*. 2016;**903**:142–8.

[43] Toubanaki D.K., Margaroni M., Karagouni E. 'Nanoparticle-based lateral flow biosensor for visual detection of fish nervous necrosis virus amplification products'. *Molecular and Cellular Probes*. 2015;**29**:158–66.

[44] Khunthong S., Jaroenram W., Arunrut N., Suebsing R., Mungsantisuk I., Kiatpathomchai W. 'Rapid and sensitive detection of shrimp yellow head virus by loop-mediated isothermal amplification combined with a lateral flow dipstick'. *Journal of Virological Methods*. 2013;**188**:51–6.

[45] Cork J., Jones R.M., Sawyer J. 'Low cost, disposable biosensors allow detection of antibodies with results equivalent to ELISA in 15 min'. *Journal of Immunological Methods*. 2013;**387**:140–6.

[46] Tarasov A., Gray D.W., Tsai M.Y., *et al.* 'A potentiometric biosensor for rapid on-site disease diagnostics'. *Biosensors and Bioelectronics.* 2016;**79**:669–78.
[47] Raaperi K., Orro T., Viltrop A. 'Epidemiology and control of bovine herpesvirus 1 infection in Europe'. *Veterinary Journal.* 2014;**201**:249–56.
[48] Gokce G., Erdem A., Ceylan C., Akgöz M. 'Voltammetric detection of sequence-selective DNA hybridization related to *Toxoplasma gondii* in PCR amplicons'. *Talanta.* 2016;**149**:244–9.
[49] Wei B., Li F., Yang H., *et al.* 'Magnetic beads-based enzymatic spectrofluorometric assay for rapid and sensitive detection of antibody against ApxIVA of *Actinobacillus pleuropneumoniae*'. *Biosensors and Bioelectronics.* 2012;**35**:390–3.
[50] Linhares I., Raposo T., Rodrigues A., Almeida A. 'Incidence and diversity of antimicrobial multidrug resistance profiles of uropathogenic bacteria'. *Biomedical Research International.* 2015;**2015**:354084.
[51] Zuo M.Y., Chen L.J., Jiang H., Tan L., Luo Z.F., Wang Y.M. 'Detecting endotoxin with a flow cytometry-based magnetic aptasensor'. *Analytical Biochemistry.* 2014;**466**:38–43.
[52] Queiros R.B., de-los-Santos-Álvarez N., Noronha J.P., Sales M.G.F. 'A label-free DNA aptamer-based impedance biosensor for the detection of *E. coli* outer membrane proteins'. *Sensors and Actuators B: Chemical.* 2013;**181**:766–72.
[53] Nieradka K., Kapczynska K., Rybka J., *et al.* 'Microcantilever array biosensors for detection and recognition of Gram-negative bacterial endotoxins'. *Sensors and Actuators B: Chemical.* 2014;**198**:114–24.
[54] Kumar M.S., Ghosh S., Nayak S., Das A.P. 'Recent advances in biosensor based diagnosis of urinary tract infection'. *Biosensors and Bioelectronics.* 2016;**80**:497–510.
[55] Das A.P., Kumar P.S., Swain S. 'Recent advances in biosensor based endotoxin detection'. *Biosensors and Bioelectronics.* 2014;**51**:62–75.
[56] Bayn A., Nol P., Tisch U., Rhyan J., Ellis C.K., Haick H. 'Detection of volatile organic compounds in *Brucella abortus*-seropositive bison'. *Analytical Chemistry.* 2013;**85**:11146–52.
[57] Andersson L. 'Subclinical ketosis in dairy cows'. *Veterinary Clinics of North America: Food Animal Practice.* 1988;**4**:233–51.
[58] Krogh M.A., Toft N., Enevoldsen C. 'Latent class evaluation of a milk test, a urine test, and the fat-to-protein percentage ratio in milk to diagnose ketosis in dairy cows'. *Journal of Dairy Science.* 2011;**94**:2360–7.
[59] McArt J.A., Nydam D.V., Oetzel G.R. 'Epidemiology of subclinical ketosis in early lactation dairy cattle'. *Journal of Dairy Science.* 2012;**95**:5056–66.
[60] Iwersen M., Falkenberg U., Voigtsberger R., Forderung D., Heuwieser W. 'Evaluation of an electronic cow side test to detect subclinical ketosis in dairy cows'. *Journal of Dairy Science.* 2009;**92**:2618–24.
[61] Voyvoda H., Erdogan H. 'Use of a hand-held meter for detecting subclinical ketosis in dairy cows'. *Research in Veterinary Science.* 2010;**89**:344–51.

[62] Weng X., Zhao W., Neethirajan S., Duffield T. 'Microfluidic biosensor for β-hydroxybutyrate (βHBA) determination of subclinical ketosis diagnosis'. *Journal of Nanobiotechnology*. 2015;**13**:13.

[63] Weng X., Chen L., Neethirajan S., Duffield T. 'Development of quantum dots-based biosensor towards on-farm detection of subclinical ketosis'. *Biosensors and Bioelectronics*. 2015;**72**:140–7.

[64] Garverick H.A., Harris M.N., Vogel-Bluel R., *et al.* 'Concentrations of nonesterified fatty acids and glucose in blood of periparturient dairy cows are indicative of pregnancy success at first insemination'. *Journal of Dairy Science*. 2013;**96**:181–8.

[65] Wako Diagnostics. Available from http://www.wakodiagnostics.com/pi/pi_hr_seriese_nefa-hr(2).pdf. HR Series NEFA-HR(2) (2012). Accessed 4 August 2016.

[66] Veerapandian M., Hunter R., Neethirajan S. 'Lipoxygenase-modified Ru-bpy/graphene oxide: electrochemical biosensor for on-farm monitoring of non-esterified fatty acid'. *Biosensors and Bioelectronics*. 2016;**78**:253–8.

[67] Johnson C.R., Ayers M.W., Ahmadzadeh A., *et al.* 'Short communication: characterization of early postpartum estrous behavior and ovulation in lactating dairy cows using radiotelemetry'. *Journal of Dairy Science*. 2012;**95**:5085–8.

[68] Contreras Jiménez G., Eissa S., Ng A., Alhadrami H., Zourob M., Siaj M. 'Aptamer-based label-free impedimetric biosensor for detection of progesterone'. *Analytical Chemistry*. 2015;**87**:1075–82.

[69] Arévalo F.J., Messina G.A., Molina P.G., Zón M.A., Raba J., Fernández H. 'Determination of progesterone (P4) from bovine serum samples using a microfluidic immunosensor system'. *Talanta*. 2010;**80**:1986–92.

[70] Bahadır E.B., Sezgintürk M.K. 'Electrochemical biosensors for hormone analyses'. *Biosensors and Bioelectronics*. 2015;**68**:62–71.

[71] Gumus A., Lee S., Karlsson K., Gabrielson R., Winkler D.W., Erickson D. 'Real-time in vivo uric acid biosensor system for biophysical monitoring of birds'. *Analyst*. 2014;**139**:742–8.

[72] Gumus A., Lee S., Ahsan S.S., *et al.* 'Lab-on-a-bird: biophysical monitoring of flying birds'. *PLoS ONE*. 2015;**10**:e0123947.

[73] Rand J. 'Pathogenesis of diabetes mellitus in the cat'. *European Journal of Companion Animal Practice*. 2004;**14**:131–4.

[74] Somogyi M., Kirstein M. 'Insulin as a cause of extreme hyperglycemia and instability'. *Weekly Bulletin of the St Louis Medical Society*. 1938;**32**:498–510.

[75] Reusch C.E., Kley S., Casella M. 'Home monitoring of the diabetic cat'. *Journal of Feline Medicine and Surgery*. 2006;**8**:119–27.

[76] Nel M., Lobetti R.G., Keller N., Thompson P.N. 'Prognostic value of blood lactate, blood glucose, and hematocrit in canine babesiosis'. *Journal of Veterinary Internal Medicine*. 2004;**18**:471–6.

[77] Acierno M.J., Johnson M.E., Eddleman L.A., Mitchell M.A. 'Measuring statistical agreement between four point of care (POC) lactate meters and a laboratory blood analyzer in cats'. *Journal of Feline Medicine and Surgery*. 2008;**10**:110–14.

[78] Waner T., Nachum-Biala Y., Harrus S. 'Evaluation of a commercial in-clinic point-of-care polymerase chain reaction test for *Ehrlichiacanis* DNA in artificially infected dogs'. *The Veterinary Journal*. 2014;**202**:618–21.

[79] Srividya G., Kulshrestha A., Singh R., Salotra P. 'Diagnosis of visceral leishmaniasis: developments over the last decade'. *Parasitology Research*. 2012;**110**:1065–78.

[80] Larson L.J., Schultz R.D. 'Three-year serologic immunity against canine parvovirus type 2 and canine adenovirus type 2 in dogs vaccinated with a canine combination vaccine'. *Research in Applied Veterinary Medicine*. 2007;**8**:305–10.

[81] Kim Y.K., Lim S.I., Choi S., Cho I.S., Park E.H., An D.J. 'A novel assay for detecting canine parvovirus using a quartz crystal microbalance biosensor'. *Journal of Virological Methods*. 2015;**219**:23–7.

[82] Reperant L.A., Brown I.H., Haenen O.L., *et al*. 'Companion animals as a source of viruses for human beings and food production animals'. *Journal of Comparative Pathology*. 2016;**155**(1):S41–53.

[83] Servat A., Feyssaguet M., Blanchard I., *et al*. 'A quantitative indirect ELISA to monitor the effectiveness of rabies vaccination in domestic and wild carnivores'. *Journal of Immunological Methods*. 2007;**318**:1–10.

[84] Hnaien M., Diouani M.F., Helali S., *et al*. 'Immobilization of specific antibody on SAM functionalized gold electrode for rabies virus detection by electrochemical impedance spectroscopy'. *Biochemical Engineering Journal*. 2008;**39**:443–9.

[85] Jeon J.W., Seo S.M., Kim H.S., *et al*. 'ELISA-on-a-chip for on-site, rapid determination of anti-rabies virus antibodies in canine serum'. *Sensors and Actuators B: Chemical*. 2012;**171–172**:278–86.

Part IV

Commercialization

Chapter 14
Commercialized point-of-care technologies

G. P. Kanakaris[1], C. Sotiropoulos[2] and L.G. Alexopoulos[1]

14.1 Introduction

Point-of-care (PoC) diagnostics have grown to become both academically and commercially a very challenging but also rewarding field of research, development, innovation, and commercialization. PoC systems can be found in the ambulance, the hospital, the clinic, a doctor's office, a pharmacy, a patient's home or the bedside. For all of these settings, there is a growing need for easier, less costly and more reliable diagnostic solutions for a wide range of diagnostic, prognostic, and predictive targets. PoC systems aspire to cover that need. The systems and technologies envisioned by researchers as future PoC in-vitro-diagnostic (IVD) candidate solutions are numerous, often very exciting as to their potential impact on clinical care and have been thoroughly reviewed in literature [1–5]. However, it is true that the staggering number of proposed technologies for PoC seen in literature is disproportionate to the ones that are eventually adapted in commercial systems. This disconnect has been attributed to the fact that academia has for long been focused at methods that are not easily transferable to industry [6], whereas it is often the case that a scientific achievement of the field might not present with a clinical impact that justifies the effort or cost of the implementation. In recent years, a number of excellent review articles have been published [7–9] regarding PoC systems commercialization. There are different viewpoints that can be used to describe this subject, notably market size, state-of-the-art technological applications or promising research findings with potential impact on commercial IVD systems. In this work, we shall attempt to review the status of PoC IVD as can be seen through commercialized applications alone, with a focus on handheld and small benchtop systems. The specific viewpoint is based on the technological and diagnostic target decisions of 104 PoC manufacturers that were selected on the basis of site visits on international trade fairs [10], PoC conferences, and market

[1]Systems Bioengineering Group, Department of Mechanical Engineering, National Technical University of Athens, Athens, Greece

[2]Hematology Division, Department of Internal Medicine, University of Patras Medical School, Rion of Patras, Greece

reports [11] for the PoC industry. The selection process for trade fair participation was the inclusion of any company offering a PoC solution on settings ranging from clinic to what is referred to as extreme PoC [12]. The scope of the selection is to get a comprehensive image of the commercialized PoC landscape, while minimizing bias in the inclusion process. Subsequently, we proceed to categorize the technologies used by each of these companies. We identify five major groups: (1) lateral-flow assays, (2) centrifugal microfluidics systems, (3) electrochemical systems, (4) nucleic acid testing systems, and (5) blood gas analyzers. Those five technological categories encompass 97 out of 104 companies, whereas 14 systems cannot be grouped into these categories. The technological groups are not strictly related to the sensing approach or the fluidic manipulation approach, rather in key design similarities. However, PoC systems are at the epicenter of a very active research field with new principles being applied constantly [13,14] that may very well render this categorization obsolete in the near future. In this sense, this article can be viewed as a snapshot of the commercial PoC landscape at the time of writing that is sure to evolve in coming years. Using as a starting point the previously mentioned 104 PoC firms, this review outlines the most prevalent technologies in the field and provides some insight on their functions, attributes, and limitations. Furthermore, using the same starting point, we present a thorough list of biomarkers that are currently the diagnostic objective of systems reviewed in this work.

14.2 Commercialized point-of-care systems—technology categorization

See Table 14.1.

14.2.1 Lateral-flow assays

Lateral-flow immunoassays or LFIAs for short, constitute by far the most represented technology in terms of number or companies as well as diversity of analytes commercially available for PoC settings. It is indicative that 81 out of the 104 companies reviewed in this work deal partially or exclusively with LFIA and LFIA readers. Furthermore, 143 out of 173 biomarkers reported here are being detected also or exclusively using LFIA technology alone, or coupled with a detection system at various levels of sensitivity, resolution, and multiplexibility. The apparent domination of these capillary-driven disposable tests be can traced into their inherent merits [6,15], which are compactness, simplicity in both manufacturing and operation, compatibility with small sample volumes, and affordability. Closely modeled to ELISA, this technology is not only rapidly growing on its current capabilities [3,16] but is also emerging as a potential alternative in the league of quantitative biomarker detection [17]. Still the innate simplicity of traditional LFIA comes with the cost of incompatibility with automated multistep protocols, detection limits that are often incompatible with low abundance targets, and the need to translate ELISA assays into their one-step counterpart for each individual test, which can pose severe challenges both in porous media design and in reagent selection.

Table 14.1 Main attributes of the discreet technological approaches for PoC systems

	Lateral-flow tests and systems	**Lab-on-a-Disk**	**Electrochemical sensing systems**	**Nucleic acid detection**	**Blood gas and electrolyte systems**	**Other**
Number of companies	81	4	10	6	4	15
Number of biomarkers	143	36	30	12	13	59
Signal detection	Visual, colorimetric, absorbance, fluorescence	Absorbance, fluorescence	Amperometry, potentiometry, conductimetry, impedance	Fluorescence, impedance	Amperometry, potentiometry, fluorescence	Physical, optical
Fluid manipulation	Capillary	Centrifugal/ capillary	Capillary/active	Capillary/active	Active	Various
Device size	No device, pocket size, benchtop small	Benchtop, small	Pocket size, benchtop small	Pocket size, benchtop small and large	Benchtop large	Various
Main applications	Infectious disease, cardiac stat testing, pregnancy	Clinical biochemistry, hematology	Glucose, coagulation, hematocrit, blood gas, electrolytes, cardiac stat testing	Infectious disease, mutation	Blood gas, electrolytes, metabolites	Clinical biochemistry, hematology, coagulation
Quantitative	Usually no, semiquantitative, quantitative when coupled with detection device	Yes	Yes	Yes	Yes	Yes
Mutliplexing	Multiple test lines (limited), multiple strips	Multiple test chambers	Multiple electrodes	Multiple fluorescent probes, multiple electrodes	Multiple electrodes	Yes

14.2.1.1 LFIA principle of operation

The driving force of LFIAs is a capillary flow in porous pads. Figure 14.1 shows the main components of a typical lateral-flow test format. The strip assembly comprises four consecutive absorbing pads into which the sample will flow in-line. There are two main architectures commonly used, the sandwich format and the competitive inhibition format [18].

The sandwich format is primarily used when the target analyte is a large compound with multiple antigenic sites. In this case, the conjugate pad (Figure 14.1(b)) embeds an air dried or lyophilized solution of a label (typically latex or gold nanoparticles) [19] conjugated with a recognition antibody or antigen specific to the target biomarker. The reaction membrane (Figure 14.1(c)) carries the test and the control line (Figure 14.1(d/e)), which are bound to its matrix. The test line is a capture antibody or antigen specific to a targeted biomarker epitope. The control line is usually a species-specific anti-immunoglobulin [20] that will bind the initial recognition conjugate stored in the conjugate pad. The assay process starts with the introduction of the sample and often an added buffer solution on the sample pad (Figure 14.1(a)). Capillary forces drive the sample inside the pad in which a pre-treatment may take place for pH adjustment, retention of interfering components or removal of particulates, and cells. Consequently, the assay-ready sample moves on to the conjugate pad in which the analyte, if present, will bind to the label conjugate and move on to the reaction membrane. Once the sample reaches the test line, the analyte/label conjugates complex will bind to the immobilized capture molecules and accrue, thus forming a visible band. As the sample solution continues to flow on, excess or unbound conjugates reach the control line and form a second visible band, that is, an accumulation of immobilized label conjugates to confirm

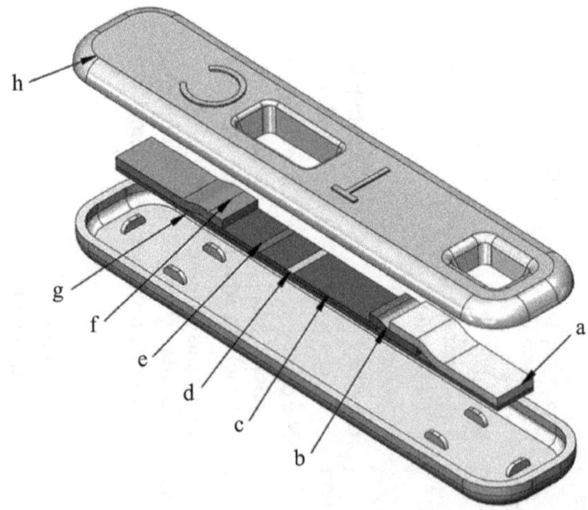

Figure 14.1 Inside a typical lateral flow test: (a) sample pad, (b) conjugate pad, (c) reaction membrane, (d) test line, (e) control line, (f) wick pad, (g) backing strip, and (h) cassette

successful assay completion. The last pad (Figure 14.1(f)) will gather excess solution and labels thus clearing the view from unbound label conjugates to the reaction matrix. Typical tests that utilize the sandwich approach are HIV, human chorionic gonadotropin (hCG), Dengue immunoglobulin G/immunoglobulin M (IgG/IgM) and Ag NS1, Troponin I, and many others.

The competitive inhibition format is less common and is primarily used when the target analyte is a small molecule with a single available antigenic site. In this case, the test line is drawn with an immobilized antigen that competes for the same antibody epitope of the label conjugate as the analyte itself. This way, once the sample reaches the conjugate pad, any present analyte will bind the label conjugate to epitope depletion, thus preventing the downstream binding that would occur on the test line. As such, in competitive assays the formation of a test line is negatively correlated to the presence of the biomarker. The control line does not compete for the same binding site, so it forms upon successful completion of the assay. In some cases, the analyte may be incubated with the label conjugate externally prior to strip usage [21]. A typical test that utilizes competitive inhibition is the detection of microalbumin in human urine specimens for the diagnosis of renal dysfunction [22].

14.2.1.2 Design, materials, and reagents in LFIA

LFIAs are mostly found in the encased format (Figure 14.1) or the dipstick format, a usually exposed strip assembly that is dipped inside a sample container. Dipstick formats are more often used when there is a need for larger sample quantities or a pretreated sample and may not include all of the pads seen in the previous paragraph. Encased formats include the serial arrangement of pads that starts with the sample pad. The sample pad is typically fabricated from cotton linter, glass fiber, or cellulose [18] strips or sheets and serves the purpose of pretreatment and smooth and homogeneous transportation of the sample to the conjugate pad. Filtration of particulates, pH adjustment, and leakage prevention are key design parameters. The conjugate pad contains the recognition molecule bound on a label. Conjugate pads are commonly made from glass fibers, polyester, or rayon [6] and are impregnated or sprayed with the label conjugate solution and eventually dried or lyophilized to store the reagent. Design parameters here include the reproducible volumetric priming of the membrane with the reagent and consistent release of conjugates during use. Attached to the conjugate pad lies the reaction matrix, a long strip of porous membrane typically made from nitrocellulose onto which the capture and control molecules are immobilized in discrete lines. This hydrophobic component by nature needs to be treated with surfactants but at the same time retain the ability to electrostatically adsorb antibodies [23]. But the most important design parameter of the reaction matrix is the capillary flow time—the time it takes for a sample to travel through the membrane laterally, as it directly affects the time available for antibody–antigen interaction both for capture and recognition molecules. Faster membranes make for faster tests but put stress both on quantity and association rate in the reagent selection. Slower membranes allow for slow association rates and less reagents but increase the waiting time of the user. Connected to the reaction matrix is the wick, the end station of the sample. The wick is typically made from cotton linter or high density cellulose and is used to collect the sample after it goes over the control line and keep the flow running. All the porous components are joined together and supported with a plastic backing strip whose purpose is to keep everything together for easy manufacturing and assembly (Figure 14.1(g)). This final, processed, and supported strip is placed within a plastic cassette (Figure 14.1(h)) that exposes only the inlet and the test lines to the user and protects the components from damage.

Labels are key components in the LFIA industry. Although a variety of different components is proposed in the literature [18,24], the ones applied commercially are colloidal gold nanoparticles, monodisperse latex colored particles, and fluorescent particles. Gold nanoparticles are low cost, available in a variety of sizes and present high sensitivity compared to latex. However, the quantification of a colloidal gold assay is limited, it is a sensitive label to pH and salts, may not be suitable for multiplex systems, and does not offer control over the orientation of the bound antibody/antigen or covalent conjugation, rather passive adsorption [16,21]. Latex on the other hand is also cost effective, comes at a variety of colors, is amenable to a variety of surface chemistries and covalent conjugation of recognition molecules, but it requires the use of surfactants as it is hydrophobic in its untreated form, and it generates a weaker signal; thus, it is not as sensitive. Fluorescent labels are used in device-based LFIAs to further increase sensitivity or when the goal is quantification and can be organic or quantum dots.

Although all of the above are essential components of an LFIA, antibodies are in the core of immunoassay technologies [16]. Aside from specificity and affinity, parameters such as the association rate constant and long-term storage become critical. Absent a sample incubation step, the performance of recognition element binding for an antibody with a given association rate constant is directly linked to the capillary flow time of the porous components. The same applies for the test line, in which incubation is not an option at all. Slow kinetics will result in unbound but also bound labels bypassing the test line for this reason. The available binding time is specified in literature [25] as 1–6 s for the test line and 10–20 s for the conjugation pad. Both monoclonal and polyclonal antibodies are used in commercial LFIAs, often in combination. Monoclonal Abs ensure consistency among product batches, are specific, and can achieve higher immobilization surface densities, however they are expensive to develop and produce and may not achieve high affinities. Polyclonal Abs are inexpensive in comparison and can achieve high affinities but are less specific and present variability between batches. A third option that is currently available from reagent vendors are aptamers, single strands of DNA or RNA that can bind to target molecules, such as proteins and peptides. Aptamers present great advantages over antibodies [26]: they have increased stability and tolerance to temperature and pH, they are produced using chemical or enzymatic synthesis and as such are highly reproducible, they can be more economical and they can achieve high specificity and affinity. To the authors' knowledge, at the time of writing, aptamer technology has not been applied commercially in LFIAs. However, it is the epicenter of a lot of discussion and research [27], and it is considered by some a good candidate for LFIA applications [26,28].

14.2.1.3 Aspects of sensitivity and quantification in LFIAs

LFIAs can be described as a really smart way to greatly simplify an ELISA assay into a low cost, one step, rapid test. There is however a tradeoff to all this inherent simplicity, and it lies in test sensitivity, quantification of the results, and the dynamic range of the test. Unlike its multistep counterpart, commercial LFIA technology is mostly qualitative, giving a yes/no answer to a threshold-related question. Part of this problem is implicit to the LFIA architecture, and part of it is user related. LFIA is typically a one-step assay that relies on a homogeneous distribution and flow of reagents at all phases including reconstitution, mixing, and binding to the capture molecules. In addition, many parameters such as membrane speed, capture Abs, label conjugate reagents, and sample quantity are subject to error due to manufacturing and fabrication tolerances of components and equipment, storage or environmental conditions during use. The test line itself is typically visible to ~10-μm depth [29] exposing less than 1/10th of the bound label conjugates to the user thus making the test susceptible to errors arising from wetting irregularities and batch to batch variations [23]. The user-related limitation arises from the fact that even a perfectly correlated signal with an assorted color chart is susceptible to subjective interpretation, perception of color, visual impairment, and psychological factors. More so, while adding an additional step to the rapid assay may improve the sensitivity of the test, for example a sample pretreatment in a competitive assay or an enzymatic enhancement of the signal generation, it also makes the test result susceptible to human error and deviations from the assay protocol. As such, LFIAs have been mostly designated for cases in which a binary answer will suffice, such as infectious and venereal disease diagnosis. Independent studies have been conducted to evaluate the performance of LFIAs in such diagnostic scenarios. HIV rapid testing is considered to be one of the most successful implementations of lateral-flow technology with performance reaching that of laboratory testing. For HIV antibody testing, high-sensitivity values are being reported [30–32] ranging from ~97.5 percent to 100 percent for more than ten LFIA brands with the lower values being reported for patients taking antiretroviral medication. Specificity values (percentage of true negatives) are being reported to be over 99 percent in two of the studies and 86 percent to 100 percent in the third with an outlier at 75 percent. In a different application [33], four LFIA tests for the detection of norovirus in fecal samples were evaluated for three virus genogroups. The results show specificity equal to 100 percent for all manufacturers, but the sensitivity ranges from 17 percent to 52 percent for the genogroup I, 39 percent to 64 for genogroup II, and 59 percent to 78 percent for genogroup GII.4 with the latter being the most common. As such, it was concluded that LFIA alone is not an acceptable solution if the goal is to rule out this condition. Another diagnostic target for which LFIAs have been developed is visceral leishmaniasis. A global comparative evaluation [33] of five different brands showed sensitivity of 92.8–100 percent and specificity of 96–100 percent for panels performed using archived samples from the Indian subcontinent. However, these values were significantly lower for samples from Brazil and East Africa (sensitivity 61.5–91 percent, specificity 36.8–87.2 percent). For noninfectious disease applications, a common LFIA product marketed by manufacturers is a rapid troponin I test. One study [34] examines four different LFIA tests for cardiac troponin I and reports positive predictive values (PPV) reaching 100 percent for three of the tests with negative predictive values (NPV) ranging from 57.6 percent to 91.6 percent for the same tests. One test performed differently with NPV over 99 percent and PPV less than 60 percent. Regarding sensitivity, out of the four tests only one gave a positive result for Tp-I less than 0.5 ng mL^{-1}, whereas the rest gave negative results for Tp-I less than 1.6 ng mL^{-1}. Many other studies exist, both independent and ones conducted by manufacturers, that demonstrate performance ranging from poor to as good as lab testing, depending on the diagnostic target, the manufacturer, and the setting. However, yes/no results cannot fully describe every clinical condition. For example, quantifying the b-hCG value can assist in ruling out ectopic pregnancy, whereas quantification of C reactive protein (CRP) can assist in the differential diagnosis between a viral or microbial infection. For this reason, quantitative LFIA technologies have been developed.

There are both instrumented and stand-alone methods to render LFIAs quantitative or semiquantitative. The ones that are frequently seen in commercial applications are color comparison charts, ladder bar assays, colorimetric strip readers, and fluorescent systems. Color comparison charts are found in the packages of urinalysis strips, drug, adulteration, and alcohol test dipsticks, which do not share the definition of lateral-flow tests. Less commonly this method can be found in lateral-flow tests for cholesterol (Prima® hometest) [35]. This method is most susceptible to human error and thus not very frequently used. As such, for biomarkers such as C-Reactive protein and fecal calprotectin in which screening and rapid testing is desirable, but clinical intervention may vary depending on the concentration range, a ladder bar is introduced [36]. This method is based on the careful deposition of specific concentration and volume of capture Ab on two or more subsequent test lines. This way, in lower concentrations of the analyte, the bound conjugate will be depleted before the final test line, so if the test includes two test lines, only one will become visible. In larger concentrations, the conjugate will be enough to bind both test lines, and both lines will appear. As lateral flow tests lack the flexibility of analytical devices in detector adjustments, the dynamic range of the measurement is often quite limited. As such, ladder assays exist for various ranges. For example, the diagnostics company Preventis® offers two different versions of a calprotectin rapid test, with cutoffs at 60 and 200 μg g^{-1} of fecal sample, respectively. Another method with multiple bars is the deposition of reference lines along with the test line (ABON™ CRP). This way, the intensity of the test line is compared to the intensity of two reference lines (high and low) in order to evaluate whether the sample concentration is above, below, of between the two reference concentrations. Many tests based on variations of this principle are marketed as semi-quantitative tests with one (PreventID® CRP 1/3) or two reference lines in sandwich or competitive (Cortez Diagnostics Inc. CRP) formats. However, these noninstrumental approaches could not cover the market demands for more sensitive, user-independent quantitative tests. For this reason, device-based and device-compatible LFIAs as well as LFIA-compatible devices have been introduced to the market by a number of manufacturers, usually at a low cost. Among the 81 LFIA rapid test companies examined in this study, 17 produce some sort of chromatographic imaging device and nine offer a urinalysis strip reader, whereas companies such as Skannex, Qiagen, Bioscitec, Cellmic, and Axxin deal exclusively with imaging technologies for LFIAs. The typical strip reader [37–39] includes a controlled illumination source (light emitting diode (LED)/Laser diode), a detector that can be a charged coupled device (CCD), complementary metal-oxide-semiconductor (CMOS), or photodiode, and suitable optics for colorimetric, fluorescent (Trinity Biotech—Meritas®, Micropoint ezLabs®), or refractometric (ReaScan®) analysis. The advantages of using such a system are numerous: Subjective interpretation of the results is replaced with a machine interpretation of the test line, either numeric or in the yes/no format that leaves no room for ambiguity. As such, the user-sourced errors are ruled out, and it becomes possible to acquire quantitative results based on the test line coloration/intensity. Also, in multi-pixel or scanning schemes, averaging over a nonuniform test line and white balance/background removal assist in more robust quantifications/interpretations of the results. Typical detection ranges for colorimetric systems are 0.5–50 and 0.2–20 ng mL^{-1} for Troponin I (Concile® Ω100, Humasis HUB-QUANpro) and typical ranges for fluorescent systems are 50–10,000 ng mL^{-1} for D-Dimer (Micropoint mLabs®) and 0.012–30 ng mL^{-1} [40] for Troponin I (Trinity Biotech—Meritas®, Figure 14.2(d)). Considering that the sensitivity of the plain immunochromatographic lateral-flow test for Tn-I is usually 1 ng mL^{-1} (Teco Diagnostics TnI, PreventID TnI and many others) and that the diagnostic limit is set at 0.04 ng mL^{-1} [41], one can see the value of using colorimetric and fluorescent systems coupled with LFIAs. The downside in using such systems is losing the inherent simplicity of these stand-alone disposable tests. Analytical devices even at this level require calibration, standards, and even maintenance, although much less than the typical clinical analytical instrument. An interesting line of products for the digitization of an LFIA result are smartphone-based systems, either devices that actually embed a smartphone (Skannex Skansmart) or adapters that allow the user's own smartphone to be used as an imaging and image processing device for lateral flow tests (CellMic—HRDR 200). Since most modern smartphones include quality CMOS sensors and illumination LEDs, they make a good fit for colorimetric applications at low cost. Cellmic has even introduced a smartphone-compatible multiwavelength fluorescence reader (CellMic HRDR-300). These solutions remove the need for an elaborate optical system putting the pressure on the analysis software and making the technology ideal for e-health applications. However, they do suffer from platform obsolescence [16] due to the constant evolution of smartphone hardware and software.

14.2.1.4 Multiplex analyses in LFIAs

The need for multiple measurements in a single disposable strip/cartridge is a constant challenge in PoC systems design. A typical example is the cardiac triple test (Troponin I, Myoglobin, CK-MB), in which different combinations of even qualitative (threshold) results indicate different stages of myocardial infraction and

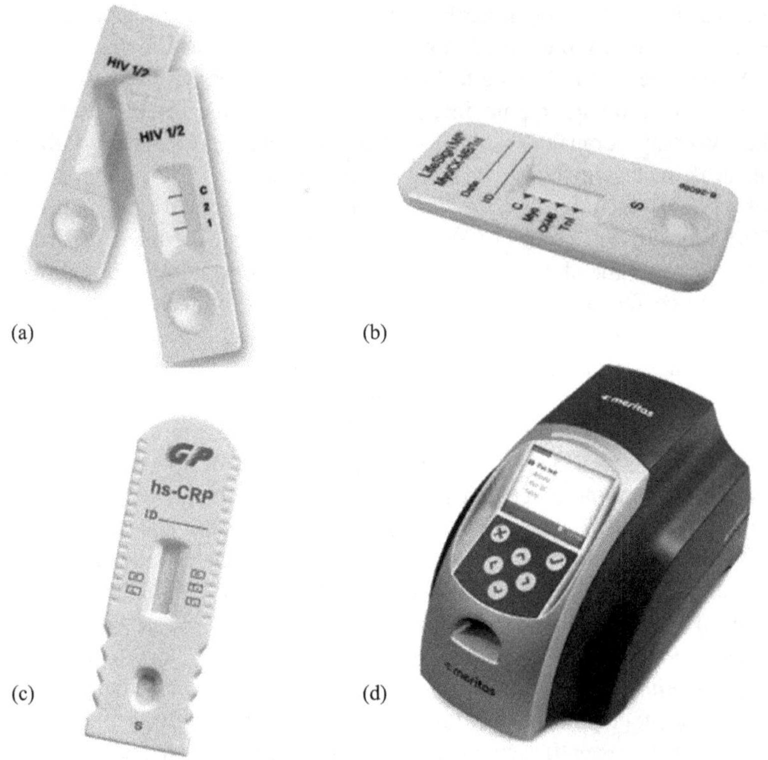

Figure 14.2 Lateral-flow rapid tests and system: (a) an HIV rapid test, (b) a multiplex rapid test for Troponin I, Myoglobin, and Creatine Kinase-MB, (c) a semiquantitative rapid test for CRP-based on reference lines, (d) Meritas by Trinity Biotech: a device-based LFIA with fluorescence detection for quantitative testing of Troponin I and Brain Natriuretic Peptide

cell necrosis. Other examples include the differentiation between influenza types A or B, malaria types Pf or Pv, and HIV types 1 or 2. In all those cases, detection of multiple analytes can assist in the identification of the subtype of the disease, the selection of a suitable treatment, and/or the prognosis of the disease outcome. In lateral-flow tests, there are two main strategies to detect multiple analytes: The first method is simply the attachment side by side of two or more strips in a common cassette. This technique is common in ToRCH rapid combo tests in which new-borns and pregnant women are screened for Toxoplasmosis, Rubella, Cytomega-lovirus, and Herpes simplex virus antibodies. This method does not share the classical definition of multiplex as a number of different samples are introduced to the same number of different strips resulting in more steps needed by the user and more samples requested from the patient. However, it does have the advantage of being as robust and reliable as the individual tests. The second method—that is common in cardiac triple, HIV, and influenza multiplex tests—involves the deposition of multiple test lines and the corresponding detection Abs in a single strip. The advantage in this "true multiplex" method is that a single drop of sample can be used for two or more tests, which is compatible with finger-prick sampling, involves only one step, and does not suffer from sample aliquot variations. The disadvantage is that just like in multiplex ELISA, nonspecific binding and cross-reactivity raise the bar for the selection of good quality antibodies [42], whereas all assays must be compatible with a single selection of membrane, sample pretreatment, and incubation times that are dependent on the positioning of the test lines. A third strategy, not frequently seen in commercial products, is the implementation of a single sample pad that feeds multiple strips. Although such a technique introduces parallelization into the assay, it may not be compatible with conventional LFIA production equipment.

14.2.2 Centrifugal point-of-care systems

Centrifugal PoCs, also known as lab-on-a-disk or LOAD, are a group of analytical microfluidic systems that share common fluid manipulation techniques based on the centrifugation of a disk-like consumable microfluidic chip. In these systems, samples and reagents are transferred and mixed in a network or microfluidic channels and chambers within a rotating disk to perform diagnostic assays. This technology is only employed by four out of 104 companies examined here; however, it is unique in many ways and also quite new in the field of commercial PoC systems with increasing interest from diagnostics companies [43]. The main benefit of centrifugal microfluidics is that traditionally troublesome aspects of lab-on-a-chip such as pumping, aliquoting, bubble formation, plasma/serum extraction, and peripheral equipment necessities suddenly become a lot easier to tackle with simple exploits of geometry, spin, and wetting properties of the substrate. More so, LOAD can be designed in a way that fluid motion is discrete, from one position to another with no stable intermediate states, which makes it conceptually and operationally preferable to systems in which everything is continuous and thus in need of carefully designed closed loop control. Having strong affiliations with the optical disk industry, this technology has been commercially applied so far in quantitative PoC devices for blood chemistry, hematology, and limited immunoassay tests; however, it possesses an inherent universality and parallelization capabilities that suggest there is more to come. That being said, these systems are by default hard to miniaturize to a handheld level and they are more costly than LFIAs for low-throughput settings.

14.2.2.1 LOAD principle of operation

LOAD operations rely on four different forces: centrifugal, capillary, Euler, and Coriolis [44]. Centrifugal force is used to pump fluid outward. Capillary force is used for the initial sample priming and to move fluid in sufficiently small channels in which capillary effects are prevalent. Euler force is used to mix components when inertial forces are large enough. Coriolis force is mostly used experimentally

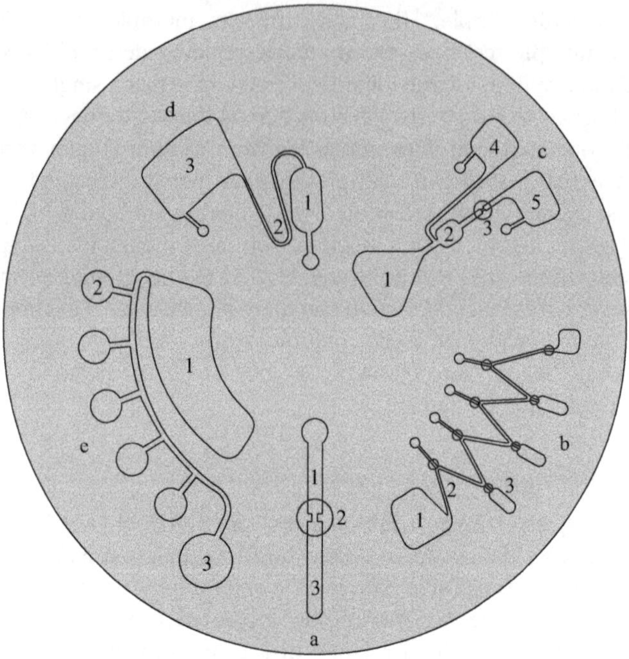

Figure 14.3 Operation-specific blocks that are found inside a LOAD disposable: (a) hydrophobic valving block, (b) aliquoting block with vents, (c) metering block, (d) siphon structure, and (e) overflow aliquoting block

in flow direction switching schemes [45]. Aside from these four (pseudo)forces, the enabling technology for LOAD systems is the local hydrophobic patching of flow channels that allows on/off valving dependent on rotational speed of the disk. However, unlike LFIAs, LOAD does not have a single operating sequence adopted in all systems rather consists of design blocks that perform a specific task and whose interconnection results in a protocol-specific chip. Figure 14.3 shows the typical blocks found in a LOAD consumable. Figure 14.3(a) is a simplified example of hydrophobic valving [46], in which hydrophilic channels Figure 14.3(a1) and (a3) are connected through a smaller and hydrophobically treated section (Figure 14.3(a2), enclosed in a red circle). Liquid that is initially present in Figure 14.3(a1) will not be able to move to Figure 14.3(a3) when the disk is motionless because of the contact angle difference at the intersection with the hydrophobic patch. If the disk starts spinning, the liquid volume will experience a centrifugal force pushing it against the hydrophobic barrier. Should this centrifugal force become large enough, the exerted pressure on the fluid front will counter the superficial force at the air/fluid interface, and the hydrophobic valve will open, that is it will be wetted allowing the liquid to flow to Figure 14.3(a3). The structures are often vented so that air pressure buildup will not interfere with pumping.

Figure 14.3(b) is a method of aliquoting of sample and reagents patented by Gyros [47], a company that pioneered the field of commercial centrifugal microfluidics. In Figure 14.3(b), a sample originally stored at Figure 14.3(b1) flows inside the meander Figure 14.3(b2) by capillary force alone, whereas the disk is stationary. Hydrophobic patches (encircled sections) allow venting of the channel but are not wetted by the inflowing liquid. Once the meander has been primed, the disk spins, outer hydrophobic patches burst, and the liquid gets distributed at the respective chambers (Figure 14.3(b3)). Figure 14.3(c) shows a metering chamber. At low rotational speed, sample stored in chamber Figure 14.3(c1) flows outward, filling chamber Figure 14.3(c2) but stopping at Figure 14.3(c3) due to the hydrophobic patch. Once Figure 14.3(c2) is full, all excess liquid will flow toward Figure 14.3(c4). Once Figure 14.3(c1) is empty, the disk is accelerated up to a point when Figure 14.3(c3) bursts, and the metered liquid flows in chamber Figure 14.3(c5). In Figure 14.3(d), a siphoning block in depicted. In this arrangement, while the disk is spinning, fluid within Figure 14.3(d1) will flow due to capillary and centrifugal force in channel Figure 14.3(d2) until the radial level in Figure 14.3(d1) matches the radial level in the first turn of Figure 14.3(d2). If the disk stops, then the flow will continue up to the entrance of Figure 14.3(d3) in which it will stop absent capillary force. However, as the entrance of Figure 14.3(d3) is outward in comparison to the exit of Figure 14.3(d1), upon spinning the disk again, Figure 14.3(d1) will empty. Figure 14.3(e) is again an aliquoting arrangement used by most commercial LOAD systems. In this block, when the disk is spinning, liquid stored in Figure 14.3(e1) flows outward priming each chamber consequently starting with Figure 14.3(e2). The last chamber Figure 14.3(e3) is used to gather excess liquid. Sufficiently large channels or vents are used to allow air to escape. Mixing can be achieved in chambers of suitable size and shape by abruptly slowing down or stopping and then reaccelerating the disk. Most systems are designed using variations and combinations of these blocks along with mixing chambers, sample, and reagent ports or liquid containing pouches embedded in the disk. Centrifugation adds the great benefit of plasma/serum separation within the chip so typically these systems provide sample to result functionality without the need for external separation steps. The result of the test is acquired using photometric or fluorescence sensing in detection chambers [48].

14.2.2.2 Design, materials, and reagents in LOAD

Centrifugal microfluidic design, while mostly based on the same operating principles, are quite diverse. Significantly, each LOAD system researched in this article employs its own unique and innovative techniques of liquid handling and detection. LOAD solutions dealing with blood chemistry (Abaxis—Piccolo Xpress [49–54], Cobas® b 101 [55–58]) typically use a spectrophotometric method of analyte detection, much like in a standard laboratory clinical analyzer. In this type of measurement, the sample is diluted within the chip and then mixed with a detection reagent for analyte quantification. The final fluid absorbs light at specific wavelengths in correlation to the concentration of the sample/reagent reaction product. In order to produce measurable absorbance, the fluid needs to be accumulated in a

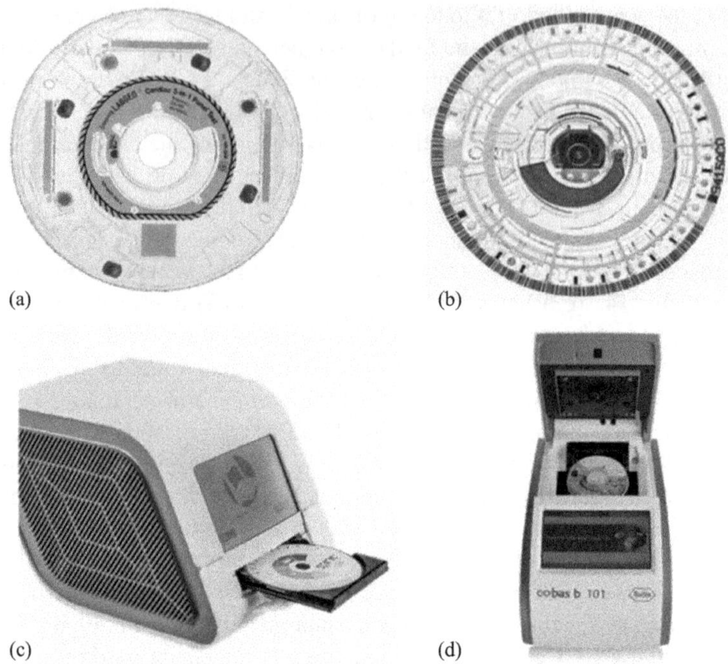

Figure 14.4 Lab-on-a-disk systems and consumables: (a) the disk used in Samsung LabGeo IB10, (b) the disk used in Abaxis' Piccolo Xpress, (c) the system Spinit from Biosurfit, and (d) the system b 101 from Cobas

cuvette-like chamber measuring a few millimeters in layer thickness, much like in clinical analyzers. For this reason, these disks are a lot thicker than typical microfluidic structures, embedding chambers, and channels that are well within the mesoscale. Since in centrifugal systems the fluid handling force acts on the entire mass of the liquid, larger structures along with microchannels do not pose a problem. Cobas b 101 (Figure 14.4(d)) takes advantage of this thick design by incorporating a lateral sampling port protected by a hinged lid. Upon closing this lid, buffer/reagent solution is released in the chip and thus the problem of reagent storage/reconstitution is solved. Similarly, Abaxis (Figure 14.4(b)) stores the buffer solution in a separate container at the center of the disk that opens by mechanical interaction when the consumable is inserted in the machine. Biosurfit (Figure 14.4(c)) with their system Spinit [59–65] uses a different approach that allows for a thinner design. The incident light is forced to undergo multiple internal reflections within the detection chamber using a diffractive reflective layer at the bottom of the chamber. This gold layer traps specific biomarkers using surface immobilized antibodies. The refractive index of the biomarker layer is measured as light reflects out of the disk and is used to quantify the bound analyte [66]. A different issue that LOAD designers need to address is that when the disk keeps spinning, the radial "gravitational" field keeps fluids secure against a wall. However, if the disk stops, then capillary forces once again become predominant and in a half full mixing/detection chamber this can be a problem. For this reason, it is often a requirement to do detection while the disk is spinning. This can only be done stroboscopically. Abaxis uses a Xenon arc lamp to counter this problem. The source is only open for 5 μs with a 50-kW radiant power output. The light travels through the sample and apertures and ends up in a series of dichroic mirrors and interference filters in which it is broken down into wavelengths and measured by individual light detectors. This arrangement allows multiple measurements at 1,200 rpm to acquire an average absorption per cuvette in the desired wavelengths. Biosurfit has patented plain LED source illumination with CCD/CMOS detection for a 200-μm sample layer. This 2D approach, in conjunction with the internal reflection format and carefully designed apertures, allows simultaneous measurements that correspond to different optical path lengths within the sample. A special case in the family of commercial LOAD is that of Samsung-Nexus and their systems LABGEO [67–69] IB10 (Figure 14.4(a)) and PA20. In this format, the disk is a carrier for up to three LFIA strips, performing fully automated quantitative immunoassays for cardiac markers from whole blood. It is mentioned in literature [70] that this system, at least initially, used fluorescence with lanthanides to provide high sensitivity with inexpensive optics. The typical materials used for centrifugal disk microfluidics are polycarbonate and polymethyl methacrylate because of their good optical characteristics, low cost, and compatibility with mass production methods such as injection molding. The hydrophobic treatment [71–75] can be done using silane derivatives or polytetrafluoroethylene in spray or plasma deposition, whereas superhydrophobic treatment is also possible by using surface micropatterning [76] prior to treatment. Apart from the fluidic aspects of LOAD, a key consideration is the storage of the reagents within the chip. All four companies investigated here use dry reagents which are reconstituted during usage of the disks. The drying can be conventional (e.g. atmospheric or hot air) or by lyophilization when the required stability cannot be achieved using other methods or when a porous structure of the pellet or spot is necessary for rapid reconstitution and mixing. Taking into consideration all of the above, one can observe that centrifugal microfluidics constitute a good canvas for a variety of design approaches and seem to be able to perform in different types of tests such as immunoassays and clinical chemistry.

14.2.2.3 Aspects of sensitivity and precision in LOAD

Commercialized centrifugal microfluidics, as it has been described in the previous paragraph, utilize standard and nonstandard optical arrangements to acquire measurements indicative of the quantity or absence of an analyte. A big benefit of this technology is the inclusion of control spots into the disposables to calibrate the device and assess the state of the reagents [49,51,52]. These sophisticated systems clearly do not aim for the rough estimation in an extreme PoC setting [12] rather aspire to become mobile or benchtop, low cost alternatives to clinical analyzers. Abaxis offers 26 different clinical chemistry tests in 17 individual disks/panels. Independent studies [77–80] produced generally favorable results with most measurements conforming with laboratory standards. For example, liver function panel intra-assay precision, linearity, and accuracy have been found to be acceptable for clinical usage [77,79] with the exception of total bilirubin. Also, lipid panel results have been found to be within the limits set by the National Cholesterol Education Program with high-density lipoproteins (HDL) falling out of the limit in one study [81]. Abaxis reports [82] linear ranges for its assays with a wide dynamic range but less sensitivity than a clinical analyzer [83], especially for electrolyte panels [84], although within the limits of diagnostic significance. Cobas offers a lipid panel and glycated hemoglobin measurement, with independent studies [85–88] for the latter reporting CVs ranging from less than 3 percent to more than 7 percent. Biosurfit offers complete blood count and a CRP immunoassay. For the complete blood count (CBC), the company reports [89] CVs ranging from 4.8 percent for Hematocrit to 13.5 percent for Monocyte differentiation in white blood cell count. For CRP, the analytical range of the measurement is reported [90] to be 4–180 mg L^{-1} with a coefficient of variation (CV) of 6.6 percent against a standard. Samsung offers immunoassays for cardiac and cardiac STAT testing, thyroid function, pregnancy, and procalcitonin measurement. For cardiac triple measurements, Samsung's LabGeo is reported [91] to have high sensitivity (0.05–30 ng mL^{-1} TpI, 2–60 ng mL^{-1} CK-MB, 30–500 ng mL^{-1} Myoglobin) with a total CV in the range of 11 percent. It can be deduced from all of the above that these devices present high sensitivity in most cases but sometimes lack in repeatability when compared to standard clinical analyzers. However, their compactness, cost, and ease of use might outweigh these drawbacks when there is no alternative for an in situ measurement, such as a doctor's office, an ambulance, or any other emergency or decentralized setting.

14.2.2.4 Aspects of multiple measurements in LOAD

Centrifugal microfluidics is inherently a multiple measurement technology, since parallelization and aliquoting are easily accomplished using standard techniques. These multiple measurements can be used for multiplexing, for real time optics calibration, for disk and reagent stability assessment, and for control measurements. The disks manufactured by Abaxis and Biosurfit are prime examples of multiple measurements using a single disk. Panels like CBC, cardiac triple, and clinical chemistry analytes have already been implemented in centrifugal technology systems. However, a multistep protocol, such as ELISA, has yet to be realized in commercial PoC LOAD systems and is perhaps a natural next step for this technology that would also presumably have a positive effect on the coefficients of variation of the devices.

14.2.3 Electrochemical sensing systems

Along with LFIAs, electrochemical sensing systems constitute the most established and applied technologies for PoC settings. Ten out of 104 companies examined here deal with purely electrochemical or hybrid technology that may also include optical detection. In these systems, the presence of an analyte is correlated to an electrical signal at the electrodes of a disposable strip. With blood glucose portable meters being the most prominent of applications, electrochemical sensors have been embedded into a wide range of commercial PoC systems covering blood gases, electrolytes, metabolites, and other analytes. Typically in less the size of a smartphone, these systems are solid state, low cost, highly sensitive, and robust. These merits, along with the fact that they cover analytical targets incompatible with LFIAs, have established them in a number of settings, such as home monitoring, and emergency testing for specific biomarkers. However, there are drawbacks to consider, including limited specificity from the electrochemical activity of species other than the analyte, limited shelf life, and nonspecific binding [92,93]. Nonetheless, electrochemical sensing systems' impact can be appreciated when considering that millions of people rely on them every day to regulate medication intake of insulin and oral anticoagulants.

14.2.3.1 Electrochemical sensing systems principles of operation

Commercial PoC systems with an embedded electrochemical sensor are quite diverse in terms of operation and analytical targets. Figure 14.5 shows a simplified disposable electrode strip similar to the ones used for glucose monitoring. In this embodiment, the sample, which can be a drop of blood from finger prick, is metered in a vented sample chamber. A volumetric electrode verifies that the chamber is full while a working and a reference electrode are used to acquire the measurement. Electrochemical sensing PoC applications can be subdivided into different categories as per their recognition element and the method of measurement. In regards to the recognition element, the two main categories are biocatalytic and affinity biosensors [92]. In biocatalytic sensors, the recognition element is

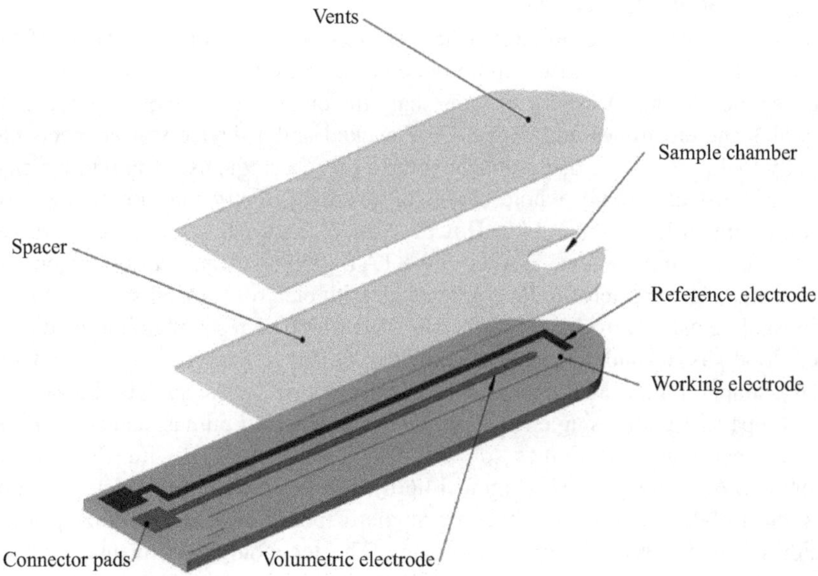

Figure 14.5 A simplified design of an electrochemical sensing disposable strip

an enzyme that catalyzes the formation of an electroactive product when the sample is introduced to the cell. In affinity biosensors, the recognition molecule is most commonly an antibody that captures the antigen biomarker whereas a secondary antibody, labeled with an enzyme, will complete the sandwich format. This enzyme label is responsible for the formation of the electroactive product in affinity biosensors. As per the method of measurement, the four main approaches are amperometry, potentiometry, conductometry, and impedance spectroscopy. In amperometry [92,93], the analyte and the sensor reagents are involved in reduction–oxidation reactions at inert metal electrodes. During this process, the electrode gains electrons from the analyte. This results in a measurable current that is linearly proportional to the concentration of the analyte. This technique is used for the quantification of metabolites, such as glucose, cholesterol, and lactate. In potentiometry, a potential difference described by the Nernst equation is measured between two electrodes with minimal current drawn. The measurement has a logarithmic dependence to the analyte concentration. Potentiometry is commonly used with ion selective electrodes for electrolyte (K^+, Na^-, and Cl^-) quantification. Conductometry is the measurement of the sample's conductivity that is in turn related to the concentration of electrolytes. Since red blood cells have insulating properties due to their lipid bilayer membrane, this method is used for hematocrit measurements. Impedance spectroscopy is primarily used in affinity biosensors, in which a sinusoidal excitation signal with varying frequency is used to acquire the impedance spectrum of the test zone. The measurement is proportional to the real-time binding activity on the test zone [94].

14.2.3.2 Design, materials, and reagents in electrochemical sensing systems

PoC devices that embed an electrochemical sensor come in a wide range of footprints and designs, the most common one being pocket size instruments for blood glucose monitoring. All these systems share in common the use of disposable strip in which the electrodes and reagents are packed and a device that connects electrically to this strip. The disposable is in most cases a single use strip into which the sample, most commonly whole blood, is inserted directly into ametering/testing chamber much like Figure 14.5. However, more elaborate approaches have been successfully commercialized. Abbot's i-STAT [95,96] (Figure 14.6(b)) combines active and passive microfluidic components with electrochemical sensing within a disposable plastic cartridge to perform a wide range of tests including cardiac triple, blood gas, coagulation, and electrolytes. Different cartridges utilize potentiometry, amperometry, and affinity methods depending on the analyte; however, all are compatible with a single handheld device containing pumps, actuators, heater, and all necessary electronics to perform measurements in situ. In a similar approach, Alere's Epoc [97] (Figure 14.6(d)) can perform 11 tests including blood gas, electrolytes, and metabolites in a single disposable cartridge utilizing microfluidics and different sensor technologies. The technological advantage of such systems is the ability to perform multiplex assays with sample pretreatment and on-chip calibration. Another notable approach is that of mobile phone integration.

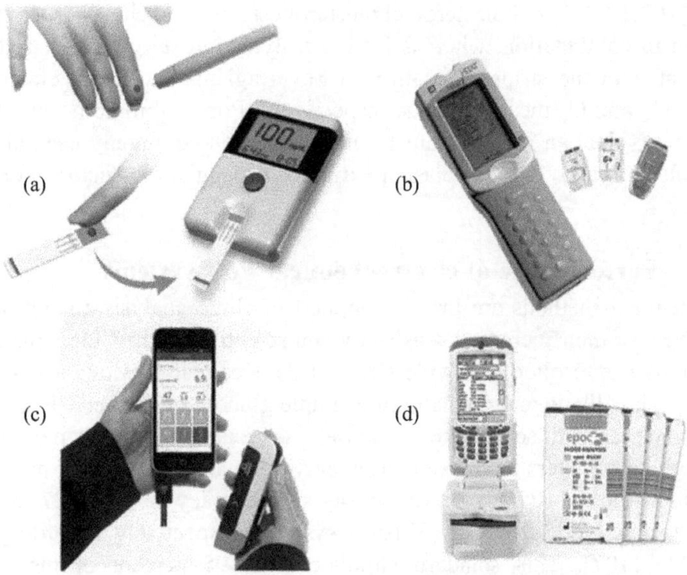

Figure 14.6 Electrochemical biosensing systems: (a) typical glucose meter, (b) i-STAT handheld blood analyzer, Abbott Point of Care Inc. (c) Dario's smartphone compatible glucose meter, and (d) Epoc PoC system by Alere

Just like in LFIA readers, the smartphone race to provide high quality consumer goods has provided the technological basis that is compatible with some PoC specifications [98,99]. For example, Dario [100] (Figure 14.6(c)) offers glucose meters that utilize the audio jack of smartphones, an I/O port able to both record and generate signals that can be used for electrochemical sensing. Regarding the electrode design in electrochemical sensor systems, configurations include the use of at least two electrodes: a reference electrode and a working electrode, while commonly an auxiliary electrode is included to function as a source or sink of electrons. Additional probes can be used for volumetric verification or to compensate for working electrode fouling or nonspecific binding. Electrodes are fabricated using either screen printing or laser ablation patterning on thin layers of metal made with vapor deposition [101]. Materials include carbon, platinum, gold, silver, and palladium. In biocatalytic sensors [92], the functionalization of the electrode is achieved by making sure that the targeted enzyme and the electrode surface are in contact and that the enzyme remains functional. Methods to achieve this include entrapment of the enzyme between membranes on the electrode surface, covalent bonding, encapsulation, inclusion in a gel or paste, adsorption, and biotinylation of the enzyme to form a biotin–avidin complex with the electrode. The enzymes used depend on the application. Glucose meters use glucose oxidase along with a redox mediator often in the form of immobilized enzyme in a redox polymer [102,103]. For cholesterol, cholesterol esterase and cholesterol oxidase are used often in combination, whereas for electrolytes ion selective electrodes [104] are integrated in the strips. Miniaturized Severinghaus and Clark electrodes are used for CO_2 and O_2 measurements, respectively. For prothrombin time (PT/INR) measurements, human recombinant thromboplastin is commonly used along with buffer solution and human plasma-extracted coagulation factors for control purposes.

14.2.3.3 Performance of electrochemical PoC systems

Electrochemical methods are broadly applied in clinic analysis among other reasons because of their increased sensitivity and selectivity. Their implementation in PoC systems has resulted in a wide range of devices that provided access to measurements typically done in the lab, for example glucose or lipid panels, or replaced testing principles with solid-state technology, for example PT/INR measurements. Portable glucose meters have been stringently evaluated as per their performance. Independent studies [105,106] have reported the accuracy of 43 and 27 commercial self-monitoring blood glucose (SMBG) systems respectively according to ISO 15197:2003 [107]. This standard stipulates that 95 percent of measurements below 75 mg dL^{-1} must fall between a $\pm$15 mg dL^{-1} zone from laboratory results, whereas measurements over 75 mg dL^{-1} must fall between $\pm$20 percent from laboratory results to acquire a CE mark. In these studies, approximately 21 percent and 41 percent of the product did not perform according to these standards. It is worth mentioning that the 2013 revision of ISO 15197 sets tighter standards for the regulation of SMBG systems, notably by requiring that 99 percent of measurements should fall within the error margin. Interfering factors in SMBG have been widely investigated [108,109]. Temperature, altitude, certain drugs, such as acetaminophen, and substances such as maltose can significantly affect the reliability of the measurement. In PT/INR monitoring, where once again frequent testing is used for the regulation of drug dosage, electrochemical systems have partly replaced classical methods based on magnetic/mechanical principles to detect clotting. With companies generally abiding by an allowable $\pm$0.3 INR (International Normalized Ratio for prothrombin time) bias in comparison to a reference system, we examine the coefficients of variation published by manufacturers for their systems. Alere's INRatio [110] is reported by the manufacturer to have an 8 percent CV for normal and 7.7 percent CV for therapeutic samples in a wide range of 0.7–7.5 INR values. Siemens, with Xprecia Stride [111], in a similar range of 0.8–7 INR reports CVs of 5.9 percent for INR < 2, 4.1–4.2 percent for INR ranging from 2 to 4.5, and 3.6 percent for INR > 4.6 up to 8. Cobas with Coaguchek [112] reports a CV of 2.6 percent for venous blood and 3.5 percent for capillary blood for samples with INR < 4.5. Abbott, with i-STAT [113], reports a CV of 4.7 percent for venous blood with mean INR = 2.4 and CV of 4.6 percent with mean INR = 2.5. The ISO standard 17593:2007 defines that for samples with INR < 2 in a reference system, 90 percent of the allowable differences between the PoC and the reference system must be $\pm$0.5 INR. For samples with INR from 2 to 4.5, this becomes $\pm$30 percent with an allowable bias of ±0.3 INR. For values larger than those, no performance criteria are set [114].

14.2.4 Nucleic acid testing systems

Nucleic acid testing (NAT), as a diagnostic tool at the PoC, is a rapidly evolving field with potentially great impact. Out of 104 companies reviewed in this work, four offer diagnostic tools based on NAT with company provided assays, whereas two offer general use amplification/detection systems that are claimed to be directed for PoC applications. In NAT PoC systems, the diagnostic targets are most commonly (1) DNA/RNA segments of bacteria/viruses or (2) human genetic material. In the first case, human body fluids such as plasma, sputum, or nasal swab are tested in an attempt to diagnose an infectious disease and differentiate between possible subtypes that would alter the therapeutic intervention. In the second case, the clinical significance lies in the detection of mutations or oncogenes, with the most widespread application being cancer treatment or prevention. One key advantage of molecular diagnostic systems lies in their ability to amplify their target sequence, a trait that is unique in comparison to other diagnostic approaches. This allows for extremely high sensitivity and specificity even at early disease stages in which immunological PoC methods fail to report a positive result due to lack of sufficient analyte. As per the target analytes, nucleic acid methods ultimately look for the most primary pieces of information that describe pathogenesis, and in this sense they can deliver high quality information to the PoC. These can be in regard to, for example, the specific subtype of an influenza infection, or a mutation in a tumor suppressor gene. On the downside, nucleic acid detection systems can only confirm the presence of a target sequence within the sample, without providing any information on whether, for example, a microorganism is alive or dead, or whether there is colonization without disease [115]. In addition, polymerase chain reaction (PCR), the most widely adopted method employed in NAT, is a slow process that is hard to implement in PoC and has led to the introduction of alternative amplification strategies [116]. Finally, the nucleic acid testing industry faces tremendous financial pressure from the technological strides of the whole-genome sequencing technology and is thus limited to scenarios in which a centralized health structure is not present.

14.2.4.1 NAT principles of operation

Nucleic acid detection presents significant diversity as to the methods applied to make it compatible with diagnostics [116–119]. This is true for both amplification and detection of the amplicons. The first step in all these techniques is releasing the DNA/RNA into a buffer in order to perform amplification. This is most commonly achieved using chemical or enzymatic agents; however, other methods, such as sonication and mechanical lysis, have been employed [118]. This step is commonly followed by purification and removal of particulates. At this point, the analyte, if present, can be at very low numbers, so it is necessary to amplify, that is multiply the segments of interest. If the target is a DNA segment, PCR is the most widely

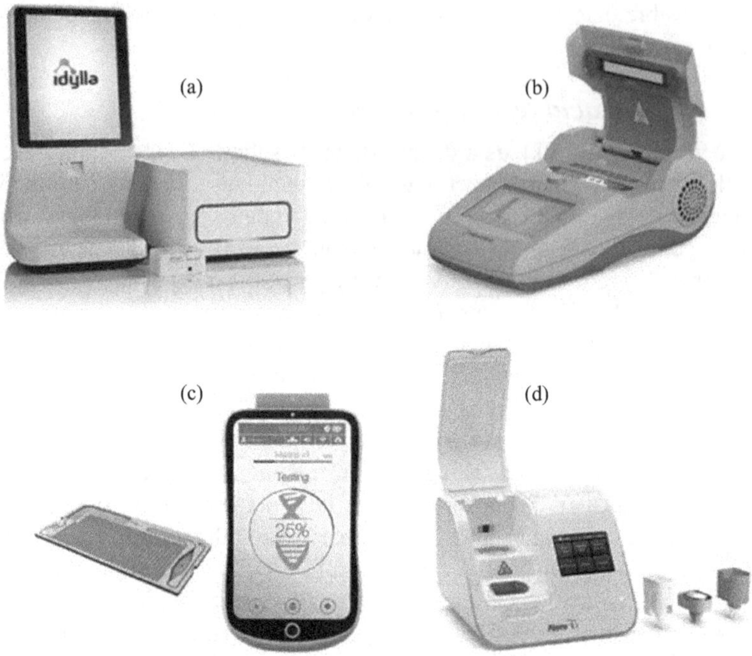

Figure 14.7 Nucleic acid testing PoC systems: (a) Idylla system from Biocartis, (b) Genie III system from Optigene, (c) Q-poc system from Quantum DX, and (d) Alere-I system from Alere

adopted approach. In PCR, the double DNA helix is denatured at high temperatures (~98 °C) to allow separation of the two strands. Once this step is complete, the temperature is lowered to allow chemically synthesized oligonucleotides called primers to pair with the separated strands on the location of interest. After this, the enzyme polymerase is used to extend the primer with complementarity to the preexisting strand in a way that it forms a copy of the previously separated strand, at least for the length of the sequence of interest. The process is usually repeated for up to 30–35 times to acquire exponentially increasing number of copies. If the target is RNA, a variation of this technique is used in which another enzyme, reverse transcriptase, is first used to create complementary DNA to the RNA of interest. Biocartis with their system Idylla (Figure 14.7(a)), Quantum DX (Figure 14.7(c)) with Q-POC, and NanoBiosys all use PCR as their amplification strategy. Another approach to amplification is loop-mediated isothermal amplification, or LAMP [120]. In this method, the temperature is constant at ~65 °C, and amplification takes place using four to six primers. An enzyme initiates synthesis by strand displacement and two of the primers form loop structures to facilitate subsequent rounds of amplification [121]. This technique is used by HiberGene in their systems HG Swift and by Optigene in Genie (Figure 14.7(b)). A third approach that is utilized by Alere in their system Alere-i (Figure 14.7(d)) is called

Nicking Enzyme Amplification Reaction. NEAR employs a nicking enzyme along with polymerase to exponentially amplify DNA at constant temperature. Both NEAR and LAMP are faster than PCR and easier to integrate in PoC systems since no thermocycling is required. Regarding detection, five out of six companies reviewed here use fluorescence in two or more channels, out of which four use a real time method, meaning that the amplification and the detection of amplicons are done simultaneously. Quantum DX uses nanowires with DNA probes whose impedance changes as DNA sequences hybridize onto their surface. Although not included in the list of companies reviewed in this article, Cepheid [122] is considered to pioneer the field of molecular PoC systems. The company has developed GeneXpert, a range of PoC benchtop devices that use a disposable cartridge to perform 6-plex real-time PCR for infectious and venereal disease, oncology, and nosocomial infections.

14.2.4.2 NAT PoC systems performance

NAT systems have the inherent advantage of amplification, which allows for very high sensitivity. Alere-i is being used to test nasal swab samples for the diagnosis of influenza A and B. Alere reports [123] sensitivity of 97.9 percent with specificity of 89.2 percent for influenza A against viral culture in a clinical study of 571 patients, whereas the same values are calculated for influenza B at 92.5 percent and 96.5 percent respectively. Results regarding the limit of detection for this assay are also presented, with values ranging from 1.88×10^5 $TCID_{50}$ mL^{-1} for subtype A/H1N1 to 5.55×10^2 $TCID_{50}$ mL^{-1} for subtype B/Yamamata lineage for direct swab testing. The same device is used for the detection of *Streptococcus pyogenes* in throat swab. Alere reports [124] 95.9 percent sensitivity and 94.6 percent from a clinical study of 481 patients against bacterial culture. The system HG-Swift by HiberGene is used for the detection of Meningococcus in a variety of samples. The company reports [125] sensitivity and specificity equal to 100 percent, however using as a reference a laboratory PCR assay for a total of 137 samples. In addition, the limit of detection of the assay is claimed to be 1.4 copies μL^{-1} for whole blood and 1.9 copies μL^{-1} for respiratory swab. Idylla by Biocartis is used for oncology assays but also for a respiratory panel to identify influenza virus subtypes and RSV. The company claims to be able to provide high sensitivity; however, the clinical trials are ongoing at the time of writing of this paper [126].

14.2.5 Blood gas/electrolyte benchtop systems

Blood gas and electrolyte testing is a requirement in critical care and emergency room testing to diagnose acute conditions relating among others to respiratory problems, lung disease, kidney disease, dehydration, or heart condition. In this sense, centralized testing often needs to be circumvented in order to get stat measurements at the PoC [127]. Various companies have developed benchtop systems with larger footprint compared to the devices that have been presented so far to cover the needs of such settings. The specifications of these systems usually include compatibility with capillary blood which is extremely useful in neonatal care, cartridge-based reagents and controls, reliability, operating simplicity, and the ability to integrate into a broad data management system. These devices are usually based on electrochemical sensing; however, fluorescent-based systems have been developed [128]. In this review, four out of 104 companies have developed one or more blood gas/electrolyte PoC analyzers. Gem 4000 [129] from Instrumentation Laboratory, Rapid Point 500 [130] from Siemens, Cobas b 123 [131] as well as Cobas b 221 [132] from Roche, and OPTI CCA-TS2 [128] from OPTIMedical perform blood gas and electrolyte testing, whereas some also include a metabolic module, a co-oximetry module, and hemoglobin measurement. Blood gas and electrolyte analyzers are in general highly automated sophisticated devices that have become pretty much standard equipment in modern clinical facilities.

14.2.6 Other technologies

14.2.6.1 Physical determination of biomarkers (coagulation)

Apart from electrochemical approaches, coagulation disorders can be diagnosed and monitored using physical determination relating to the change of viscosity in a blood sample in the presence of clotting reagents. Accriva diagnostics have developed a whole-blood microcoagulation system called Hemochron Signature Elite [133]. In this system, the blood is introduced in a disposable cartridge which in turn is inserted in a pocket-size PoC device. Using a pumping system embedded in the device, the sample is mixed with coagulation reagents and forced into a reciprocating motion within the cartridge. This motion is directly affected by the formation of clots and is monitored optically within the device [134]. The cessation of movement marks a clinically important clotting time and is used to provide a series of measurements, such as PT/INR, activated partial thromboplastin time (APTT), citrated APTT, and activated clotting time. Similarly, Coagusense has developed a PT/INR monitoring system called Coag-Sense [135]. In this system, a rotating spoke wheel is operating inside a mix of sample and recombinant rabbit thromboplastin. When the sample clots, the clot is caught up in the spokes and interrupts a beam/sensor couple. Physical methods are promoted by their manufacturers on the basis that alternative systems, such as electrochemical sensors, are affected by the concentration of red blood cells or other substances in the sample.

14.2.6.2 Optical detection systems

Optical detection in clinical testing is probably the most applied technology of all for the detection of biomarkers [136]. Optical techniques such as fluorescence, turbidimetry, nephelometry, spectrophotometry, chemiluminescence, and others are routinely used in clinical settings. In fact, apart from electrochemical systems, most other automated PoC systems mentioned in this article fall into this sensing category. However, the viewpoint of this article is the categorization of devices based on discreet set of design choices that go beyond the sensing or the fluidic manipulation method alone. In this sense, the single unifying principle of devices presented in this paragraph is an optical detection system around which the system is built.

Immunofluorimetric [137] methods have been employed widely in clinical testing. Siemens has developed the Stratus CS200 [138] that utilizes immunofluorimetry to perform STAT cardiac tests, notably CK-MB, Tp-I, Myoglobin, and D-Dimer. This device accepts whole blood in a test tube with anticoagulant. Test reagents are inserted in individual single-use cartridges. Up to four different cartridges can be used each time to create the emergency panel that is deemed most useful at the time of measurement. The device contains a centrifuge and fluidic samplers and components to perform the automated assay. This system has a comparably large footprint and a level of automation complexity that exceeds what is commonly attributed to PoC systems as they have been described so far. However, it represents a different category of PoC systems, the ones that are placed inside a clinic or emergency room to provide fast and reliable information outside the normal pipeline of biochemical testing. In this sense, such devices are autonomous and easy to use, provide lab-quality results, and have low sample to result times. As such their intended use justifies a higher design sophistication and cost.

Siemens has also developed the DCA Vantage analyzer that uses a colorimetric method along with a multipoint absorbance measurement from a competitive latex agglutination assay [139]. The device is used to monitor glycemic control and detect early kidney disease by measuring a glycated hemoglobin ratio and urine albumin and creatinine in self-contained immunoassay cartridges. In a similar approach by Alere with the system Affinion [140], glycated hemoglobin is measured using a boronate affinity assay. A spectral reflectance method is used to measure different colors that correlate to the glycated and total hemoglobin in the sample, and whose ratio is proportional to the required measurement. The same device with a different cartridge can be used to quantify albumin and creatinine in human urine. These devices are simple to use, table top systems that are compatible with finger-prick sample acquisition and are directed toward decentralized diabetes/hypertension management.

In a different application, Diagon with Coag S [141,142] use optical turbidimetry to quantify cloudiness generated by a reaction between the sample and recombinant thromboplastin. A disposable cuvette that includes lyophilized reagents, and a steel ball is introduced with the sample (whole blood) after being placed in the PoC device in a controlled temperature position. The sphere is moved using a rotating magnetic field, and clotting is monitored using turbidimetry.

A common application of optical detection is the quantitative determination of hemoglobin using a disposable chip and a tabletop small footprint device. Alere's Hemopoint [143,144] is an example of such systems. In this device, a disposable microcuvette is primed with finger-prick blood and placed in the system. Reagents stored in the microcuvette hemolyze the sample. A modified azide methemoglobin method is utilized, and the colored product which is proportional to the blood hemoglobin concentration is quantified using an absorption method at 570 nm with an additional wavelength at 880 nm for turbidity compensation. Alere has also developed a lipid panel PoC system called Cholestech [145–148]. This device is used to measure total cholesterol, high-density lipoprotein cholesterol, triglycerides, and glucose in a single-disposable cartridge from finger-prick blood collected with a heparin coated capillary tube. The device uses reflectance photometry to quantify quinoneimine dye produced by a series of reactions [149] for each individual analyte.

Of special interest to critical care testing is the measurement of ammonia in blood in order to assist in the diagnosis of liver disease, kidney failure or a urea cycle disorder. Menarini diagnostics [150] is one of the companies from this list that produces a portable ammonia meter. These devices typically utilize a strip into which ammonium ions are separated from whole blood vertically to react with a reagent producing a color change. This change is detected using reflectance measurements at an appropriate wavelength typically provided using LED technology.

Finally, in a different application, Diesse has developed Chorus [151,152], a cartridge-based ELISA benchtop device for a wide range of immunoassay tests ranging from infectious disease to autoimmunity markers. This system uses self-contained reagent cartridges and performs full ELISA protocols with washing steps to acquire results photometrically. A carousel-type design allows for up to 30 different cartridges to be analyzed in one run with integrated fluidic components handling all protocol steps.

14.3 Commercialized point-of-care systems—biomarkers

Selection of biomarkers/conditions has a great impact on the commercialization success of a PoC system. On the first part of this work, different technological approaches applied in commercial PoC systems have been described. This second part will address the targeted biomarkers of these systems. Although review articles referenced throughout this work present notable applications of their reviewed technology, to the best knowledge of the authors, an extended list of the conditions/biomarkers currently being tested using PoC technology has not been published. The purpose of the second part of this work is to present a thorough list of biomarkers that are currently the diagnostic, prognostic or predictive objective of the systems that have been investigated in this work. Although it would fall outside of the scope of this review to investigate each biomarker separately, it is the hope of the authors that this tabulated form of analytes with their respective technologies could assist researches into identifying technological and diagnostic gaps in PoC applications.

Table 14.2 includes 173 targets categorized on the basis of their type and the sample that is used to perform the PoC test. For each target, all relevant tests/devices from companies investigated in this work are cited and categorized on the basis of their technology. Companies that deal exclusively with glucose monitoring and pregnancy testing and applications dealing with allergen identification and veterinary testing have not been included in this table. Hematology panels are grouped into one entry due to the low number of PoC systems currently performing them. The devices have been categorized on the basis of the first section of this article. Table 14.3 includes the references to the system presented in Table 14.2.

Table 14.2 An extended list of conditions/biomarkers tested by PoC technology and the companies investigated here that have developed relevant systems organized by technology: Lateral-flow immunoassays (LFIA), lab-on-a-disk (LOAD), electrochemical systems (EC), blood gas (BG), and electrolyte PoC systems and other systems

Test category	**Sample type**	**Diagnostic target**	**LFIA**	**LOAD**	**EC**	**NAT**	**BG**	**Other**
Clinical biochemistry								
Lipid panel	S/P/WB	CHOL: Cholesterol	94,119	121,122	27,32,33			7,10,20
Lipid panel	S/P/WB	HDL (Direct)	119	121,122	32			7,10,20
Lipid panel	S/P/WB	LDL		122				
Lipid panel	S/P/WB	TRIG: Triglyceride	94,119	121,122	27,32,33			7,10,20
Lipid panel	WB	Lipid panel (cholesterol, LDL, HDL, triglycerides)		122	27			
Atherosclerosis marker	WB	sPLA2-IIA (atherosclerosis)	57					
Cardiac panel	S/P/WB	BNP: Brain natriuretic peptide	41,57,72,111	124,125				10,19
Cardiac panel	S/P/WB	CK-MB	40,41,44,47,49,54, 64,67,69,72,75, 76,88,93,96, 107,111,114	124,125	29			10,19
Cardiac panel	S/P/WB	CK: Creatine kinase	119	121				20
Cardiac panel	S/P/WB	h-FABP	57,67,72,76,96					
Cardiac panel	S/P/WB	Myoglobin	40,41,44,47,49,54, 64,67,69,72,75,76, 77,88,93,96,107, 111,114	124,125	29			10,19

(Continues)

Table 14.2 (*Continued*)

Test category	Sample type	Diagnostic target	LFIA	LOAD	EC	NAT	BG	Other
Cardiac panel	S/P/WB	Troponin I	40,41,44,47,49,52, 53,54,57,64,67,69, 70,72,74,75,76,77, 82,83,88,93,96,99, 107,108,111,114	124,125	29			10,19
Liver panel	S/P/WB	ALT: Alanine aminotransferase	119	121				7,20
Liver panel	S/P/WB	AST: Aspartate aminotransferase	119	121				7,20
Liver panel	S/P/WB	Gamma GT						7,20
Liver panel	S/P/WB	ALP: Alkaline phosphatase	119	121				7,20
Liver panel/anemia marker/body fluid analysis	S/P/WB	LDH: Lactate dehydrogenase		121				20
Liver panel/anemia marker	S/P/WB	TBIL: Total bilirubin	119	121			3,4,5	7,20
Liver panel/hemolytic anemia marker	S/P/WB	DBIL: Direct bilirubin		121				7
Liver panel/renal panel/nutritional status	S/P/WB	ALB: Albumin	42,64	121				7,20
Liver panel/nutritional status marker	S/P/WB	TP: Total protein		121				7,20
Liver disease/kidney failure/Reye syndrome	WB	Ammonia						20
Renal panel	S/P/WB	BUN: Blood urea nitrogen	119	121	29		1,3	7,20

Renal panel	S/P/WB	CREA: Creatinine	119	121	29,35,36		7,13,20
Renal panel	S/P/WB	Cystatine C	72				
Gout/kidney disease/ monitoring cancers treatment marker	S/P/WB	UA: Uric acid	119		38		7,20
Electrolytes/renal panel/acid–base balance marker	S/P/WB	K+: Potassium	119	121	29,35	1,2,3, 4,5,6	7
Electrolytes/renal panel/acid–base balance marker	S/P/WB	NA+: Sodium		121	29,35	1,2,3, 4,5,6	7
Electrolytes/acid–base balance marker	S/P/WB	Cl−: Chloride		121	29,35	1,2,3, 4,5,6	7
Electrolytes/renal panel/parathyroid function marker	S/P/WB	Ca: Calcium		121	35	1,2,3, 4,5,6	7,20
Electrolytes/renal panel	S/P/WB	iCa: ionized calcium			29		
Electrolytes/renal panel/parathyroid function marker	S/P/WB	PHOS: Phosphorus, inorganic		121			20
Electrolytes/renal disorders	S/P/WB	Magnesium		121			20
Pancreatic function marker	S/P/WB	AMY: Amylase	119	121			20
Pancreatic function marker	Stool	Pancreas Elastase 1	60				
Diabetes disease/ pancreatic function marker	S/P/WB	Insulin	72				
Diabetes disease marker	S/P/WB	Fructosamine					20
Diabetes disease marker	WB	Glycated hemoglobin		122	27,32,37		7,11,13

(Continues)

Table 14.2 (Continued)

Test category	Sample type	Diagnostic target	LFIA	LOAD	EC	NAT	BG	Other
Diabetes disease marker	Urine	Ketoacidosis (β-hydroxybutyrate)	114		36,37			
Diabetes disease marker	Urine	Microalbumin	44,54,69,70,71, 72,114		39			
Diabetes disease marker	S/P/WB	GLU: Glucose	94,119	121	27,29,32, 33,35,36,37		1,3,4, 5,6	7,10,20
Acid–base balance marker	S/P/WB	LAC: Lactate		121	29,33,35, 36,37		1,3,4, 5,6	
Blood gases/acid–base balance marker	S/P/WB	pO_2			29,35		1,2,3, 4,5,6	
Blood gases/acid–base balance marker	S/P/WB	pCO_2			29,35		1,2,3, 4,5,6	
Blood gases	S/P/WB	sO_2			29		1,2	
Blood gases/acid–base balance marker	S/P/WB	tCO_2: Total carbon dioxide		121				7
Thyroid function marker	S/P	TSH	40,57,64,70, 72,75,77,94					
Thyroid function marker	S/P	Thyroxine free	57					
Sex hormone (female)	Urine	Luteinizing hormone (ovulation)	40,44,45,47,49,54, 57,64,67,70,72,73, 74,76,77,80,82,83, 84,85,88,93,94,97, 99,107,114					
Sex hormone (female)	Urine	FSH: Follicle stimulating hormone	44,47,54,57,64,67, 70,72,73,80,84, 93,94,99,107					

Sex hormone/pregnancy marker	Urine	b-hCG	40,44,45,47,48,49, 50,53,54,57,64,67, 69,70,71,72,73,74, 76,77,80,82,83,84, 85,88,92,93,94,96, 97,99,102,104,107, 112,114	29		19
Sex hormone	S/P	Testosterone	40,57			
Tumor markers	S/P	AFP	40,47,49,54,57,64, 67,72,73,74,75,76, 77,85,93,99,114			
Tumor markers	S/P	CA-125	57,72			
Tumor markers	S/P	CA-15-3	57			
Tumor markers	S/P	CA-19-9	57			
Tumor markers	S/P	CEA	40,47,49,54,57,64, 67,72,73,74,75, 76,77,85,93,114			
Tumor markers	Stool	M2-PK	60			
Tumor markers	S/P/WB	PSA	40,54,64,67,72, 74,75,76,77,85, 94,96,99			
Tumor markers	S/P	PSA	47,49,52,54,72, 73,93,96,114			
Tumor markers	Urine	PSA	83			
Tumor markers	Urine	Urinary bladder cancer Ag	44,57,70,75,104			
Metastatic cancer marker	P	ctBRAF mutation			25	
Urinalysis	Urine	Urinalysis strip test	44,46,53,54,69,71, 72,73,74,75,77,82, 85,93,94,99,107, 114,115	38		

(Continues)

Table 14.2 (Continued)

Test category	Sample type	Diagnostic target	LFIA	LOAD	EC	NAT	BG	Other
Other	Urine/WB	Drugs/alcohol/ adulterants	44,47,49,54,61,64, 65,71,72,73,74,75, 77,83,85,88,91,93, 94,98,100,107,116					10
Hematology								
Complete blood count/anemia marker	WB	Hgb: Blood hemoglobin	72,107,119		27,37,39		1,2,5	12
Complete blood count/anemia marker	WB	Hct: Hematocrit		123	29,35,37,38		2,3,4,5	
Complete blood count/inflammation marker	WB	WBC: White blood cell count		123	39			
Complete blood count	WB	Hematology panels		123,124				8,9
Anemia marker	WB	Ferritin	64,72,94					
Hemolytic anemia marker	WB	G6PD deficiency	63					
Other	WB	Blood type classification	55					
Coagulation								
Hemostasis status/ coagulation disorders	P/WB	INR PT			28,29,30,34			17,18
Coagulation disorders marker	P/WB	D DIMER	41,42,44,57,64,70, 72,75,76,106,111	124				10,19
Inflammation								
Inflammatory marker	S/P/WB	CRP: C reactive protein	42,44,51,54,57,64, 70,72,75,97,104	121,123				19
Inflammatory marker	S/P	Neopterin	57					
Inflammatory marker	S/P/WB	Procalcitonin	47,72					

Inflammatory marker	Tear	MMP-9 (dry eye disease)	51	
Inflammatory bowel disease (IBD) marker	Stool	Calprotectin	57,58,68, 70,78,86	14
Inflammatory bowel disease (IBD) marker	Stool	Lactoferrin	47,68,86,112	
Oxidative stress marker	Urine	Malondialdehyde (ROS indicator)	114	
Rheumatoid arthritis marker	S/P/WB	Anti-CCP	59	14
Rheumatoid arthritis marker	S/P/WB	Anti-MCV	62,70	
Rheumatoid arthritis/ autoimmune diseases marker	S	Rheumatoid factor	75,97	14
Infections differential diagnosis tests				
Bacterial serology tests	S/P/WB	*Borrelia burgdorferi* IgG Ab	72	14
Bacterial serology tests	S/P/WB	*B. burgdorferi* IgM Ab	72,75	14
Bacterial serology tests	S/P	*Brucella* spp. Abs	73	
Bacterial antigen detection tests	Stool	*Campylobacter* spp. Ag	56,68,86	
Bacterial antigen detection tests	Stool	*Vibrio cholerae* spp. (serotypes O1 & O139) Ag	75,93,96	
Bacterial antigen detection tests	Stool	*Clostridium difficile* Ag	56,66,68,72, 86,108,112	
Bacterial antigen detection tests	Stool	*Clostridium difficile* Toxin A/B	50,56,68,72,75, 86,108,112	
Bacterial antigen detection tests	Stool	*Clostridium perfingens* Ag	68	

(Continues)

Table 14.2 (Continued)

Test category	Sample type	Diagnostic target	LFIA	LOAD	EC	NAT	BG	Other
Bacterial antigen detection tests	Stool	*E. coli* (serotype O157) Ag	47,56,68, 75,86,112					
Bacterial serology tests	S/P/WB	*Helicobacter pylori* IgG Ab	40,44,45,47,49,52, 53,54,61,64,67,69, 70,71,72,73,74,75, 76,88,92,93,94,99, 102,107,112,114					14
Bacterial antigen detection tests	Stool	*Helicobacter pylori* Ag	40,44,47,50,54,56, 57,58,61,66,68,69, 70,71,72,73,74,75, 78,85,86,88,93,96, 104,108					
Bacterial infection marker	Biopsy sample	*Helicobacter pylori* (liquid urease test)	109					
Bacterial antigen detection tests	Urine	*Legionella pneumophila* Ag	49,54,56,57,72,73, 75,80,86,93,104, 108,112					14
Parasite antigen detection tests	S/P	*Leishmania* spp. (Kala-Azar) Ag	47,54,66,73, 75,96,101					
Bacterial serology tests	S/P/WB	*Leptospira* spp. IgG/IgM Abs	47,48,52,64, 75,79,96					
Bacterial antigen detection tests	Stool	*Listeria monocytogenes* Ag	68,86					
Bacterial antigen detection tests	WB, CSF, Nasophagal swab	*Neisseria meningitidis* (serogroups A, B, C, 29E, W135, X, Y, Z) Ag				21		

Bacterial serology tests	WB	*Orientia tsutsugamushi* IgM Ab	63		
Bacterial antigen detection tests	(Stool)/S/P	*S. typhi/S. paratyphi* Ag	47,49,52,53, 64,68,74		
Bacterial antigen detection tests	Stool/S/P	*S. typhi* Ag	44,47,49,54,56, 64,68,73,74,86		
Bacterial serology tests	S/P/WB	*S. typhi* IgG/IgM Abs	44,47,49,53,73, 74,79,93,96		
Bacterial serology tests	S/P/WB	*Rickettsia rickettsii* IgG/IgM Abs	44,49		
Bacterial antigen detection tests	Stool	*Shigella* spp. Ag	56,68,86		
Bacterial antigen detection tests	Throat swab	*Streptococcus group A* (*S. pyogenes*) Ag	44,45,47,49,50,54, 57,64,68,69,70,71, 72,73,74,75,80,86, 88,92,93,96,102, 103,104,107,112	26	14
Bacterial antigen detection tests	Throat swab	*Streptococcus group B* (*S. agalactiae*) Ag	47,50,64, 69,70,71		
Bacterial antigen detection tests	Urine	*Streptococcus group B* (*S. agalactiae*) Ag	44		
Bacterial antigen detection tests	Nasal & Rectal swab	*Streptococcus group B* (*S. agalactiae*) Ag	72	21	
Bacterial antigen detection tests	Urine	*Streptococcus pneumoniae* Ag	57,72,75, 86,108,112		
Bacterial antigen detection tests	WB	*Mycobacterium tuberculosis* IgG Ab	44,49,70,72		

(Continues)

Table 14.2 (Continued)

Test category	Sample type	Diagnostic target	LFIA	LOAD	EC	NAT	BG	Other
Bacterial antigen detection tests	WB	*Mycobacterium tuberculosis* IgM/IgG Abs	44,75,85, 93,96					
Bacterial serology tests	S/P/WB	*Mycobacterium tuberculosis* Abs	40,52,64, 73,83,88			23		
Bacterial antigen detection tests	Stool	*Yersinia enterocolitica* (serotypes O:3) Ag	68					
Bacterial antigen detection tests	Vaginal swab	*Chlamydia* spp. Ag	50,54,64,67,69, 71,72,74,77,88, 92,93,96			23		14
Parasite antigen detection & serology tests	WB	*Plasmodium* spp. (malaria Pan) Ag & Abs	40,44,47,48,49,54, 64,66,67,72,73,74, 75,79,82,88,93,96, 101,104,112			23		
Parasite antigen detection & serology tests	WB	*Plasmodium falciparum* (malaria Pf) Ag & Abs	40,44,47,48,49,52, 53,54,63,64,67,72, 73,74,75,76,79,82, 83,88,93,96,104, 105,112,114			23		
Parasite antigen detection & serology tests	WB	*Plasmodium vivax* (malaria Pv) Ag & Abs	40,44,47,49,52,53, 54,63,64,67,73,74, 75,76,79,82,83,93, 96,105,114			23		
Parasite serology tests	S/P	*Trypanosoma cruzi* Ab	54,64,72, 73,75,96					
Parasite antigen detection tests	Stool	*Entamoeba* spp. Ag	68,75					

Parasite serology tests	S/P/WB	*Wuchereria bancrofti*, *Brugia malayi* or *B. timori* IgG/IgM (Filariasis)	47,75		
Parasite antigen detection tests	Stool	*Giardia lamblia* Ag	47,50,58,61,68,71, 72,75,86,108,112		
Parasite antigen detection tests	Urine	*Schistosoma* spp. (*mansoni*, *japonicum*) Ag	73,75,95		
Parasite antigen detection tests	Stool	*Cryptosporidium parvum* Ag	50,58,68,75, 86,108,112		
Parasite serology tests/ToRCH	S/P/WB	*Toxoplasma gondii* IgG/IgM Abs	47,53,73, 93,96		14
ToRCH panel	S/P/WB	ToRCH panel (TOX, CMV, RUB, HSV-1,HSV-2) Abs	47,53,64, 73,74,93		
Sexual transmitted diseases (STDs) panel	Vaginal/ throat swab	*Neisseria gonorrhoeae* Ag	47,54,64,73, 77,88,93,96	23	
Sexual transmitted diseases (STDs) panel	S/P/WB	*Treponema pallidum* IgG/IgM/IgA Abs	40,44,47,48,49,52, 54,61,67,72,73,74, 76,82,83,85,88,90, 93,96,108,114		14
Sexual transmitted diseases (STDs) panel	Vaginal swab	*Trichomonas vaginalis* Ag	45		
Virus antigen detection tests	Tear	Adenovirus Ag	51		
Virus antigen detection tests	Stool	Adenovirus Ag	40,47,49,50,54,57, 58,61,64,66,68,72, 74,75,78,80,86,87, 88,93,97,104		

(Continues)

Table 14.2 (Continued)

Test category	Sample type	Diagnostic target	LFIA	LOAD	EC	NAT	BG	Other
Virus antigen detection tests	Nasal/ Throat swab	Adenovirus Ag	45,68,86					
Virus serology tests	Stool	Astrovirus Ag	44,68,75,86					
Virus serology tests	S/P/WB	Chikungunya virus IgG/IgM Abs	47,49,64, 75,79,82					
Virus serology tests/ToRCH	S/P/WB	Cytomegalovirus (CMV) IgG/IgM Abs	47,53,73, 93,96					14
Virus serology tests	S	Dobrava–Hantaan virus IgM Ab	110					
Virus serology tests	S/P/WB	Dengue virus IgG/IgM Abs	40,44,47,48,49,52, 54,63,64,74,75,79, 82,88,93,96,101,114					
Virus antigen detection tests	S/P/WB	Dengue virus Ag (NS1)	40,54,63,75,79, 82,93,101					
Virus serology tests	S/P/WB	Ebola virus IgG/IgM Abs	47,91					
Virus antigen detection tests	Stool	Enterovirus Ag (VP1 Peptide)	68,86					
Virus serology tests	S/P	Epstein Barr virus VCA-IgG/IgM Abs	45,54,64,69,71, 72,74,75,88,92, 101,102,107,112					14
Virus serology tests	WB	Hantavirus IgG/IgM Abs	49,64					
Virus serology tests/viral liver disease	S/P	Hepatitis A virus IgM Ab	53,64,73, 93,96					

Virus serology tests/viral liver disease	S/P/WB	Hepatitis C virus Ab (anti-HCV)	40,44,47,52,53,54, 64,67,69,71,72,73, 75,76,79,82,83, 88,91,93,114			
Virus antigen detection tests/viral liver disease	S/P	Hepatitis B virus HBsAg	40,44,47,48,50,52, 53,54,64,67,69,71, 72,73,74,75,76,82, 85,88,93,96,114			
Virus serology tests/viral liver disease	S/P/WB	Hepatitis B virus panel (HBsAg, HBsAb, HBeAg, HBeAb, HBcAb)	42,47,64,72,73, 75,79,83,104			
Virus serology tests/ToRCH/sexual transmitted diseases (STDs)	S/P/WB	HSV 1 IgG/IgM Abs	40,47,73,80,93			14
Virus antigen detection tests	Nasal Swab	Influenza A + B (differentiates) Ag	40,44,45,47,49,54, 67,68,70,72,74,75, 76,77,78,80,86,88, 91,92,96,97,103, 104,112,114		25,26	14
Viral infection marker	S/P/WB	Myxovirus resistance A protein	51			
Virus antigen detection tests	Stool	Norovirus Ag	40,47,68,75, 86,87			
Virus serology tests	S/P/WB	PUUMALA virus IgM Ab	110			
Virus antigen detection tests	Stool	Rotavirus Ag	40,44,47,50,54,57, 58,61,64,66,68,72, 74,75,78,80,86,87, 88,93,97,104			

(Continues)

Table 14.2 (Continued)

Test category	Sample type	Diagnostic target	LFIA	LOAD	EC	NAT	BG	Other
Virus antigen detection tests	Nasal swab	RSV virus Ag	45,47,50,57,61,67, 68,72,75,80,86, 92,103,112			25		14
Virus serology tests/ToRCH	S/P/WB	Rubella virus IgG/IgM Abs	47,73,93					14
Virus serology tests	S	Tickborne encephalitis virus IgM Ab	110					
Virus serology tests	S/P/WB	Zika virus IgG/IgM Abs	49					
Virus serology tests/ToRCH/sexual transmitted diseases (STDs) panel	S/P/WB	HSV 2 virus IgG/IgM Abs	47,73					14
HIV treatment marker	WB	HIV (CD4 Cell count for treatment determination)	48					
Sexual transmitted diseases (STDs)/HIV marker	S/P/WB	HIV 1/2 Abs	40,44,47,48,49, 50,54,64,67,69, 71,72,73,74,75, 76,77,79,82,83, 85,88,90,91,93, 96,101,104,105, 108,114					

Sexual transmitted diseases (STDs)/HIV marker	S/P/WB	HIV 1/2 Tri-Line Abs	40,47,53,64,73, 74,75,79,82, 88,93,96,101		
Sexual transmitted diseases (STDs)/HIV marker	S/P/WB	HIV 4th generation tests (Ag & Abs)	47,75,79, 96,112		
Fungus antigen detection tests	Vaginal swab	*Candida* spp. Ag	50		
Other	WB	*Clostridium tetani* anti-toxoid Ab (immunization status)	72,75,94		14
Other	Vaginal swab	Sialidase activity (bacterial vaginosis)	45		
Miscellaneous					
Gastrointestinal bleeding marker	Stool	Transferrin	47,68,93		
Gastrointestinal bleeding marker	Stool	Occult blood (Hb)	40,45,47,54,57, 58,60,64,67,68, 69,70,71,72,73, 74,75,76,77,84, 85,92,93,94,96, 99,102,104, 107,114		16
Gastrointestinal bleeding marker	Stool	Occult blood (Hb/Hp)	44,61,66,70, 75,86		
Other	Breast milk	Creamatocrit (optical)		37	

Table 14.3 The list of 125 systems/companies investigated in this work. The reference number corresponds to the biomarker list in Table 14.2

Reference number	Company/product name	Category	Reference number	Company/product name	Category
1	Optimedical	Blood gas/electrolyte analyzer	64	Amgenix	LFIA
2	Cobas-Roche b 121	Blood gas/electrolyte analyzer	65	BioGMS	LFIA
3	Cobas-Roche b 221	Blood gas/electrolyte analyzer	66	Apacor	LFIA
4	Cobas-Roche b 123	Blood gas/electrolyte analyzer	67	Humasis	LFIA
5	Instrumentation Lab	Blood gas/electrolyte analyzer	68	Certest Biotech	LFIA
6	Siemens RAPIDPoint 500	Blood gas/electrolyte analyzer	69	Futura System	LFIA
7	Nexus—Samsung LABGEO PT10	Other—clinical chemistry	70	Preventis	LFIA
8	Nexus—Samsung LABGEO HC10	Other—hematology	71	Sanymed Diagnostics	LFIA
9	Norma Diagnostica	Other—hematology	72	VedaLab	LFIA
10	Alere Cholestech	Other	73	Atlas Link	LFIA
11	Alere Afinion	Other—diabetes	74	Dialab	LFIA
12	Alere Hemopoint	Other—hemoglobin	75	Nal von minden	LFIA
13	Siemens DCA Vantage	Other—diabetes	76	HBI 21	LFIA
14	DIESSE	Other—immunoassays	77	MH Medical	LFIA
15	OJ Bio	Other	78	Generic Assays	LFIA
16	Accumetrics	Other—coagulation	79	Jmitra. Co	LFIA
17	Coagusense	Other—coagulation	80	SAS Scientific	LFIA
18	Diagon	Other—coagulation	81	Rapigen	LFIA
19	Siemens Stratus CS 200	Other	82	BHAT biotech	LFIA
20	Menarini Diagnostics	Other	83	Intecasi	LFIA
21	Hibergene	Nucleic acid testing	84	Lobeck	LFIA
22	NanoBioSys	Nucleic acid testing	85	Victorch	LFIA
23	QuantumDX	Nucleic acid testing	86	Vidia	LFIA
24	Optigene	Nucleic acid testing	87	Rimco	LFIA
25	BioCartis	Nucleic acid testing	88	Prometheus Bio	LFIA
26	Alere—Alere i	Nucleic acid testing	89	Hemaprompt	LFIA
27	BioSys	Electrochemical	90	Biolytical	LFIA
28	Alere INRatio	Electrochemical	91	Orasure	LFIA

29	Abbott i-stat	Electrochemical	92	Quidel	LFIA
30	Siemens Xprecia Stride	Electrochemical	93	Aluxbio	LFIA
31	MagellanDX	Electrochemical	94	Primahometest	LFIA
32	PTSdiagnostics	Electrochemical	95	Rapid Medical	LFIA
33	Cobas-Roche Accutrend	Electrochemical	96	Audit Diagnostics	LFIA
34	Cobas-Roche Coagucheck	Electrochemical	97	EVL	LFIA
35	Alere epoc	Electrochemical	98	Drugcheck	LFIA
36	NovaBiomedical	Electrochemical	99	Teco Diagnostics	LFIA
37	EKF Diagnostics	Electrochemical	100	DST Diagnostics	LFIA
38	Urit	Electrochemical	101	BioRad	LFIA
39	Hemocue	Electrochemical	102	Beckman Coulter	LFIA
40	Nanoentek	LFIA	103	Becton Dickinson	LFIA
41	Alere Triage	LFIA	104	Biomerieux	LFIA
42	Technoclone	LFIA	105	Atomo Diagnostics	LFIA
43	Siemens Urinalysis	LFIA	106	Micropointbio	LFIA
44	Dutch Diagnostics	LFIA	107	Clarity Diagnostics	LFIA
45	Sekisui Diagnostics	LFIA	108	IMMCO	LFIA
46	YD Diagnostics	LFIA	109	AB Analytica	LFIA
47	Biocan Diagnostics	LFIA	110	Reagena	LFIA
48	Omega Diagnostics	LFIA	111	Cobas-Roche h 232	LFIA
49	Lumiquick Diagnostics	LFIA	112	Alere Rapid tests	LFIA
50	Sanyon Diagnostics	LFIA	113	Sugentech	LFIA
51	RPS Diagnostics	LFIA	114	DFI Care	LFIA
52	LabCare Diagnostics	LFIA	115	77 Elektronika	LFIA
53	Spectrum Diagnostics	LFIA	116	Cellmic	LFIA—Devices
54	Cortez Diagnostics	LFIA	117	Abingdon Health	LFIA—Devices
55	MTC InVitro	LFIA	118	Skannex	LFIA—Devices
56	Pro-Lab Diagnostics	LFIA	119	Cobas-Roche Reflotron	LFIA
57	Concile	LFIA	120	Bioscitec	LFIA—Devices
58	Epitope Diagnostics	LFIA	121	ABAXIS—Piccolo Xpress	Lab on a Disk
59	Eurodiagnostica	LFIA	122	Cobas-Roche b101	Lab on a Disk
60	Schebo Biotech	LFIA	123	Biosurfit Spinit	Lab on a Disk
61	Biomaxima	LFIA	124	Nexus—Samsung LabGeo IB10	Lab on a Disk
62	Orgentech	LFIA	125	Nexus—Samsung LabGeo PA20	Lab on a Disk
63	Accessbio	LFIA			

14.4 Conclusions

In this review, a total of 104 PoC companies have been reviewed as per their technology, design, and target analytes/applications. We identified five distinct technological approaches relating to the PoC design configurations and sensing decisions. These are (1) lateral-flow tests with or without a reader system, (2) centrifugal systems (LOAD) with optical sensing, (3) handheld or small benchtop electrochemical systems, (4) handheld or small benchtop nucleic acid testing systems, and (5) benchtop blood gas and electrolyte analysis systems. One thing that immediately becomes evident is that simpler approaches such as standalone LFIA tests or direct electrochemical strip-based systems greatly outnumber more sophisticated systems such as lab-on-a-disk or cartridge-based microfluidic devices, presumably because it is a well-established, accessible technology with off the shelf-production equipment available for most applications. This is also valid for the number of analytes being detected by such systems. However, clinical biochemistry and hematology measurements, but also all measurements for which the objective is high quality quantitative testing in panels, are mostly designated to more complex systems. It is also evident that from a large pool of fluidic and sensing methodologies reviewed in literature, only a few seem to make their way into commercial applications. Centrifugal microfluidics, capillary microfluidics, and active pump-based systems seem to have become the gold standards when it comes to reagent and sample handling. Electrochemical sensing is the prominent detection method for blood gases, electrolytes, and metabolites. Immunochromatography, absorbance, reflectance, and fluorescence are mostly applied for immunoassays; however, there are targets for which both methods are used. In addition, the implementation of user-friendly embellishments (device integrated lancets, modified protocols to allow sample pretreatment in the device) defines a focus point of commercial systems that is seldom seen in research efforts: usability might outweigh technology refinement or even new, groundbreaking principles of operation. Moreover, it is clear that no single approach is better than all others for all settings and all applications. Quantitation, detection limits, and repeatability are important specifications, although it is often the case that an extremely sensitive system with unprecedented resolution might not justify its development costs when examined against its impact in clinical practice. In the specific field of immunoassays, we identify two trends, the lateral flow technology and the automated assay in a device approach. In LFIA, it is pretty much established that the vision is a one-step assay for all applications. As such, efforts are being directed into improving the antibody performance, the labels, and the instrumented quantification methods. In automated device-based immunoassay PoC systems, the efforts are also directed into assimilating multistep and multiplex lab protocols, with a successful example being the LOAD. Nucleic acid testing PoC systems are emerging as a highly sensitive and specific method for infectious disease differential diagnosis with the focus being on reagents, the amplification, and the detection methodology. The targeted biomarkers of PoC systems are numerous and each has its own impact in a wide range of diagnostic settings: β-hCG, Tp-I/CK-MB/Myoglobin, fecal occult blood, *Helicobacter pylori*, influenza A/B, HIV 1/2, Malaria species, and glucose are the most prominent diagnostic targets of PoC. As a concluding remark, it is the authors' viewpoint that innovative and applicable ideas in the field of PoC systems engineering have yet a lot to offer to this field as long as their implementation is realistic, and their merits outweigh the effort and cost of their development and production.

References

[1] Tang M., Wang G., Kong S.K., Ho H.P. 'A review of biomedical centrifugal microfluidic platforms'. *Micromachines*. 2016;**7**(2):26.

[2] Tokel O., Inci F., Demirci U. 'Advances in plasmonic technologies for point of care applications'. *Chemical Reviews*. 2014;**114**(11):5728–52.

[3] Vashist S.K., Luppa P.B., Yeo L.Y., Ozcan A., Luong J.H.T. 'Emerging technologies for next-generation point-of-care testing'. *Trends in Biotechnology*. 2015;**33**(11):692–705.

[4] Warsinke A. 'Point-of-care testing of proteins'. *Analytical and Bioanalytical Chemistry*. 2009;**393**(5):1393–405.

[5] Wu L., Qu X. 'Cancer biomarker detection: recent achievements and challenges'. *Chemical Society Reviews*. 2015;**44**:2963–97.

[6] Yetisen A.K., Akram M.S., Lowe C.R. 'Paper-based microfluidic point-of-care diagnostic devices'. *Lab on a Chip*. 2013;**13**(12):2210–51.

[7] Kim L. 'Overview of the microfluidic diagnostics commercial landscape'. *Methods in Molecular Biology*. 2013;**949**:65–83.

[8] Volpatti L.R., Yetisen A.K. 'Commercialization of microfluidic devices'. *Trends in Biotechnology*. 2014;**32**(7):347–50.

[9] Haber C. 'Microfluidics in commercial applications; an industry perspective'. *Lab on a Chip*. 2006;**6**(9):1118–21.

[10] Medica Tradefair. 2015. Available from http://www.medica-tradefair.com/

[11] Point-Of-Care Diagnostic Market worth $27.5 Billion by 2018. 2014. Available from http://www.marketsandmarkets.com/PressReleases/point-of-care-diagnostic.asp

[12] Michael I., Kim T.H., Sunkara V., Cho Y.K. 'Challenges and opportunities of centrifugal microfluidics for extreme point-of-care testing'. *Micromachines*. 2016;**7**(2):32.

[13] Andreasen S.Z., Kwasny D., Amato L., *et al*. 'Integrating electrochemical detection with centrifugal microfluidics for real-time and fully automated sample testing'. *RSC Advances*. 2015;**5**(22):17187–93.

[14] Kim T.H., Abi-Samra K., Sunkara V., *et al*. 'Flow-enhanced electrochemical immunosensors on centrifugal microfluidic platforms'. *Lab on a Chip*. 2013;**13**(18):3747.

[15] Posthuma-Trumpie G.A., Korf J., van Amerongen A. 'Lateral flow (immuno) assay: its strengths, weaknesses, opportunities and threats. A literature survey'. *Analytical and Bioanalytical Chemistry*. 2009;**393**(2):569–82.

[16] O'Farrell B. 'Lateral flow immunoassay systems: evolution from the current state of the art to the next generation of highly sensitive, quantitative rapid assays', in Wild D. (ed.). *The Immunoassay Handbook*. Amsterdam: Elsevier Ltd.; 2013. pp. 89–107.

[17] Mak W.C., Beni V., Turner, A.P.F. 'Lateral-flow technology: from visual to instrumental'. *Trends in Analytical Chemistry*. 2016;**79**:297–305.

[18] Sajid M., Kawde A.N., Daud M. 'Designs, formats and applications of lateral flow assay: a literature review'. *Journal of Saudi Chemical Society*. 2015;**19**(6):689–705.

[19] Chun P. 'Colloidal gold and other labels for lateral flow immunoassays'. In: *Lateral Flow Immunoassay*. Totowa, NJ: Humana Press; 2009. pp. 1–19.

[20] Bangs Laboratories. Lateral Flow Tests. 2013. Available from https://www.google.gr/url?sa=t&rct=j&q=&esrc=s&source=web&cd=1&ved=0ahUKEwj8te6a9M_MAhXCC5oKHTGVAHkQFggbMAA&url=https%3A%2F%2Fwww.researchgate.net%2Ffile.PostFileLoader.html%3Fid%3D549e3a43d5a3f202388b4576%26assetKey%3DAS%253A273658700861477%2540144225672

[21] Eltzov E., Guttel S., Kei A.L.Y., Sinawang P.D., Ionescu R.E., Marks R.S. 'Lateral flow immunoassays—from paper strip to smartphone technology'. *Electroanalysis*. 2015;**27**(9):2116–30.

[22] Pfab T., Franz U., Herfeld F., Lun A., Armbruster F., Hocher B. 'Rapid immunochromatographic strip test for the detection of albuminuria and brief literature review on albuminuria screening'. *European Journal of Medical Research*. 2006;**11**(1):3–6.

[23] Mansfield M.A. 'The use of nitrocellulose membranes in lateral-flow assays', in Wong R.C., Tse H.Y. (eds.). *Forensic Science and Medicine: Drugs of Abuse: Body Fluid Testing*. Totowa, NJ: Humana Press Inc.; 2005. pp. 71–85.

[24] Linares E.M., Kubota L.T., Michaelis J., Thalhammer, S. 'Enhancement of the detection limit for lateral flow immunoassays: evaluation and comparison of bioconjugates'. *Journal of Immunological Methods*. 2012;**375**(1–2):264–70.

[25] Brown M.C. 'Antibodies: key to a robust lateral flow immunoassay'. In: *Lateral Flow Immunoassay*. Totowa, NJ: Humana Press; 2009. pp. 1–16.

[26] Jayasena S.D. 'Aptamers: an emerging class of molecules that rival antibodies in diagnostics'. *Clinical Chemistry*. 1999;**45**(9):1628–50.

[27] Bruno J. 'Predicting the uncertain future of aptamer-based diagnostics and therapeutics'. *Molecules*. 2015;**20**(4):6866–87.

[28] Chen A., Yang S. 'Replacing antibodies with aptamers in lateral flow immunoassay'. *Biosensors and Bioelectronics*. 2015;**71**:230–42.

[29] Millipore. Rapid Lateral Flow Test Strips: Considerations for Product Development. 2008. Available from papers3://publication/uuid/F7A438B7-0365-4DA7-83B5-4375A96FD0C4

[30] Juarez S.I., Nuñez A.E., Aranda M.M., Mojica D., Kim A.A., Parekh B. 'Field evaluation of 4 rapid tests for diagnosis of HIV infection in Panama'. *Journal of Clinical Microbiology*. 2016;**54**(4):1127–9.

[31] Delaney K.P., Branson B.M., Uniyal A., *et al.* 'Evaluation of the performance characteristics of 6 rapid HIV antibody tests'. *Clinical Infectious Diseases*. 2011;**52**(2):257–63.

[32] Stürenburg E., Junker R. 'Point-of-care testing in microbiology: the advantages and disadvantages of immunochromatographic test strips'. *Deutsches Ärzteblatt International*. 2009;**106**(4):48–54.

[33] Cunningham J., Hasker E., Das P., *et al.* 'A global comparative evaluation of commercial immunochromatographic rapid diagnostic tests for visceral leishmaniasis'. *Clinical Infectious Diseases*. 2012;**55**(10):1312–19.

[34] Ramparany L., Ramirez J., Nizou J.-Y., Le Saux D., Richard V., Talarmin A. 'Evaluation of four rapid immunochromatographic tests for the detection of cardiac troponin I'. *Clinical and Vaccine Immunology*. 2011;**18**(3): 414–17.

[35] Monti A. PRIMA LAB SA: www.primahometest.com. Easy to Use Rapid Self-Test for the Detection of Total Cholesterol Level in Blood. Instructions for Use. Available from https://primahometest.com/products/1/pht_ifu_chol_en.pdf. Accessed July 4, 2016.

[36] Byrnes S., Thiessen G., Fu E. 'Progress in the development of paper-based diagnostics for low-resource point-of-care settings'. *Bioanalysis*. 2013; **5**(22):2821–36.

[37] Chaumat P. *Method for diagnosing and differentiating HIV-2 infections*. Patent WO2013178737 A1, 2013.

[38] Esfandiari J. *Dual path immunoassay device*. Patent US7189522 B2, 2006.

[39] Anderson E V., Lapointe J, Martinez, R, *et al. Point of care diagnostic systems*. Patent US6394952 B1, 1998.

[40] Eriksson M. *Initial Evaluation of the Meritas Troponin I test for Measurement of Equine Cardiac Troponin I*. Report, 2014.

[41] Mahajan V.S., Jarolim P. 'How to interpret elevated cardiac troponin levels'. *Circulation*. 2011;**124**(21):2350–4.

[42] Yen C.W., de Puig H., Tam J.O., Gómez-Márquez J. 'Multicolored silver nanoparticles for multiplexed disease diagnostics: distinguishing dengue, yellow fever, and Ebola viruses'. *Lab on a Chip*. 2015;**15**(7):1638–41.

[43] Strohmeier O., Keller M., Schwemmer F., *et al.* 'Centrifugal microfluidic platforms: advanced unit operations and applications'. *Chemical Society Reviews*. 2015;**44**:6187–229.

[44] Ducrée J. 'Centrifugal microfluidics'. In *Encyclopedia of Microfluidics and Nanofluidics SE—203*. 2008;(Figure 1):234–245. doi:10.1007/978-0-387-48998-8_203.

[45] Brenner T., Glatzel T., Zengerle R., Ducree J. 'A flow switch based on coriolis force'. Presented at *Seventh International Conference on Miniaturized Chemical and Blochemlcal Analysts Systems*; Squaw Valley, CA, 2003.

[46] Feng Y., Zhou Z., Ye X., Xiong J. 'Passive valves based on hydrophobic microfluidics'. *Sensors Actuators A: Physical*. 2003;**108**(1):138–43.

[47] Andersson P., Ekstrand G. *Retaining microfluidic microcavity and other microfluidic structures*. Patent US7300199 B2, 2003.

[48] Burd T.L., Schembri C.T. *Reagent container for analytical rotor*. Patent US5304348, 1994.

[49] U.S. Pat. Nos. 5,061,381; 5,173,193; 5,186,844; 5,122,284.

[50] Schembri C.T., Burd T.L., Kopf-Sill A.R., Shea L.R., Braynin B. 'Centrifugation and capillarity integrated into a multiple analyte whole blood analyser'. *Journal of Automatic Chemistry*. **17**(3):99–104.

[51] Schembri C.T. *Analytical rotor with dye mixing chamber*. Patent US5472603 A, 1995.

[52] Bernstein D.M., Lingane P.J., Nagle R., Ostoich V.E. *Methods for photometric analysis*. Patent US5478750 A, 1995.

[53] Schembri C.T., Ostoich V., Lingane P.J., Burd T.L., Buhl S.N. 'Portable simultaneous multiple analyte whole-blood analyzer for point-of-care testing'. *Clinical Chemistry*. 1992;**38**(9):1665–70.

[54] Abaxis—Piccolo Xpress. Available from http://www.piccoloxpress.com/

[55] Boehm C., Lutz S., Spinke J. *Rotatable cartridge with a metering chamber for analyzing a biological sample*. Patent WO2015185763 A1, 2015.

[56] Boehm C., Lutz S., Dolbinow T. *Rotatable cartridge for measuring a property of a biological sample*. Patent WO2015173097 A1, 2015.

[57] Boehm C., Lutz S. *Rotatable cartridge for analyzing a biological sample*. Patent WO2015185226 A1, 2015.

[58] Cobas b 101. Available from http://www.cobas.com/home/product/point-of-care-testing/cobas-b-101-poc-system.html

[59] Pedro Santos Manso Côrte-Real J., Correia de Matos Nolasco Lamas F. *Focusing method*. Patent WO2016050767 A1, 2016.

[60] Correia de Matos Nolasco Lamas F., Manuel de Oliveira Garcia da Fonseca J., Do Rosário Magalhães A., Santos Manso Côrte-Real J.P. *Cell counting*. Patent WO2016050755, 2016.

[61] Burger R., Reis N., da Fonseca J.G., Ducree J. 'Plasma extraction by centrifugo-pneumatically induced gating of flow'. *Journal of Micromechanics and Microengineering*. 2013;**23**(3):035035.

[62] da Fonseca J.G., Reis Nuno A.E. *Liquid distribution and metering*. Patent WO2011122972 A3, 2011.

[63] Manuel de Oliveira Garcia da Fonseca J., Reis Nuno A.E., Burger R. *Analytical rotors and methods for analysis of biological fluids*. Patent US8440147 B2, 2013.

[64] Manuel de Oliveira Garcia da Fonseca J. *Photometric device and method*. Patent WO2013079619, 2013.

[65] Biosurfit—Spinit. Available from http://biosurfit.com/solution/

[66] Point Grey Application Example—Biosurfit Spinit. Available from http://www.ptgrey.com/case-study/id/10414

[67] St John A., Price C.P. 'Existing and emerging technologies for point-of-care testing'. *Clinical Biochemist Reviews*. 2014;**35**(3):155–67.

[68] Yoo J.C. *Bio-disc reading apparatus and assay method using same*. Patent EP2241893 A2, 2010.

[69] Samsung LabGeo IB 10. Available from http://www.samsung.com/global/business/healthcare/healthcare/in-vitro-diagnostics/BCA-IB10/DE

[70] Egan R.L. Better and Faster: Improving Rapid Point-of-Care Testing. IVD Technology; 2009.

[71] Feng Y., Zhou Z., Ye X., Xiong J. 'Passive valves based on hydrophobic microfluidics'. *Sensors Actuators A: Physical*. 2003;**108**(1–3):138–43.

[72] Chew M., Teo W., Xie L., *et al*. 'Study of a capillary force driven passive valve for a microfluidic package'. In *Electronics Packaging Technology Conference*. Singapore: Pan Pacific Hotel; 2006:448–53.

[73] Avram M., Avram A.M., Bragaru A., Ghiu A., Iliescu C. 'Plasma surface modification for selective hydrophobic control'. *Romanian Journal of Information Science and Technology*. 2008;**11**(4):409–22.

[74] Arkles B. 'Hydrophobicity, hydrophilicity and silanes'. *Paint and Coatings Industry*. 2006;**22**(10):114.

[75] Andersson H., Van der Wijngaart W., Griss P., Niklaus F., Stemme G. 'Hydrophobic valves of plasma deposited octafluorocyclobutane in DRIE channels'. *Sensors Actuators B: Chemical*. 2001;**75**(1–2):136–41.

[76] Ellinas K., Tsougeni K., Petrou P.S., Kakabakos S.E., Tserepi A., Gogolides E. 'Control of flow and protein adsorption on plasma nanotextured polymeric microfluidics'. *Proceedings of the Third European Conference on Microfluidics*; Heidelberg, December 2012.

[77] Schnabl K.L., Tarek M., Cursio C., Yip P.M. 'Evaluation of the Piccolo Xpress chemistry analyzer for point of care testing of plasma liver function markers, electrolytes and metabolites'. *Clinical Biochemistry*. 2008;**41**(14):1284–5.

[78] Owen W.E., Caron J.E., Genzen J.R. 'Liver function testing on the Abaxis Piccolo Xpress: use in Ebola virus disease protocols'. *Clinica Chimica Acta*. 2015;**446**:119–27.

[79] Akbas N., González G., Edwards R., Devaraj S. 'Assessment of liver function tests on Piccolo Xpress point of care chemistry analyzer in a pediatric hospital'. *Practical Laboratory Medicine*. 2015;**3**:1–7.

[80] van Gammeren A.J., van Gool N., de Groot M.J.M., Cobbaert C.M. 'Analytical performance evaluation of the Cobas 6000 analyzer—special emphasis on trueness verification'. *Clinical Chemistry and Laboratory Medicine*. 2008;**46**(6):863–71.

[81] Simpson P., Jolly L., Tirimacco R., Gill J., Tideman P. 'Evaluation of the lipid panel on the abaxis Piccolo Xpress to determine its potential as a point-of-care instrument in a nonlaboratory setting'. *Point of Care The Journal of Near-Patient Testing and Technology*. 2010;**9**(1):32–5.

[82] Piccolo Xpress product inserts. Available from http://www.piccoloxpress.com/products/panels/pls-msds/

[83] Beckman Coulter—Alanine Aminotransferase Product insert. Available from https://www.beckmancoulter.com/wsrportal/techdocs?docname=/cis/BAOSR6x07A/%25%25/EN_ALANINEAMINOTRANSFERASE(ALT).pdf. Accessed June 3, 2016.

[84] 510(k) Substantial equivalence determination decision summary assay only template. Available from http://www.accessdata.fda.gov/cdrh_docs/reviews/K071211.pdf. Accessed June 3, 2016.

[85] Criel M., Jonckheere S., Langlois M. 'Evaluation of three hemoglobin A1c point-of-care instruments'. *Clinical Laboratory*. 2016;**62**(3):285–91.

[86] Lenters-Westra E., Slingerland R.J. 'Three of 7 hemoglobin A1c point-of-care instruments do not meet generally accepted analytical performance criteria'. *Clinical Chemistry*. 2014;**60**(8):1062–72.

[87] Sreenan S., Tormey W. 'American Diabetes Association recommendations on haemoglobin A1c use in diabetes diagnosis: time to include point-of-care devices?' *Annals of Clinical Biochemistry*. 2016;**53**(5):620–4.

[88] Jonckheere S., Langlois M. 'Evaluation of the Roche cobas b101 point-of-care instrument for glycated hemoglobin testing in an ambulant hospital setting'. *Conference paper*. 2013.

[89] Biosurfit Spinit-BC, package insert. PL-002-05, Revision 4, 1/2016.

[90] Biosurfit—CRP product insert. PL-001-05, Revision 11, 3/2015.

[91] *Samsung LabGeo IB10 Brochure*. CT-LABGEO IB10 4.5-151125-EN.

[92] Ronkainen N.J., Halsall H.B., Heineman W.R. 'Electrochemical biosensors'. *Chemical Society Reviews*. 2010;**39**: 1747–1763.

[93] Wang Y., Xu H., Zhang J., Li G. 'Electrochemical sensors for clinic analysis'. *Sensors*. 2008;**8**(4):2043–81.

[94] Daniels J.S., Pourmand N. 'Label-free impedance biosensors: opportunities and challenges'. *Electroanalysis*. 2007;**19**(12):1239–57.

[95] Lauks I.R. 'Microfabricated biosensors and microanalytical systems for blood analysis'. *Accounts of Chemical Research*. 1998;**31**(5):317–24.

[96] Martin C.L. 'i-STAT—combining chemistry and haematology in PoCT'. *Clinical Biochemist Reviews*. 2010;**31**(3):81–4.

[97] Lauks I., Maczuszenko A. *Diagnostic devices incorporating fluidics and methods of manufacture*. Patent US7842234 B2, 2010.

[98] Choi S. 'Powering point-of-care diagnostic devices'. *Biotechnology Advances*. 2016;**34**(3):321–30.

[99] Wang X., Gartia M.R., Jiang J., *et al*. 'Audio jack based miniaturized mobile phone electrochemical sensing platform'. *Sensors Actuators B: Chemical*. 2015;**209**:677–85.

[100] Dario Smart Meter. Available from http://mydario.co.uk/solution/

[101] Turner A.P.F. 'Biosensors: sense and sensibility'. *Chemical Society Reviews*. 2013;**42**(8):3184–96.

[102] Ohara T.J., Rajagopalan R., Heller A. 'Glucose electrodes based on cross-linked [Os(bpy)2Cl]+/2+ complexed poly(1-vinylimidazole) films'. *Analytical Chemistry*. 1993;**65**(23):3512–17.

[103] Gorton L., Lindgren A., Larsson T., Munteanu F.D., Ruzgas T., Gazaryan I. 'Direct electron transfer between heme-containing enzymes and electrodes as basis for third generation biosensors'. *Analytica Chimica Acta*. 1999; **400**(1):91–108.

[104] Bakker E., Qin Y. 'Electrochemical sensors'. *Analytical Chemistry*. 2006;**78**(12):3965–84.

[105] Freckmann G., Baumstark A., Jendrike N., *et al*. 'System accuracy evaluation of 27 blood glucose'. *Diabetes Technology and Therapeutics*. 2010;**12**(3):221–31.

[106] Freckmann G., Schmid C., Baumstark A., Pleus S., Link M., Haug C. 'System accuracy evaluation of 43 blood glucose monitoring systems for self-monitoring of blood glucose according to DIN EN ISO 15197'. *Journal of Diabetes Science and Technology*. 2012;**6**(5):1060–75.

[107] ISO 15197:2003. 2003. Available from http://www.diabetes.co.uk/blood-glucose-meters/iso-accuracy-standards.html

[108] Heinemann L. 'Quality of glucose measurement with blood glucose meters at the point-of-care: relevance of interfering factors'. *Diabetes Technology and Therapeutics*. 2010;**12**(11):847–57.

[109] Hellman R. 'Glucose meter inaccuracy and the impact on the care of patients'. *Diabetes/Metabolism Research and Reviews*. 2012;**28**(3):207–9.

[110] *Alere IN Ratio Report*. 2013. Available from http://www.alere.com/en/home/product-details/inratio-pt-inr-monitoring-systems-us.html?c=US

[111] Fernando R., Ak J., Kennedy S., *et al*. *PT/INR Test Performance of the Xprecia Stride Coagulation Analyzer on Capillary Blood Is Equivalent to a Reference Laboratory Hemostasis System PT/INR Test Performance of the Xprecia Stride Coagulation Analyzer on Capillary Blood Is Equivalent to a Re*.; 2015.

[112] Cobas—Roche : CoaguChek XS Plus & Pro Systems. 2014, Roche.

[113] Abbott I-STAT : PROTHROMBIN TIME/(PT/INR). Article 715236-01M, Revision Date: 01-Jul-13.

[114] Alere—Clinical Guidelines for PT/INR Comparisons; 2007.

[115] García-Arroyo L., Prim N., Martí N., Roig M.C., Navarro F., Rabella N. 'Benefits and drawbacks of molecular techniques for diagnosis of viral respiratory infections. Experience with two multiplex PCR assays'. *Journal of Medical Virology*. 2016;**88**(1):45–50.

[116] Zanoli L.M., Spoto G. 'Isothermal amplification methods for the detection of nucleic acids in microfluidic devices'. *Biosensors*. 2013;**3**(1):18–43.

[117] Holland C.A., Kiechle F.L. 'Point-of-care molecular diagnostic systems—past, present and future'. *Current Opinion in Microbiology*. 2005;**8**(5):504–9.

[118] Niemz A., Ferguson T.M., Boyle D.S. 'Point-of-care nucleic acid testing for infectious diseases'. *Trends in Biotechnology*. 2011;**29**(5):240–50.

[119] Yan L., Zhou J., Zheng Y., *et al*. 'Isothermal amplified detection of DNA and RNA'. *Molecular Biosystems*. 2014;**10**(5):970–1003.

[120] Mori Y., Notomi T. 'Loop-mediated isothermal amplification (LAMP): a rapid, accurate, and cost-effective diagnostic method for infectious diseases'. *Journal of Infection and Chemotherapy*. 2009;**15**(2):62–9.

[121] New England Biolabs—Isothermal amplification. 2015. Available from https://www.neb.com/applications/dna-amplification-and-pcr/isothermal-amplification

[122] Cepheid – molecular diagnostics. Available from http://www.cepheid.com/en/
[123] *Alere – Influenza A & B Package Insert*; 2014.
[124] Alere—Strep A Package Insert. IN733000 Rev. 2, 09/2015.
[125] Fairley D.J., Fairley D., McKenna J., *et al*. Rapid and accurate detection of *Neisseria meningitidis* DNA in clinical specimens using the Hibergene HG Meningococcus LAMP assay. Available from http://www.hibergene.com/wp-content/uploads/ECCMID_2016_ePoster_EP0217.pdf. Accessed July 1, 2016.
[126] Idylla IFV-RSV Panel Clinical Testing (RP1). 2016. Available from https://clinicaltrials.gov/ct2/show/study/NCT02786381
[127] González A.L., Waddell L.S. 'Blood gas analyzers'. *Topics in Companion Animal Medicine*. 2016; **31**(1):27–34.
[128] OPTIMedical—OPTI CCA-TS2. Available from http://www.optimedical.com/products-services/opti-CCA-TS2.html
[129] Instrumentation Laboratory – GEM premier 4000. Available from http://www.instrumentationlaboratory.com/products-services/critical-care-diagnostics/instruments/gem-premier-4000-with-iqm.aspx
[130] Siemens—RAPIDPoint 500 Systems. Available from http://usa.healthcare.siemens.com/point-of-care/blood-gas/rapidpoint-500-systems. Accessed July 5, 2016.
[131] Roche—Cobas b 123. Available from http://www.cobas.com/home/product/point-of-care-testing/cobas-b-123-poc-system.html
[132] Roche—Cobas b 221. Available from http://www.cobas.com/home/product/point-of-care-testing/cobas-b-221-system.html
[133] Accriva Diagnostics—Hemochron Signature Elite. Available from http://www.accumetrics.com/products/hemochron-signature-elite-whole-blood-microcoagulation-system
[134] Harris L.F., Castro-López V., Killard A.J. 'Coagulation monitoring devices: past, present, and future at the point of care'. *Trends in Analytical Chemistry*. 2013;**50**:85–95.
[135] Coagusense—Coag-Sense. Available from http://www.coagusense.com/
[136] Burtis C.A., Bruns D.E. *Tietz Fundamentals of Clinical Chemistry and Molecular Diagnostics*. Elsevier Health Sciences; 2014.
[137] U.S. Department of Health & Human Services, Save Request, Mitsubishi Chemical Mediance Co. 2012.
[138] Stratus CS 200 Acute Cardiac Care Troponin Analyzer. Available from http://www.healthcare.siemens.com/point-of-care/cardiac/stratus-cs-200-acute-care. Accessed July 4, 2016.
[139] Siemens. Siemens—DCA Vantage System Reagent Kit product insert. 2008:4.
[140] Alere Afinion HbA1c. Available from http://www.alere.com/en/home/product-details/afinion-hba1c.html
[141] Diagon—Coag S system. Available from http://www.diagon.com/en/pocreagent/coag-s-inr-test-kit

[142] Antal J., Neubrandt J., Pető M. *Procedure and measuring system for determining blood coagulation characteristics*. Patent WO2015040441 A1, 2015.

[143] *Alere HemoPoint H2—Product Insert*. 2012. Available from http://www.hmscweb.com/PDF_Files/HemoPoint/HemoPoint_H2_Microcuvettes_Package_Insert.pdf. Accessed July 4, 2016.

[144] Alere Hemopoint H2. Available from http://www.alere.com/en/home/products-services/brands/hemopoint.html

[145] Assay device for direct measurement of LDL cholesterol. 2005. Available from https://patents.google.com/patent/US7772007B2/en

[146] Method for the determination of HDL cholesterol by means of a rapid diagnostic agent with an integrated fractionating step. 2000. Available from https://patents.google.com/patent/US6171849B1/en

[147] High-density lipoprotein assay device and method. Patent US 7582484 B2.

[148] Alere Cholestech. Available from http://www.alere.com/en/home/products-services/brands/cholestech.html

[149] *Alere Cholestech LDX GLU Product Insert*. 26196en Rev. B 7/2013.

[150] Menarini Diagnostics Pocketchem BA. Available from http://www.menarinidiagnostics.com/Products/point_of_care_testing/pocketchem_ba

[151] Francesco C., Berti B., Paoli C. 'Immunoenzymatic analysis device for once-only use, and the relative analytical method field'. 2002. Patent US 2002/0164779 A1.

[152] Diesse—Chorus. Available from http://www.diesse.it/en/Instruments/id:39/

Chapter 15

Consumer diagnostics

Spyridon E. Kintzios[1]

15.1 Introduction

An extensive overview of the progress in the commercialization of portable biosensors is given in the present chapter. The survey is focused on these systems and technologies that are either already incorporated in commercial products or are very close to be commercialized, being characterized with a Technology Readiness Level value of 7 or higher. Updated information is provided in the following sectors: (i) optimization and advances in conceptual approaches and operating principles, (ii) novel applications of commercial handheld/portable biosensors, and (iii) novel biomarkers that could be used as targets for emerging biosensor platforms.

15.2 Improvements in working principles/assay concepts

15.2.1 Breath analysis biosensors

Significant research has been focused on the development of point-of-test (POT) systems for detecting volatile organic compounds (VOCs) in breath. It is self-explainable that exhaled breath gas analysis is a textbook noninvasive diagnostic procedure, even more than urine or saliva testing. The development of breath-based diagnostic systems and devices can be viewed as a natural evolution of the relatively recent use of animals as selective odor detectors, with applications ranging for early disease diagnosis (e.g., sensing of cancer or tuberculosis by dogs) to explosives identification (e.g., mapping of minefields by rats) to customized uses (e.g., truffle detection by pigs, drug screening by bees) [1–4]. From the approximately 3,000 VOCs identified in human breath, 200 have been associated with chronic disease, such as cancer and diabetes or even Alzheimer's disease [5]. Examples of these gas biomarkers are methanol, ethanol, limonene, acetone, ethylene, ethane, cyclohexanone, undecane, toluene, 2-butanone, 2,2-dimethyl-hexanal, and methylated hydrocarbons [6]. In spite of the superior sensitivity and high-throughput capacity of

[1]School of Food, Biotechnology and Development Director, Laboratory of Cell Technology, Department of Biotechnology, Agricultural University of Athens, Athens, Greece

the natural animal olfactory systems (able to detect a few molecules of a target analyte even tens of kilometers away), there are practical limitations to the everyday use of animals (including insects) as living biosensors: in particular, the requirement of extensive training as well as the dependence of the detection response on the animal's mood and breed makes the use of animals promising, but currently less attractive than artificial biosensor constructs.

Several breath biosensors, generically named "bio-sniffers" or "e-noses," have been developed [7–12]. Most of them are enzyme-based: For example, Arakawa *et al.* [11] developed a system for the direct measurement of ethanol concentration in breath in relation to the intensity of chemiluminescence resulting from luminol reaction induced by alcohol oxidase and horseradish peroxidase–luminol–hydrogen (H_2) peroxide system. Using this biosensor, the authors were able to classify breath samples according to gaseous ethanol from 30 (low level) to 400 ppm (very high level, probability of acute toxicosis). Kudo *et al.* [13] developed an optical biosensor for a nicotinamide adenine dinucleotide (NADH)-dependent formaldehyde determination. Following a similar approach, Ye *et al.* [12] built their acetone bio-sniffer system upon an NADH-dependent secondary alcohol dehydrogenase immobilized membrane, which was coupled to a fiber-optic NADH measurement system. The working range of the biosensor was 20–5,300 ppb. The authors were able to correlate the elevated acetone concentration in exhaled breath with increased lipolysis due to the exercise activity of the test subjects. Since lipid metabolism is also associated with pathological conditions, such as diabetes, an acetone bio-sniffer could probably be used as a user-friendly, patient-centered self-test. Capuano *et al.* [14] presented a novel biosensor for VOCs as biomarkers for lung cancers based on microarrays of porphyrin-coated quartz microbalances. Their methodological concept was based on mimicking, with the use of porphyrin, of cytochrome P450, a naturally occurring enzyme catalyzing the oxidation of VOCs in the diseased lung and possibly functioning as a VOC sensor, too. The novel e-nose was able to detect cancer cases with a high rate of success (93 percent) however could not discriminate between adenocarcinomas and squamous cancer cells. The success of breath biosensors is also dependent on the physical properties of the sample and the sampling process: For example, difference in the temperature between the internal and the external of the body may reduce the measured concentration of some VOCs due to condensation. It should be kept in mind that e-nose systems are not always successful, as, for example, shown in the case of detecting *Mycobacterium tuberculosis* in sputum headspace samples [15,16].

An entirely different approach is based on the expression of insect olfactory receptors in transfected mammalian cell lines, such as HEK293 [17], Sf21 [18], or *Xenopus laevis* melanophores [19]. Either the transfected cells or isolated receptors can be used as biorecognition elements in a variety of biosensor configurations, based on electrochemical [20–23] or optical working principles [24,25] for real-time monitoring of VOC distribution, even without sample precollection and concentration. Excellent reviews of the biosensory applications of the olfactory system are given by Sankaran *et al.* [26], Du *et al.* [27], and Wu *et al.* [28]. Such systems may offer superior sensitivity to enzyme-based ones, although they may also suffer from poor reproducibility due to the unpredictable pattern of cellular response against both the target VOCs and other sample components. A major effort to meet this challenge is the customized development of training of artificial intelligence software, more commonly artificial neural networks (ANNs), which can be employed to automatically recognize, classify, and interpret a response pattern associated with a given analyte, thus increasing the reliability of the olfactory biosensor at satisfactory levels, usually >90 percent [29–33]. Employment of tailored ANNs has proven to be a very powerful toll for cell-based biosensors, as it will be also shown in following for the FOODSCAN system.

15.2.2 Electrochemical and bioelectrical sensors

Several technological developments have independently contributed to further miniaturization of electrochemical biosensor systems, in parallel increasing the sensitivity, selectivity and speed of detection. A few examples will be briefly presented in following:

Kraus *et al.* [34] demonstrated an elegant bioelectrical chip design for the POC detection of anti-hepatitis C virus (anti-HCV) antibodies within just 20 min. The core of their approach was based on "Single Electrode Redox Cycling," that is, using alternating positive and negative potentials for signal detection on a single gold electrode, thus enhancing current signals up to 40-fold in comparison to simple oxidation. Detection of anti-HCV antibodies was feasible even in a whole-blood sample volume of just 2 μL. More recently, Tarasov *et al.* [35] developed a low-cost extended-gate field-effect transistor FET immunosensor for the detection of Bovine Herpes Virus 1 (BHV-1). The sensor was based on recombinant BHV-1 gE and provided results much faster than enzyme-linked immunosorbent assay (ELISA) (<10 min vs 19 h) with a limit of detection of $<2 \times 10^{-3}$, making them a promising tool that would contribute to the increase of the number of countries that have gained "BHV-1 free status" through the application of "test and cull" disease eradication programs for rapid diagnostic applications.

An increasing number of detection applications are based on electrochemical immunosensors using graphene nanostructures and/or carbon nanotubes (CNTs) and taking advantage of their large specific surface area and high electrical conductivity. Examples include assays for cancer biomarkers, such as cancer embryonic antigen [36], prostate-specific antigen (PSA) [37], and vascular endothelial growth factor [38], but also other disease markers, such as troponin [39] and cholesterol [40]. Assay times as low as 10 s have been reported using this approach, along with sub-picogram sensitivities. In spite of these promising features, graphene is still a challenging material for biosensor applications, due to the strong dependence of its conductivity on the number of active sites exposed and the surface-to-volume ratio, as well as the reduced stability of graphene-like structures in biological samples [41].

Equally advancing is the field of portable devices able to carry out Electrochemical Impedance Spectroscopy (EIS) assays. Until recently, EIS has been used only for in vitro toxicology and drug efficacy studies, in the context of high throughput albeit costly and elaborate instrumentation [42,43]. This is bound to change with the advent of portable EIS devices with dedicated analytical functions.

For example, Barton *et al.* [44] developed an impedance immunosensor for digoxin detection. Anti-digoxin antibodies were immobilized on a conductive polyaniline thin planar layer electropolymerized on the electrode surface. The limit of dioxin detection was 0.1 ng mL^{-1}. More recently, Hushegyi *et al.* [45] reported an impedimetric glycan-based biosensor for the detection of intact influenza viruses H3N2 with a very low limit of detection of 5 aM, corresponding to just 13 viral particles in a sample volume of 1 μL.

Finally, Ebrahimi and Alam [46] reported an electronic bacteria droplet sensor able to assess cell viability (and thus, bacteria concentration) through the indirect measurement of evaporation-induced stimulation of bacterial osmoregulation. The gradual evaporation of the bacteria-containing droplet causes the activation of osmoregulatory transporters in the bacterial cell membrane, leading to altered release of water and electrolytes in the surrounding medium, which is reflected on the measured system conductivity.

15.2.3 Optical biosensors

Photonic crystal slabs (PCSs) are waveguides with a periodic nanostructure in a high refractive index material. Jahns *et al.* [47] reported a handheld optical biosensor based on PCSs embedded on a disposable microfluidic chip. In a spatially multiplexed detection scheme, the binding of different blood biomarkers on the PCS surface was measured through the intensity change of reflected light.

Optical measurement of magnetic resonance, in particular proton spin resonance, in a miniaturized sensor is another recent development that may lead to more sensitive and simplified biomarker detection [48].

Joshi *et al.* [49,50] reported the development of a prototype solid-state localized surface plasmon resonance sensor based on gold nanoprisms for the label-free, amplification-less, and ultra-sensitive (at attomolar levels) detection of microRNAs (miRs) in human plasma in pancreatic cancer patients.

Advances in mobile/smartphone cameras have enabled the miniaturization of optical analytical systems (a detailed presentation of the merits of cellphone camera-assisted signal transduction is given in Chapter 1). For example, Connelly *et al.* [51] developed a handheld, disposable device (a "paper machine") that integrates paper microfluidics, and a multilayer structure allowing a central patterned paper strip to slide in and out of fluidic path, thus reducing the otherwise many pipetting steps in order to carry out sample preparation and loop-mediated isothermal amplification. Analyte detection is done with a mobile camera and a ultraviolet source. Using this prototype biosensor, the authors were able to detect *Escherichia coli* by means of *malB* gene determination with a limit of detection of just five cells.

15.2.4 DNA nanotechnology-based sensors

Existing DNA-based biosensors are distinguished for their superior selectivity and, quite often, sensitivity. These features have been proven valuable in several applications, for example the detection of pathogens in food, such as the aptamer-based, electrochemical detection of *Salmonella* sp. reported by Ma *et al.* [52]. DNA nanotechnology represents a springboard to considerably improve the use of DNA as biorecognition elements, by enabling the construction of DNA static nanostructures with designed geometry, thus resulting in dynamic DNA-devices with precisely controlled spatiochemical properties [53]. Critical for the implementation of this approach are the so-called triggers, which are overhangs with custom sequences in DNA or RNA strands able to initiate branch migration, which in turn would put the corresponding DNA nanodevice in motion (Figure 15.1). The resulting changes in geometrical configuration can be measured by an appropriate optical or electrical read-out system. The control of the geometrical changes of the nanodevices can be facilitated by molecular agents such as adenosine triphosphate or magnesium ions [54].

Prominently placed in the family of DNA nanodevices is the so-called DNA-origami. This is a rectangular-shaped platform of folded DNA allowing the creation of nonarbitrary two- and three-dimensional shapes when unfolded [55]. Applications of the origami technology include the detection of the PSA [56] and lung cancer-specific miRs [57].

DNAzymes are DNA molecules with catalytic properties. Although known for more than 20 years [58], they have only recently started to be used in biosensor-based detection applications. For example, Liu *et al.* [59] used pistol-like DNAzyme (PLDz), an oxidative DNA-cleaving DNAzyme for determining glucose in saliva and tears. For this purpose, they developed a dual enzyme biosensor operating in two stages: First, glucose is oxidized by glucose oxidase (GOx) to produce H_2 peroxide, which in turn activates PLDz to self-cleave in the presence of Mn^{2+}, Cu^{2+}, and Co^{2+} as cofactors. This truly noninvasive biosensor could be used for glucose determination at concentrations as low as 5 μM.

Yet another approach was recently reported by Aliberti *et al.* [60], who developed a novel high-sensitive fluorescence detection of small (up to 100 nt) nucleic acids based on the conjugation of double-strand probes onto polymeric poly(ethylene glycol) (PEG)-based microgels. It was possible to detect DNA sequences from HIV, HCV, and

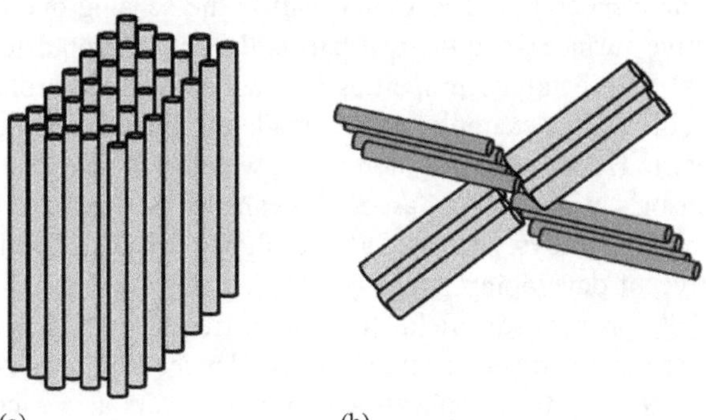

Figure 15.1 Examples of purpose-designed different geometrical configurations (a and b) of dynamic DNA devices resulting from shape-complementary static nanostructures

severe acute respiratory syndrome corona viruses as low as 1.4 fM. The key factor attributing to the success of this method is enhancement of sensitivity of detection caused by cyanine Cy5 dye fluorescence emission after confinement of the reacting parties on the microgel. Another advantageous feature is the use of PEG as a hydrophilic polymeric background able both to enhance PEG nucleic acid hybridization compared to other materials and having anti-fouling properties.

15.2.5 Ultraminiaturized and endoscopic biosensors

Eventually, progress in biosensor miniaturization will result into devices so small that can be easily implanted into the body as a tool for real-time monitoring of the distribution patterns of various biomarkers and/or drugs. Recent advances in this field allow for a projection of the realization of such microscopic, yet fully functional systems in the very near future.

A working example is given by Tahirbegi *et al.* [61] who developed a micrometric (8 mm in diameter) beryllium copper 12× microelectrode array for endoscopic measuring of ischemia in the stomach, based on the potentiometric determination of pH and nitrate ion concentration. Effective insulation of the microelectrode array from the environment was achieved by covering them with the biocompatible Epoxy 301-2 resin. Progress in implantable device technology may further benefit from advances in signal transduction outside the body, for example, by means of miniature antennas able to transmit and receive data at relatively long distances (several feet from their location) [62]. Injectable flexible semiconductor nanoscale electronic scaffolds are able to monitor neural activity represent another fascinating approach [63,64].

Even more revolutionary is the perspective of fabricating robotic microsensors with embedded micromotors. There are several reports in the literature revealing the considerable progress done in this direction. Micomotor designs are based on various materials, such as multilayered metallic nanowires or microtubules [65,66]. Of particular interest are the so-called Janus micromotors (Figure 15.2), composing of two parts ("hemispheres"): The frontal part is the sensing one (i.e., a functionalized conductive surface), and the rear part is the actual propulsion unit, usually made of a metal with catalytic properties in reaction with environmental stimuli that serve as fuel [67]. For example, motors made of zinc can reduce H_2 ions (e.g., provided by gastric HCl) to gaseous atomic H_2, which will provide the momentum for the microsensor's motion [68]. Gaseous H_2 can also be produced by the reaction of water with metallic alloys, such as Ga/Al or Pt/Mg [69–71]. Even more realistic is the perspective of developing microsensors guided with the aid of an external magnetic [72,73] or acoustic field (e.g. high-frequency ultrasound) [74,75]. Undoubtedly, here is a series of engineering challenges that need to be answered before achieving the practical application of microsensors as a routine diagnostic tool. Most of them are related to the viscosity and complex composition of biological fluids, in particular blood. On the other hand, considerations about the possible toxicity of microsensors, their components, and/or dedicated fuel (in particular, H_2O_2 at high concentrations) cannot be easily overlooked.

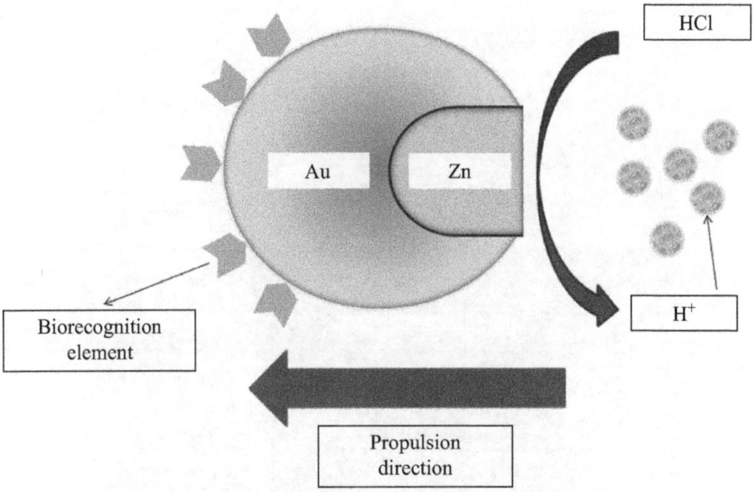

Figure 15.2 *Schematic representation of the working principle of a Janus micromotor. The frontal (left "hemisphere") part is the sensing one (i.e., an Au surface functionalized with biorecognition elements such as antibodies or aptamers) and the rear (right "hemisphere") one is the propulsion unit, made of a metal with catalytic properties using as fuel environmental trigger molecules (e.g., hydrogen ions provided by gastric HCl)*

15.3 Recent examples of commercial biosensors

15.3.1 Food safety analysis

A number of systems have emerged in the last few years promising to challenge the conventional chromatography-based analytical instrumentation for the routine determination of chemical residues and other additives in foods and drinks. Santonico *et al.* [76] reported the application of the BIONOTE (BIOsensor-based multisensorial system for mimicking Nose, Tongue, and Eyes) biosensor system for the simultaneous analysis of the vapor and liquid phase of olive oil samples as a rapid tool for the high throughput and sensitive screening for adulterants. The biosensor comprised a dual quartz microbalance (for gas analysis)/electrochemical array (for liquid analysis). By using BIONOTE, the authors were able not only to deduce the quality of each sample but also to track its origin (e.g., by establishing individual response profiles of 12 Italian extra virgin olive oils originating from different Apulian neighboring olive tree orchards). The application of biosensors to food authenticity verification process can create considerable added value to food analytics and generate incentives for adopting such novel approaches by the food industry.

Another recent system is FOODSCAN (foodscan.net) (Figure 15.3), a commercial cell-based bioelectric sensor able to detect organophosphate, carbamate, and pyrethroid pesticide residues in selected food matrices. The working principle

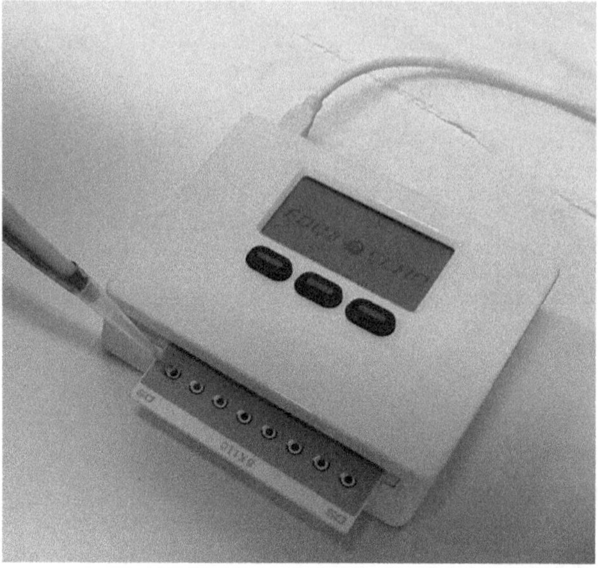

Figure 15.3 The FOODSCAN system: liquid samples are applied on the biosensing surface, comprising of arrays of cells on electrodes, which measure changes in the electrical potential of the cell membrane in a characteristic, "signature-like" pattern corresponding, each time, to the particular target analyte. Pattern classification is facilitated by a dedicated artificial neural network software. The entire analytical process is completed in less than 5–6 min

of FOODSCAN is based on the bioelectric recognition assay technology (*see also* Chapter 6), which measures characteristic, "signature-like" changes in the electrical potential of cells contained in an electro conductive probe and corresponding to cellular interactions with bioactive substances, such as pesticides and toxins [77,78]. In order to achieve a desired level of selectivity and specificity, the FOODSCAN system is supported by dedicated ANN software, which can classify the screened residues in one of the three aforementioned groups with an overall success rate of 83.6 percent [79]. By operating the system, a lay user can test up to eight food samples simultaneously and get results in 3–5 min.

In addition to FOODSCAN, various versions of electrochemical biosensors dominate the field of POT for food safety analysis. For example, various amperometric sensors have been developed for the detection of tyramine, a biogenic amine frequently associated with spoilage of fish and seafood, the most recent system being based on tyramine/monoamine oxidase immobilized on functionalized CNTs [80]. Another interesting POT product is the TCA-Safe^{TM} biosensor for the detection of 2,4,6-tricholoanisole (TCA) in cork and wine. TCA is the product of fungal cork infection and responsible for the so-called cork taint in wine, which seriously deteriorates its taste [81]. Even worse, there are considerations about a possible carcinogenic effect of the chronic exposure to even very low TCA concentrations, hence the necessity of analytical, easy-to-use and high-throughput methods to detect this substance. Conventionally, TCA determination is either done chromatographically by means of dedicated LC–MS systems, or empirically, with the aid of "testers," that is, people with olfactory senses (either through genetic predisposition or training) sensitive enough to determine the cut-off concentration of TCA in a sample (usually set at 1–2 parts per trillion (ppt), depending on the desired quality standards) [82,83]. Obviously, the second approach is advantageous over chromatography in terms of capacity (with an expert tester being able to screen more than 500 cork samples per day) but lacks the quantitative precision offered by conventional analytical methods. In either case, however, it is necessary to extract TCA from cork in ethanol, methanol, or water, in a very lengthy (overnight) process. This process is required also for nonconventional TCA analytical methods, such as immunoamperometry [84]. This is a serious limitation for the realization of a POT system for TCA screening, which could help the cork and wine industry secure a higher level of product quality and minimize the risk of recalls. The aforementioned TCA-Safe™ biosensor provides a unique solution to this problem, since it allows for detecting traces of TCA contained in cork detritus instead of applying the standard extraction protocol. This is possible due to the use of cells membrane-engineered with anti-TCA antibodies as the biorecognition elements, according to the method developed by Apostolou *et al.* [85] (details of the membrane-engineering process are given in Chapter 6). The TCA-Safe™ comprises a customized portable potentiometer measuring the engineered cells' response to the presence of TCA at concentrations as low as 0.1 ppt within just 3 min, representing a true reagent less, POT approach to cork and wine quality control.

On a rather different level, portable near-infrared (NIR) spectroscopy-based biosensors are definitely the current market trend setters. NIR spectroscopy itself has gained in popularity for food processing industry-related applications thanks to its noninvasiveness, non-requirement for sample preparation, speed and user-friendliness [86]. NIR spectra (780 nm–2.5 μm), associated with fundamental vibrations of C–H, N–H, O–H and S–H bonds in organic molecules can serve as a fingerprint of individual food composition, in particular regarding the food texture and its protein and fat content, though not providing information on detailed type and structure of the individual food components. As such, NIR spectroscopy can be very useful for the rapid and cost-efficient assessment of food freshness and/or adulteration, as well as the possible presence of allergens, in particular those with well-established NIR spectral profiles like eggs, and nuts. As a rule, NIR biosensing is facilitated with the aid of sophisticated interpretative and predictive software, supported by multivariate statistical models. There are currently a number of notable commercial portable NIR biosensors used in food quality control. All of them are composed of a dual system: a handheld NIR spectrometer which is used to scan the food sample of interest and a read-out device (usually a smartphone with an embedded app). Due to the inherent limitations of the working principle of the NIR spectroscopy, these true-to-form POT devices, although very attractive to the lay user, can only be employed as "general" food screeners (e.g., by tourists introduced to meals of doubtful origin or quality of preparation). They cannot be used for the determination of individual compounds and possible hazards, such as pesticide residues or mycotoxins. In spite of this limitation, the outlook of the commercial success of these handheld food sensors is bright, not at least thanks to an excellent marketing effort over the past two years. Two of the most known products in this category are TellSpec [87] and SCiO [88]. Even the European Commission has recently launched the FoodScanner €1 million funding initiative for the development of "an affordable and noninvasive mobile solution that will enable users to measure and analyze their food intake," particularly addressing the needs of people with conditions such as obesity, allergies, or food intolerance [89]. It remains to be seen if the winning proposal will be based on a portable NIR spectrometer or another system.

15.3.2 Glucose sensors

Contrary to the standard electrochemical mediator-based approach used in most commercial glucose biosensors [90], the new generation of POC devices for use by diabetic patients is based on direct electron transfer (DET) between the electrode surface and the active site of GOx used as the enzymatic biorecognition element. A considerable number of reports demonstrate the advantages of this approach, whereby increases in DET are concomitant with increased sensor sensitivity (to a limit of detection of almost 1 μM) and range of linear response (for a review, Song *et al.* [41]). This can be achieved by increasing the surface coverage of GOx, for example by integrating polymers with graphene and DET, even though it is questionable whether GOx retains its normal catalytic activity after adsorption on the graphene surface. A possible non-enzymatic GOx-based DET mechanism has been suggested by several authors [91].

15.3.3 Cholesterol sensors

Next to glucose, cholesterol or, more accurately, its pathologically high concentration in blood (hypercholesterolemia) is probably the most important health index, generally used to access the risk associated with unhealthy life style, in particular the incidence of cerebrovascular disease, although its validity as a biomarker is dependent on the individual genetic background [92]. Various approaches toward the development of cholesterol biosensors have been reported over the last 15 years (for a review, see Reference 93). The majority of them are based on the enzymatic assay of cholesterol using cholesterol oxidase (ChOx) as the biosensing element. Common sources for ChOx are microorganisms such as *Streptomyces hygroscopicus* and *Brevibacterium sterolicum*. Some approaches use another enzyme, cholesterol esterase (ChEt), as the biorecognition element, in combination with ChOx. ChEt catalyzes the hydrolysis of dietary cholesterol esters, triacylglycerols, and phospholipids via a serine protease mechanism [94]. In either case, the end product of the catalytic reaction using cholesterol as a substrate is H_2O_2, which can be assayed electrochemically, in a similar fashion to glucose detection using GOx (see above). For this reason, the majority of commercial cholesterol sensors utilize the same read-out format with blood sugar meters and are often distributed by the same companies (e.g., the AccuChek Instant Plus Dual Testing System by Roche Diagnostics). Modification of the electrode surface, for example using CNTs [95], layer-by-layer nano thin films of polyallylamine hydrochloride/multi-walled CNTs/gold nanoparticles/ChOx [96,97] or graphene nanostructures [98], has contributed to significant increase of the sensitivity (down to 0.02 mM) of the electrochemical detection of cholesterol, at the same time reducing the interference by compounds such as uric acid and acetaminophen. Recently, Cinti *et al.* [99] reported the construction of a highly sensitive cholesterol biosensor based on inkjet-printed Prussian blue nanoparticles (PBNPs) on screen-printed electrodes. On the other hand, systems employing an optical transducer, usually based on the change of fluorescence of ruthenium-containing dyes interacting with cholesterol have proven to be less applicable in a commercial sense. Simpler systems are more successful, for example the EnzyChrom™ (BioAssay Systems) assay kit that measures the concentration of NADH in direct quantitative relationship to the cholesterol dehydrogenase catalyzed conversion of cholesterol to cholest-4-ene-3-one. The same concept, but using a different optical test methodology, reflectance photometry, is applied as a working principle in another commercial product, CardioChek® (Polymer Technology Systems). Finally the Cholestech LDX® Analyzer (Alere), a dual cholesteror + glucose meter, also employs ChOx and GOx; however, the resulting H_2O_2 is catalyzed by horseradish peroxidase to react with 4-aminoantipyrine and *N*-ethyl-*N*-sulfohydroxypropyl-*m*-toluidine sodium salt (TOOS) to form a purple-colored quinoneimine dye proportional to the total cholesterol, high-density lipoprotein cholesterol, and glucose concentrations of the sample.

15.3.4 Wearable POC systems

By default, a biosensor that can be clad by the end-user represents an ultimate POC form. In the past few years, wearable sensing devices have attracted considerable interest as commercial gadgets for real-time, noninvasive monitoring the individual health and fitness status [100–103]. Designers of wearable systems face two major challenges: First, the devices should be based on materials flexible enough to fit the specification of attire or a gear. Second, the scope of biological information that can be retrieved without invading the body is rather narrow, mainly including physical assay indices such as respiration and heart rate, skin temperature, blood pressure, brain activity, and cardiac electric activity. These limitations give electrochemical and electrophysiological sensors precedence over other assay approaches [104,105]. At the same time, progress in sensitivity and speed of electrochemical biosensors (as illustrated in previous sections of the present chapter) has allowed for specific analyte detection in body fluids (such as sweat) locally excreted through the skin, sometimes in ultra-low volumes (under conditions of low physical activity, non-acute pathological health status or normal room temperature). These target analytes may include metabolites that are critical health biomarkers, such as glucose, lactate, or uric acid [106–111]. In addition to general health/fitness monitoring, wearable biosensors have found particular application in the assessment of wound healing, for example through the determination of hemostasis (measurement of the viscosity of coagulated blood, as well as assay of thrombin, fibrin, and prostacyclin activities), electrolyte (e.g., sodium chloride), and uric acid leakage from the wound area and the measurement of wound-specific biomarkers, such as tumor necrosis factor-alpha and C-reactive protein [112–114]. Among wearable wound-healing biosensors, electrochemical pH and or blood leakage sensors embedded in adhesive bandages probably represent the most promising commercial approach [115,116].

15.3.5 Niche consumer diagnostic POC/POT systems

When coming to new consumer-tailor applications, the sky is literally the limit. A seemingly endless variety of analytes or conditions are being targeted by an ever-increasing number of emerging novel diagnostic products. A considerable portion of POC applications are just dedicated software apps operated on smartphones, that is they are not based on actual, "hardcore" analytical principles. Some of these "exotic" niche applications are presented in Table 15.1. Although it is rather expected that many of these will fail to turn into long-term, self-sustainable commercial items, a few might actually become game-changers in their respective field.

When projecting advances in commercial POC/POT systems, one should not neglect the ongoing progress in miniaturizing conventional instrumentation. For example, Nakamura *et al.* [127] developed a micro-ion extractor for trace anion determination by ion chromatography–mass spectrometry from a single drop (25 μL) of whole blood without pretreatment. Piraino *et al.* [128] reported a multiplexed microfluidic platform for the rapid and highly sensitive detection of different (three to four) biomarkers in a single sample of only 5 μL, without the need of any sample (blood) pretreatment. The speed and simplicity of nucleic acid detection has been dramatically increased thanks to novel technologies, such as recombinase polymerase amplification, which has allowed for field-based, rapid (30–60 min) diagnosis of Ebola using portable instrumentation, including a mobile glovebox and a Diagnostics-in-a-Suitcase powered by a battery and solar panel [129]. Beyond miniaturization, drastic improvement in the reduction of assay times may come from innovative techniques for processing a reaction mixture. For example, Li *et al.* [130] recently introduced the term "cyclic solution draining and replenishing" to overcome size-associated diffusion limitations in immunoassays. Contrary to the conventional static incubation and mixing of bulk reaction solution, their approach is based on its repeated reaction solution draining and replenishing from/in the measurement cell, thus allowing for assay kinetics irrespective of the probe size and achieving the shortening of assay times from hours to minutes.

Inevitably, the trend for portable, even handheld multiplexed commercial analytical and diagnostic instrumentation will only increase in significance and market share. There are a few testimonials supporting this forecast, such as the initial breakthrough success of Theranos, a company that pioneered POC testing by developing and distributing the Edison device, able to run 229 tests on a single drop of blood. Theranos made its platform available to consumers directly through pharmacy retailers like Walgreens and CV, but also general stores like Walmart

Table 15.1 Novel/emerging biosensor products/technologies with niche applications for the end-consumer market

Product	Target analyte/condition	Application	Technology	Manufacturer	Reference
PTS Detect	Cotinine	Tobacco smoking assessment	Colorimetry	PTS Diagnostics	[117]
SmokeScreen	Cotinine	Tobacco smoking assessment	Colorimetry	GFC Diagnostics	[118]
IsoScreen	Isoniazid metabolites	Tobacco smoking assessment	Colorimetry	GFC Diagnostics	[118]
MyDX	Total pesticide residues Tetrahydrocannabinol (THC)	Cannabis quality control	Electronic nose (resistometer)	CDXLife	[119]
ResApp	Acoustic (cough analysis) lung structure profile	Lung disease assessment, in particular childhood respiratory diseases	Smartphone App	ResAppHealth	[120,121]
SpiroCall/SpiroSmart	Acoustic (blow analysis) lung structure profile	Lung health assessment	Smartphone App	University of Washington	[122]
NeuroNet	Cognitive test	Cognitive assessment	Smartphone App	Savonix	[123]
TrueAllele	Combined Probability of Inclusion (CPI) algorithm	Forensic DNA analysis	Software App	Cybergenetics	[124]
Odoreader	VOCs in urine	Prostate cancer	Gas chromatography sensor	University of Liverpool University of the West of England	[125]
-	Zika virus RNA	Zika CRISP/Ca9 sequence detector	Paper-based colorimetry	Addgene	[126]

Table 15.2 Novel biomarkers with promising application to innovative POC test development

Biomarker(s)	Type	Disease/condition	Reference
Nav 1.5 sodium channel	Protein	Breast cancer	[133]
Orexin receptor	Protein	Heart failure	[134]
Androgen receptor (AR)	Protein	Prostate cancer chemotherapy resistance	[135]
ST2	Protein	Graft rejection	[136]
Peripheral blood mononuclear cells (PBMC) metabolic profiling	Proteins	Major depressive disorder	[137]
Cyclin-D1 (*CCND1*)	Gene	Lymphoma	[138]
Transformation-related protein 53 (*TRP53*)	Gene	Cancer	[139]
Interferon regulatory factor 4 (*IRF4*)	Gene	Leukemia/Myeloma	[140]
Early growth response protein 1 (*EGR1*)	Gene	Myeloma	[141]

and Target. In essence, Theranos "democratized" diagnostics by introducing a direct test-to-consumer model. The significance of this fact cannot be diminished by the recent ruckus regarding the significant interservice variability of the Theranos panel test results [131], which has placed the company under heavy critic by the medical community [132]. In particular respect of the POC market, further progress will also depend on the identification of novel and/or reliable disease biomarkers, which could become targets of innovative assay principles, much in a similar way to companion diagnostics, where a drug is co-developed with the assay that measures the drug's effect on its particular target. A sample of most promising novel biomarkers is presented in Table 15.2.

References

[1] Teodoro-Morrison T., Diamandis E.P., Rifai N., *et al.* 'Animal olfactory detection of disease: promises and pitfalls'. *Clinical Chemistry*. 2014;**60**(12): 1473–9.

[2] Brooks S.W., Moore D.R., Marzouk E.B., Glenn F.R., Hallock R.M. 'Canine olfaction and electronic nose detection of volatile organic compounds in the detection of cancer: a review'. *Cancer Investigation*. 2015;**33**(9):411–19.

[3] Schott M., Klein B., Vilcinskas A. 'Detection of illicit drugs by trained honeybees (*Apis mellifera*)'. *PLoS ONE*. 2015;**10**(6):e0128528.

[4] Yoel U., Gopas J., Ozer J., Peleg R., Shvartzman P. 'Canine scent detection of volatile elements, characteristic of malignant cells, in cell cultures'. *The Israel Medical Association Journal*. 2015;**17**(9):567–70.

[5] Kimball B.A., Wilson D.A., Wesson D.W. 'Alterations of the volatile metabolome in mouse models of Alzheimer's disease'. *Scientific Reports*. 2016;**6**:19495.

[6] Fernández Del Río R., O'Hara M.E., Holt A., *et al.* 'Volatile biomarkers in breath associated with liver cirrhosis – comparisons of pre- and post-liver transplant breath samples'. *EBioMedicine*. 2015;**2**(9):1243–50.

[7] Peng G., Tisch U., Adams O., *et al.* 'Diagnosing lung cancer in exhaled breath using gold nanoparticles'. *Nature Nanotechnology*. 2010;**14**:669–73.

[8] Hakim M., Billan S., Tisch U., *et al.* Diagnosis of head-and-neck cancer from exhaled breath'. *British Journal of Cancer*. 2011;**104**:1649–55.

[9] Boots A.W., Berkel J.J.B.N., Dallinga J.W., Smolinska A., Wouters E.F., Schooten F.J. 'The versatile use of exhaled volatile organic compounds in human health and disease'. *Journal of Breath Research*. 2012;**6**:027108.

[10] Mazzone P.J. 'Exhaled breath volatile organic compound biomarkers in lung cancer'. *Journal of Breath Research*. 2012;**6**:027106.

[11] Arakawa T., Wang X., Kajiro T., *et al.* 'A direct gaseous ethanol imaging system for analysis of alcohol metabolism from exhaled breath'. *Sensors and Actuators B: Chemical*. 2013;**186**:27–33.

[12] Ye M., Chien P.J., Toma K., Arakawa T., Mitsubayashi K. 'An acetone bio-sniffer (gas phase biosensor) enabling assessment of lipid metabolism from exhaled breath'. *Biosensors and Bioelectronics*. 2015;**73**:208–13.

[13] Kudo H., Sawai M., Suzuki Y., *et al.* 'Fiber-optic bio-sniffer (biochemical gas sensor) for high selective monitoring of ethanol vapor using 335 nm UV-LED'. *Sensors and Actuators B: Chemical*. 2010;**132**(2):676–80.

[14] Capuano R., Santonico M., Pennazza G., *et al.* 'The lung cancer breath signature: a comparative analysis of exhaled breath and air sampled from inside the lungs'. *Scientific Reports*. 2015;**5**:16491.

[15] Kolk A., Hoelscher M., Maboko L., *et al.* 'Electronic-nose technology using sputum samples in diagnosis of patients with tuberculosis'. *Journal of Clinical Microbiology*. 2010;**48**(11):4235–8.

[16] Kolk A.H., van Berkel J.J., Claassens M.M., *et al.* 'Breath analysis as a potential diagnostic tool for tuberculosis'. *The International Journal of Tuberculosis and Lung Disease*. 2012;**16**(6):777–82.

[17] Carraher C., Dalziel J., Jordan M.D., Christie D.L., Newcomb R.D., Kralicek A. V. 'Towards an understanding of the structural basis for insect olfaction by odorant receptors'. *Insect Biochemistry and Molecular Biology*. 2015;**66**:31–41.

[18] Mitsuno H., Sakurai T., Namiki S., Mitsuhashi H., Kanzaki R. 'Novel cell-based odorant sensor elements based on insect odorant receptors'. *Biosensors and Bioelectronics*. 2014;**65**:287–94.

[19] Suska A., Ibáñez A.B., Preechaburana P., Lundström I., Berghard A. 'G protein-coupled receptor mediated sensing of TMA'. *Procedia Chemistry*. 2009;**1**(1):321–4.

[20] Goldsmith B.R., Mitala J.J., Josue J., *et al.* 'Biomimetic chemical sensors using nanoelectronic readout of olfactory receptor proteins'. *ACS Nano* 2011;**5**(7):5408–16.

[21] Lee S.H., Kwon O.S., Song H.S., *et al.* 'Mimicking the human smell sensing mechanism with an artificial nose platform'. *Biomaterials* 2012;**33**(6): 1722–9.

[22] German P.F., van der Poel S., Carraher C., Kralicek A.V., Newcomb R.D. 'Insights into subunit interactions within the insect olfactory receptor complex using FRET'. *Insect Biochemistry and Molecular Biology*. 2013;**43**(2): 138–45.

[23] Lu Y., Zhang D., Zhang Q., *et al.* 'Impedance spectroscopy analysis of human odorant binding proteins immobilized on nanopore arrays for biochemical detection'. *Biosensors and Bioelectronics*. 2016;**79**:251–7.

[24] Dacres H., Wang J., Leitch V., Horne I., Anderson A.R., Trowell S.C. 'Greatly enhanced detection of a volatile ligand at femtomolar levels using bioluminescence resonance energy transfer (BRET)'. *Biosensors and Bioelectronics*. 2011;**29**(1):119–24.

[25] Corcoran J.A., Jordan M.D., Carraher C., Newcomb R.D. 'A novel method to study insect olfactory receptor function using HEK293 cells'. *Insect Biochemistry and Molecular Biology*. 2014;**54**:22–32.

[26] Sankaran S., Khot L.R., Panigrahi S. 'Biology and applications of olfactory sensing system: a review'. *Sensors and Actuators B: Chemical*. 2012;**171–172**:1–17.

[27] Du L., Wu C., Liu Q., Huang L., Wang P. 'Recent advances in olfactory receptor-based biosensors'. *Biosensors and Bioelectronics*. 2013;**42**:570–80.

[28] Wu C., Lillehoj P.B., Wang P. 'Bioanalytical and chemical sensors using living taste, olfactory, and neural cells and tissues: a short review'. *Analyst*. 2015;**140**(21):7048–61.

[29] Siripatrawan U., Linz J.E., Harte B.R. 'Electronic sensor array coupled with artificial neural network for detection of *Salmonella typhimurium*'. *Sensors and Actuators B: Chemical*. 2006;**119**(1):64–9.

[30] Li B., Santhanam S., Schultz L., *et al.* 'Inkjet printed chemical sensor array based on polythiophene conductive polymers'. *Sensors and Actuators B: Chemical*. 2007;**123**:651–60.

[31] Panagou E.Z., Sahgal N., Magan N., Nychas G.-J.E. 'Table olives volatile fingerprints: potential of an electronic nose for quality discrimination'. *Sensors and Actuators B: Chemical*. 2008;**134**(2):902–7.

[32] Bachtiar L.R., Unsworth C.P., Newcomb R.D., Crampin E.J. 'Multilayer perceptron classification of unknown volatile chemicals from the firing rates of insect olfactory sensory neurons and its application to biosensor design'. *Neural Computation*. 2013;**25**(1):259–87.

[33] Bachtiar L.R., Unsworth C.P., Newcomb R.D. 'Artificial neural network prediction of specific VOCs and blended VOCs for various concentrations from the olfactory receptor firing rates of *Drosophila melanogaster*'. *Conference Proceedings: Annual International Conference of the IEEE Engineering in Medicine and Biology Society*; Chicago, IL: Aug 2014:3232–5.

[34] Kraus S., Kleines M., Albers J., *et al.* 'Quantitative measurement of human anti-HCV Core immunoglobulins on an electrical biochip platform'. *Biosensors and Bioelectronics*. 2011;**26**(5):1895–901.

[35] Tarasov A., Gray D.W., Tsai M.Y., *et al.* 'A potentiometric biosensor for rapid on-site disease diagnostics'. *Biosensors and Bioelectronics*. 2016;**79**:669–78.

[36] Huang J., Tian J., Zhao Y., Zhao S. 'Ag/Au nanoparticles coated graphene electrochemical sensor for ultrasensitive analysis of carcinoembryonic antigen in clinical immunoassay'. *Sensors and Actuators B: Chemical.* 2015;**206**:570–6.

[37] Zhu C., Yang G., Li H., Du D., Lin Y. 'Electrochemical sensors and biosensors based on nanomaterials and nanostructures'. *Analytical Chemistry.* 2015;**87**(1):230–49.

[38] Lin C., Wei K., Liao S., *et al.* 'A reusable magnetic graphene oxide-modified biosensor for vascular endothelial growth factor detection in cancer diagnosis'. *Biosensors and Bioelectronics.* 2015;**67**:431–7.

[39] Singal S., Srivastava A.K., Biradar A.M., Mulchandani A., Rajesh. 'Pt nanoparticles-chemical vapor deposited graphene composite based immunosensor for the detection of human cardiac troponin I'. *Sensors and Actuators B: Chemical.* 2014;**205**:363–70.

[40] Shukla S.K., Turner A.P., Tiwari A. 'Cholesterol oxidase functionalised polyaniline/carbon nanotube hybrids for an amperometric biosensor'. *Journal of Nanoscience and Nanotechnology.* 2015;**15**(5):3373–7.

[41] Song Y., Luo Y., Zhu C., Li H., Du D., Lin Y. 'Recent advances in electrochemical biosensors based on graphene two-dimensional nanomaterials'. *Biosensors and Bioelectronics.* 2016;**76**:195–212.

[42] Caviglia C., Zór K., Canepa S., *et al.* 'Interdependence of initial cell density, drug concentration and exposure time revealed by real-time impedance spectroscopic cytotoxicity assay'. *Analyst.* 2015;**140**(10):3623–9.

[43] Kasukabe T., Okabe-Kado J., Kato N., Honma Y., Kumakura S. 'Cotylenin A and arsenic trioxide cooperatively suppress cell proliferation and cell invasion activity in human breast cancer cells'. *International Journal of Oncology.* 2015;**46**(2):841–8.

[44] Barton A.C., Collyer S.D., Davis F., *et al.* 'Labeless AC impedimetric antibody-based sensors with pgmL(-1) sensitivities for point-of-care biomedical applications'. *Biosensors and Bioelectronics.* 2009;**24**(5):1090–5.

[45] Hushegyi A., Pihíková D., Bertok T., Adam V., Kizek R., Tkac J. 'Ultrasensitive detection of influenza viruses with a glycan-based impedimetric biosensor'. *Biosensors and Bioelectronics.* 2016;**79**:644–9.

[46] Ebrahimi A., Alam M.A. 'Evaporation-induced stimulation of bacterial osmoregulation for electrical assessment of cell viability'. *Proceedings of the National Academy of Sciences of the United States of America.* 2016; **113**(26):7059–64.

[47] Jahns S., Bräu M., Meyer B.O., *et al.* 'Handheld imaging photonic crystal biosensor for multiplexed, label-free protein detection'. *Biomedical Optics Express.* 2015;**6**(10):3724–36.

[48] Sauer V.T., Diao Z., Freeman M.R., Hiebert W.K. 'Wavelength-division multiplexing of nano-optomechanical doubly clamped beam systems'. *Optics Letters.* 2015;**40**(9):1948–51.

[49] Joshi G.K., Deitz-McElyea S., Johnson M., Mali S., Korc M., Sardar R. 'Highly specific plasmonic biosensors for ultrasensitive microRNA

detection in plasma from pancreatic cancer patients'. *Nano Letters*. 2014; **14**(12):6955–63.

[50] Joshi G.K., Deitz-McElyea S., Liyanage T., *et al.* 'Label-free nanoplasmonic-based short noncoding RNA sensing at attomolar concentrations allows for quantitative and highly specific assay of microRNA-10b in biological fluids and circulating exosomes'. *ACS Nano*. 2015;**9**(11):11075–89.

[51] Connelly J.T., Rolland J.P., Whitesides G.M. '"Paper Machine" for molecular diagnostics'. *Analytical Chemistry*. 2015;**87**(15):7595–601.

[52] Ma X., Jiang Y., Jia F., Yu Y., Chen J., Wang Z. 'An aptamer-based electrochemical biosensor for the detection of *Salmonella*'. *Journal of Microbiological Methods*. 2014;**98**:94–8.

[53] Chao J., Zhu D., Zhang Y., Wang L., Fan C. 'DNA nanotechnology-enabled biosensors'. *Biosensors and Bioelectronics*. 2016;**76**:68–79.

[54] Gerling T., Wagenbauer K.F., Neuner A.M., Dietz H. 'Dynamic DNA devices and assemblies formed by shape-complementary, non-base pairing 3D components'. *Science*. 2015;**347**(6229):1446–52.

[55] Veneziano R., Ratanalert S., Zhang K., *et al.* 'Designer nanoscale DNA assemblies programmed from the top down'. *Science*. 2016;**352**(6293): 1534.

[56] Yan J., Hu C., Wang P., *et al.* 'Novel rolling circle amplification and DNA origami-based DNA belt-involved signal amplification assay for highly sensitive detection of prostate-specific antigen (PSA)'. *ACS Applied Materials & Interfaces*. 2014;**6**(22):20372–7.

[57] Liu S., Su W., Li Z., Ding X. 'Electrochemical detection of lung cancer specific microRNAs using 3D DNA origami nanostructures'. *Biosensors and Bioelectronics*. 2015;**71**:57–61.

[58] Breaker R.R., Joyce G.F. 'A DNA enzyme that cleaves RNA'. *Chemistry and Biology*. 1994;**1**(4):223–9.

[59] Liu C., Sheng Y., Sun Y., *et al.* 'A glucose oxidase-coupled DNAzyme sensor for glucose detection in tears and saliva'. *Biosensors and Bioelectronics*. 2015;**70**:455–61.

[60] Aliberti A., Cusano A.M., Battista E., Causa F., Netti P.A. 'High sensitive and direct fluorescence detection of single viral DNA sequences by integration of double strand probes onto microgels particles'. *Analyst*. 2016; **141**(4):1250–6.

[61] Tahirbegi I.B., Alvira M., Mir M., Samitier J. 'Simple and fast method for fabrication of endoscopic implantable sensor arrays'. *Sensors*. 2014;**14**(7): 11416–26.

[62] Kiourti A., Lee C.W., Chae J., Volakis J.L. 'A wireless fully passive neural recording device for unobtrusive neuropotential monitoring'. *IEEE Transactions on Biomedical Engineering*. 2016;**63**(1):131–7.

[63] Hong G., Fu T.M., Zhou T., Schuhmann T.G., Huang J., Lieber C.M. 'Syringe injectable electronics: precise targeted delivery with quantitative input/output connectivity'. *Nano Letters*. 2015;**15**(10):6979–84.

[64] Xie C., Liu J., Fu T.M., Dai X., Zhou W., Lieber C.M. 'Three-dimensional macroporous nanoelectronic networks as minimally invasive brain probes'. *Nature Materials*. 2015;**14**(12):1286–92.

[65] Chałupniak A., Morales-Narváez E., Merkoçi A. 'Micro and nanomotors in diagnostics'. *Advance Drug Delivery Reviews*. 2015;**95**:104–16.

[66] Wang J. 'Self-propelled affinity biosensors: moving the receptor around the sample'. *Biosensors and Bioelectronics*. 2016;**76**:234–42.

[67] Parmar J., Ma X., Katuri J., *et al*. 'Nano and micro architectures for self-propelled motors'. *Science and Technology of Advanced Materials*. 2015;**16**:1.

[68] Gao W., Dong R., Thamphiwatana S., *et al*. 'Artificial micromotors in the mouse's stomach: a step toward in vivo use of synthetic motors'. *ACS Nano*. 2015;**9**(1):117–23.

[69] Mou F., Chen C., Ma H., Yin Y., Wu Q., Guan J. 'Self-propelled micromotors driven by the magnesium–water reaction and their hemolytic properties'. *Angewandte Chemie*. 2013;**52**(28):7208–12.

[70] Chng E.L.K., Zhao G., Pumera M. 'Towards biocompatible nano/microscale machines: self-propelled catalytic nanomotors not exhibiting acute toxicity'. *Nanoscale*. 2014;**6**(4):2119–24.

[71] Sanchez S., Soler L., Katuri J. 'Chemically powered micro- and nanomotors'. *Angewandte Chemie*. 2015;**54**(5):1414–44.

[72] Garcia M., Orozco J., Guix M., *et al*. 'Micromotor-based lab-on-chip immunoassays'. *Nanoscale*. 2013;**5**:1325–31.

[73] Mhanna R., Qiu F., Zhang L., *et al*. 'Artificial bacterial flagella for remote-controlled targeted single-cell drug delivery'. *Small*. 2014;**10**(10):1953–7.

[74] Wang W., Li S., Mair L., Ahmed S., Huang T.J., Mallouk T.E. 'Acoustic propulsion of nanorod motors inside living cells'. *Angewandte Chemie*. 2014;**53**(12):3201–4.

[75] Xu T., Soto F., Gao W., *et al*. 'Reversible swarming and separation of self-propelled chemically powered nanomotors under acoustic fields'. *Journal of the American Chemical Society*. 2015;**137**(6):2163–6.

[76] Santonico M., Grasso S., Genova F., Zompanti A., Parente F.R., Pennazza G. 'Unmasking of olive oil adulteration via a multi-sensor platform'. *Sensors*. 2015;**15**(9):21660–72.

[77] Kintzios S., Pistola E., Panagiotopoulos P., *et al*. 'Bioelectric recognition assay (BERA)'. *Biosensors and Bioelectronics*. 2001;**16**:325–36.

[78] Kintzios S., Pistola E., Konstas J., *et al*. 'Application of the bioelectric recognition assay (BERA) for the detection of human and plant viruses: definition of operational parameters'. *Biosensors and Bioelectronics*. 2001;**16**:467–80.

[79] Ferentinos K.P., Yialouris C.P., Blouchos P., Moschopoulou G., Kintzios S. 'Pesticide residue screening using a novel artificial neural network combined with a bioelectric cellular biosensor'. *BioMed Research International*. 2013;813519:1–8.

[80] Apetrei I.M., Apetrei C. 'The biocomposite screen-printed biosensor based on immobilization of tyrosinase onto the carboxyl functionalised carbon nanotube for assaying tyramine in fish products'. *Journal of Food Engineering*. 2015;**149**:1–8.

[81] Sefton M.A., Simpson R.F. 'Compounds causing cork taint and the factors affecting their transfer from natural cork closures to wine – a review'. *Australian Journal of Grape and Wine Research*. 2005;**11**(2):226–40.

[82] Sung Y.H., Li T.Y., Hyang S.D. 'Analysis of earthy and musty odors in water samples by solid-phase microextraction coupled with gas chromatography/ion trap mass spectrometry'. *Talanta*. 2005;**65**:518–24.

[83] Salemi A., Lacorte S., Bagheri H., Barceló D. 'Automated trace determination of earthy-musty odorous compounds in water samples by on-line purge-and-trap–gas chromatography–mass spectrometry'. *Journal of Chromatography A*. 2006;**1136**:170–5.

[84] Moore E., Pravda M., Guilbault G. 'Development of a biosensor for the quantitative detection of 2,4,6-trichloroanisole using screen printed electrodes'. *Analytica Chimica Acta*. 2003;**484**:15–24.

[85] Apostolou T., Pascual N., Marco M.P., *et al.* 'Extraction-less, rapid assay for the direct detection of 2,4,6-trichloroanisole (TCA) in cork samples'. *Talanta*. 2014;**125**:336–40.

[86] Porep J.U., Kammerer D.R., Carle R. 'On-line application of near infrared (NIR) spectroscopy in food production'. *Trends in Food Science & Technology*. 2015;**46**(2):211–30.

[87] *TellSpec*. 2016. Available from http://tellspec.com/en/ (accessed date 15 July 2016).

[88] *SCiO*. 2015. Available from https://www.consumerphysics.com/myscio/ (2016; access date: 7 June 2016)

[89] *FoodScanner*. Available from http://ec.europa.eu/research/horizonprize/index.cfm?prize=food-scanner (accessed date 21 July 2016).

[90] Turner A.P. 'Biosensors: sense and sensibility'. *Chemical Society Reviews*. 2013;**42**(8):3184–96.

[91] Liang B., Guo X., Fang L., *et al.* 'Study of direct electron transfer and enzyme activity of glucose oxidase on graphene surface'. *Electrochemistry Communications*. 2015;**50**:1–5.

[92] Sharifi M., Rakhit R.D., Humphries S.E., Nair D. 'Cardiovascular risk stratification in familial hypercholesterolaemia'. *Heart*. 2016;**102**(13):1003–8.

[93] Arya S.K., Datta M., Malhotra B.D. 'Recent advances in cholesterol biosensor'. *Biosensors and Bioelectronics*. 2008;**23**(7):1083–100.

[94] Chiou S.Y., Lai C.Y., Lin L.Y., Lin G. 'Probing stereoselective inhibition of the acyl binding site of cholesterol esterase with four diastereomers of 2′-*N*-alpha-methylbenzylcarbamyl-1,1′-bi-2-naphthol'. *BMC Biochemistry*. 2005;**6**:17.

[95] Yang M., Yang Y., Qu F., Lu Y., Shen G., Yu R. 'Attachment of nickel hexacyanoferrates nanoparticles on carbon nanotubes: preparation, characterization and bio-application'. *Analytica Chimica Acta*. 2006;**571**(2):211–17.

[96] Ram M.K., Bertoncello P., Ding H., Paddeu S., Nicolini C. 'Cholesterol biosensors prepared by layer-by-layer technique'. *Biosensors and Bioelectronics*. 2001;**16**(9–12):849–56.

[97] Cai X., Gao X., Wang L., Wu Q., Lin X. 'A layer-by-layer assembled and carbon nanotubes/gold nanoparticles-based bienzyme biosensor for cholesterol detection'. *Sensors and Actuators B: Chemical*. 2013;**181**: 575–83.

[98] Kuila T., Bose S., Khanra P., Mishra A.K., Kim N.H., Lee J.H. 'Recent advances in graphene-based biosensors'. *Biosensors and Bioelectronics*. 2011;**26**(12):4637–48.

[99] Cinti S., Arduini F., Moscone D., Palleschi G., Gonzalez-Macia L., Killard A.J. 'Cholesterol biosensor based on inkjet-printed Prussian blue nanoparticle-modified screen-printed electrode'. *Sensors and Actuators B: Chemical*. 2015;**221**:187–90.

[100] Windmiller J.R., Wang J. 'Wearable electrochemical sensors and biosensors: a review'. *Electroanalysis*. 2013;**25**(1):29–46.

[101] Bandodkar A.J., Wang J. 'Non-invasive wearable electrochemical sensors: a review'. *Trends in Biotechnology*. 2014;**32**(7):363–71.

[102] Bandodkar A.J., Jia W., Yardimci C., Wang X., Ramirez J., Wang J. 'Tattoo-based noninvasive glucose monitoring: a proof-of-concept study'. *Analytical Chemistry*. 2015;**87**(1):394–8.

[103] Kaewkannate K., Kim S. 'A comparison of wearable fitness devices'. *BMC Public Health*. 2016;**16**(1):433–49.

[104] Gao W., Emaminejad S., Nyein H.Y., *et al.* 'Fully integrated wearable sensor arrays for multiplexed in situ perspiration analysis'. *Nature*. 2016;**529**(7587):509–14.

[105] Liao X., Zhang Z., Liao Q., *et al.* 'Flexible and printable paper-based strain sensors for wearable and large-area green electronics'. *Nanoscale*. 2016;**8**:13025–32.

[106] Bandodkar A.J., Hung V.W., Jia W., *et al.* 'Tattoo-based potentiometric ion-selective sensors for epidermal pH monitoring'. *Analyst*. 2013;**138**(1):123–8.

[107] Jia W., Bandodkar A.J., Valdés-Ramírez G., *et al.* 'Electrochemical tattoo biosensors for real-time noninvasive lactate monitoring in human perspiration'. *Analytical Chemistry*. 2013;**85**(14):6553–60.

[108] Kim J., Imani S., de Araujo W.R., *et al.* 'Wearable salivary uric acid mouthguard biosensor with integrated wireless electronics'. *Biosensors and Bioelectronics*. 2015;**74**:1061–8.

[109] Salvo P., Dini V., Di Francesco F., Romanelli M. 'The role of biomedical sensors in wound healing'. *Wound Medicine*. 2015;**8**:15–18.

[110] Imani S., Bandodkar A.J., Mohan A.M., *et al.* 'A wearable chemical-electrophysiological hybrid biosensing system for real-time health and fitness monitoring'. *Nature Communications*. 2016;**7**:11650.

[111] Salvo P., Smajda R., Dini V., *et al.* 'A D-optimal design to model the performances of dressings and devices for negative pressure wound therapy'. *Journal of Tissue Viability*. 2016;**25**(2):83–90.

[112] Battaglia T.M., Masson J.F., Sierks M.R., *et al.* 'Quantification of cytokines involved in wound healing using surface plasmon resonance'. *Analytical Chemistry*. 2005;**77**(21):7016–23.

[113] Pache S., Angeloni S., Ischer R., Liley M., Luprano J., Voirin G. 'Wearable biosensors for monitoring wound healing'. *Advances in Science and Technology*. 2008;**57**:80–7.

[114] Kim D.H., Wang S., Keum H., *et al.* 'Thin, flexible sensors and actuators as "instrumented" surgical sutures for targeted wound monitoring and therapy'. *Small*. 2012;**8**(21):3263–8.

[115] Guionovart T., Valdés-Ramírez G., Windmiller J.R., Andrade F.J., Wang J. 'Bandage-based wearable potentiometric sensor for monitoring wound pH'. *Electroanalysis*. 2014;**26**(6):1345–53.

[116] Du Y.C., Lim B.Y., Ciou W.S., Wu M.J. 'Novel wearable device for blood leakage detection during hemodialysis using an array sensing patch'. *Sensors*. 2016;**16**(6):849.

[117] *PTS Diagnostics*. Available from http://www.ptsdiagnostics.com/ (2015)

[118] Nicolau I., Tian L., Menzies D., Ostiguy G., Pai M. 'Point-of-care urine tests for smoking status and isoniazid treatment monitoring in adult patients'. *PLoS ONE*. 2012;**7**(9):e45913.

[119] *CDXLife*. Available from http://www.cdxlife.com/mydx-analyzer-sensor/ (2016)

[120] Kosasih K., Abeyratne U.R., Swarnkar V., Triasih R. 'Wavelet augmented cough analysis for rapid childhood pneumonia diagnosis'. *IEEE Transactions on Biomedical Engineering*. 2015;**62**(4):1185–94.

[121] *ResAppHealth*. Available from http://www.resapphealth.com.au (2016)

[122] Larson E.C., Goel M., Boriello G., Heltshe S., Rosenfeld M., Patel S.N. 'SpiroSmart!: using a microphone to measure lung function on a mobile phone'. *Proceedings of the 14th International Conference on Ubiquitous Computing*, Pittsburgh, PA, Sep 2012.

[123] *Savonix*. Available from https://www.savonix.com/ (2014)

[124] Perlin M.W., Dormer K., Hornyak J., Schiermeier-Wood L., Greenspoon S. 'TrueAllele casework on Virginia DNA mixture evidence: computer and manual interpretation in 72 reported criminal cases'. *PLoS ONE*. 2014;**9**(3): e92837.

[125] Khalid T., Aggio R., White P., *et al.* 'Urinary volatile organic compounds for the detection of prostate cancer'. *PLoS ONE*. 2015;**10**(11):e0143283.

[126] Pardee K., Green A.A., Takahashi M.K., *et al.* 'Rapid, low-cost detection of Zika virus using programmable biomolecular components'. *Cell*. 2016; **165**(5):1255–66.

[127] Nakamura Y., Maeda S., Nishiyama H., Ohira S., Dasgupta P.K., Toda K. 'Micro ion extractor for single drop whole blood analysis'. *Analytical Chemistry*. 2015;**87**(13):6483–6.

[128] Piraino F., Volpetti F., Watson C., Maerkl S.J. 'A digital-analog microfluidic platform for patient-centric multiplexed biomarker diagnostics of ultralow volume samples'. *ACS Nano*. 2016;**10**(1):1699–710.

[129] Faye O., Faye O., Soropogui B., *et al.* 'Development and deployment of a rapid recombinase polymerase amplification Ebola virus detection assay in Guinea in 2015'. *Euro Surveill.* 2015;**20**(44):10–18.

[130] Li J., Zrazhevskiy P., Gao X. 'Eliminating size-associated diffusion constraints for rapid on-surface bioassays with nanoparticle probes'. *Small.* 2016;**12**(8):1035–43.

[131] Kidd B.A., Hoffman G., Zimmerman N., *et al.* 'Evaluation of direct-to-consumer low-volume lab tests in healthy adults'. *The Journal of Clinical Investigation*. 2016;**126**(5):1734–44.

[132] Li M., Diamandis E.P. 'Theranos promises a new era of preventive health care – but where's the physician?'. *Clinical Biochemistry*. 2015;**48** (16–17):1027.

[133] Fairhurst C., Watt I., Martin F., Bland M., Brackenbury W.J. 'Exposure to sodium channel-inhibiting drugs and cancer survival: protocol for a cohort study using the QResearch primary care database'. *BMJ Open*. 2014;**4**(11): e006604.

[134] Perez M.V., Pavlovic A., Shang C., *et al.* 'Systems genomics identifies a key role for hypocretin/orexin receptor-2 in human heart failure'. *Journal of the American College of Cardiology*. 2015;**66**(22):2522–33.

[135] Beltran H., Antonarakis E.S., Morris M.J., Attard G. 'Emerging molecular biomarkers in advanced prostate cancer: translation to the clinic'. *American Society of Clinical Oncology Educational Book*. 2016;**35**:131–41.

[136] Paczesny S., Abu Zaid M. 'CXCL10: most consistent cGVHD biomarker?'. *Blood*. 2016;**127**(24):2950–1.

[137] Zheng P., Fang Z., Xu X.J., *et al.* 'Metabolite signature for diagnosing major depressive disorder in peripheral blood mononuclear cells'. *Journal of Affective Disorders*. 2016;**195**:75–81.

[138] Wang Y., Ma S. 'Risk factors for etiology and prognosis of mantle cell lymphoma'. *Expert Review of Hematology*. 2014;**7**(2):233–43.

[139] Walters B., Thompson S.R. 'Cap-independent translational control of carcinogenesis'. *Frontiers in Oncology*. 2016;**6**:128.

[140] Hough K.P., Chisolm D.A., Weinmann A.S. 'Transcriptional regulation of T cell metabolism'. *Molecular Immunology*. 2015;**68**(2 Pt C):520–6.

[141] Tian J., Li Z., Han Y., Jiang T., Song X., Jiang G. 'The progress of early growth response factor 1 and leukemia'. *Intractable and Rare Diseases Research*. 2016;**5**(2):76–82.

Chapter 16

A market case report: point-of-care infusion management and intelligent patient monitoring

Alexandre Tsoukalis[1]

16.1 Infusion and infusion management at point-of-care today

Infusion, compared to injection, is a longer time medication delivery, from a few minutes to a week. It can be done by several routes, IV (parenteral), epidural, subcutaneous, perineural (regional analgesia), intrathecal and enteral. A needle is delivering the medication to the body in most cases, but in some others a catheter is transferring the medication deeper in the body to a specific location such as duodenum or vein.

Infusion at point-of-care (POC) is meant infusions in hospital setting and also infusions at home care. General infusion of fluids at both sites in majority is performed by gravity infusion sets (Figure 16.1), and only drugs are infused by LVPs (large volumetric pumps), syringe pumps very popular in Europe or ambulatory pumps. In home care, elastomeric pumps are very popular, but their larger market chemotherapy is lately changing back to ambulatory pumps.

Reported complications are also known for gravity infusions, especially for parenteral nutrition very popular as gravity delivery in Germany. Complications with air in line, bad control of flow rate and non-delivery are described in the literature.

16.2 Infusion pump basics

An infusion pump has a pumping mechanism that is a syringe pushed by a plunger or a peristaltic mechanism with a resilient tube squeezed by fingers or rollers, or an elastomeric material inflated by the drug and expelled in controlled slow way by a tubing bottleneck, or many other techniques that have been developed for Bolus Injectors market, that is for quick less than an hour infusions. A pump has most cases a battery, a buzzer for alarms, a display and keyboard, and motor with a resolver for controlling its speed.

[1]Micrel Medical Devices S.A., Gerakas, Greece.

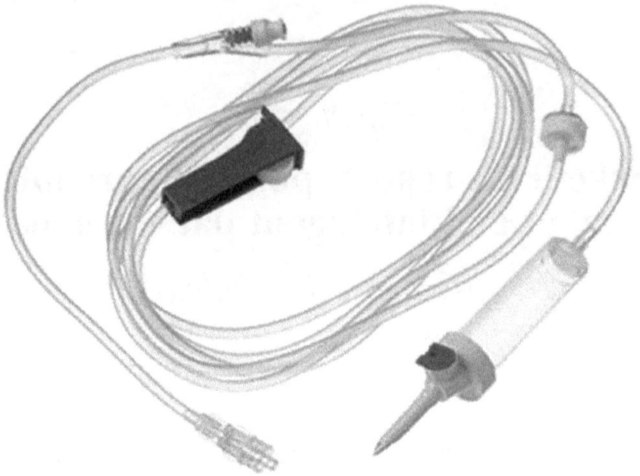

Figure 16.1 Gravity infusion set

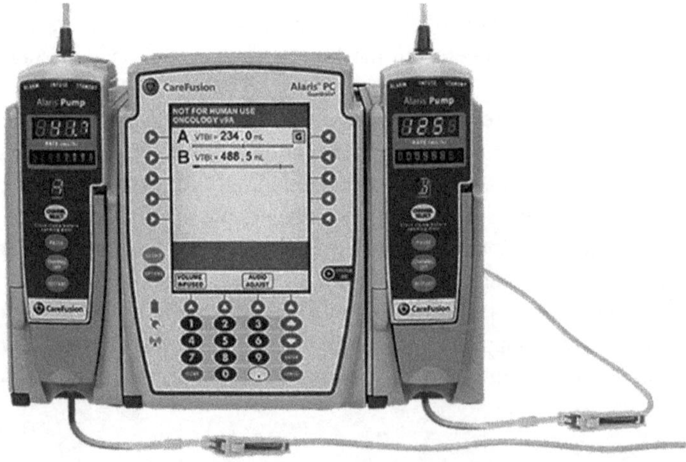

Figure 16.2 LVP pump (CareFusion)

Infusion pumps called 'Bedside' or LVP (Figures 16.2 and 16.3) by far have linear peristaltic mechanisms, that is a cam shaft sequentially pushing fingers against rubber or plastic tubing so drug flows in one direction that of the 'wave' of the fingers. Linear peristaltic achieves 5 per cent accuracy, acceptable in most infusion cases; the problem is that nominal accuracy is not achievable in practice, as most types of tubing are deformed after prolonged use and nominal accuracy is not true at the end of infusion, so it is often that under-infusion is happening by

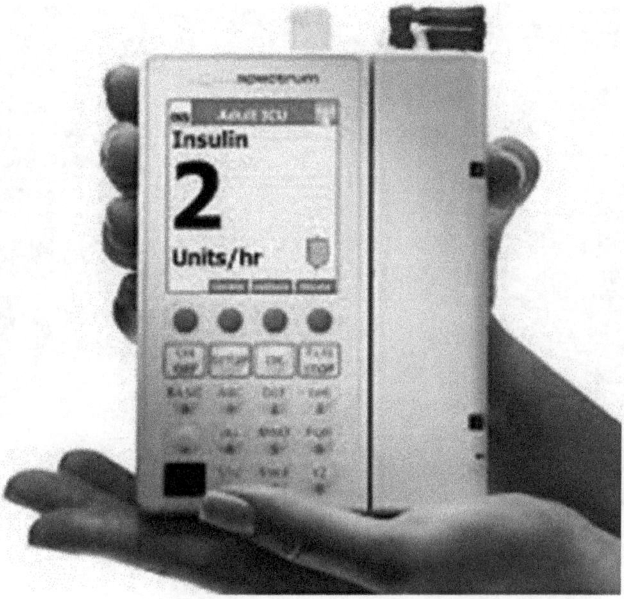

Figure 16.3 LVP pump (Baxter)

some percentage. Silicone tubing is accurate in long periods of infusion, but being more expensive is not preferred material from most pump manufacturers.

Infusion protocols need in simplest form an infusion rate in mL h^{-1}, usually from 0.1 to 1,200 mL h^{-1}.

More complex protocols use two of the three parameters (rate, volume to be infused (VTBI), time) to determine the third one as Rate = VTBI/Time. Most common is to define Rate and VTBI and get time of infusion, or pump to calculate Rate from Volume and Time definition.

Rate can be in mL h^{-1} or if a known concentration of drug is used in mg h^{-1} or mcg h^{-1}, or if body weight is used in mg kg^{-1} min^{-1}, so the pump asks the patient's weight to calculate the rate in mL h^{-1}.

More complex protocols use ramps for starting and ending the infusion, such as in parenteral nutrition to allow the body to control its glucose levels without hyperglycaemia/hypoglycaemia.

Infusion pumps have many alarm functions on any possible infusion or setup complication and a reporting system to hospital computer systems; they can also get from e-prescription/electronic medical record the infusion protocols in paper-less therapy systems. For this, there are Interoperability standards that allow devices like pumps to talk to therapy systems used by hospital like the ones of EPIC, CERNER and others and download drug libraries edited by a Hospital Care Area/Point of Care.

LVP pump biggest manufacturers are CareFusion/Becton Dickinson (BD), BBraun, Baxter, Fresenius and Hospira/Pfizer. All of them have branded safety IT

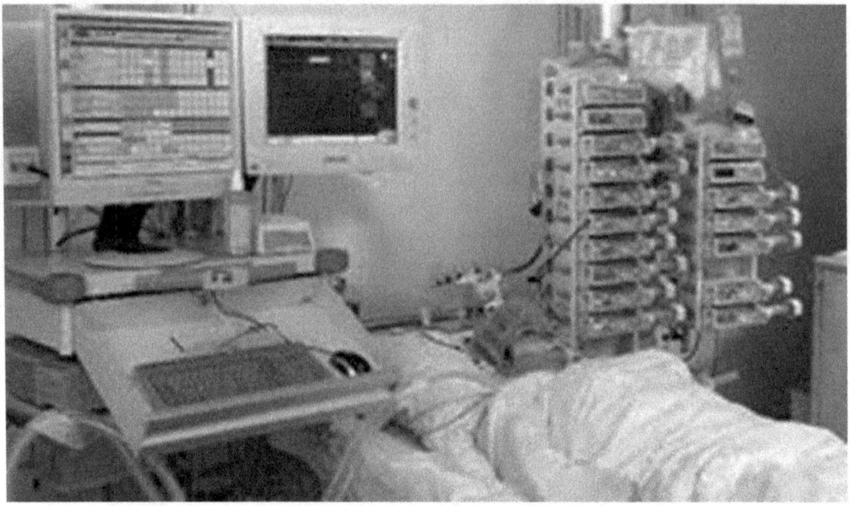

Figure 16.4 Multiple syringe pumps (B. Brown)

Figure 16.5 Elastomeric pumps

systems based on Drug Libraries and bar code scanning with central screen for multiple pump control.

Most of them have also syringe pumps that are working alone or on a pump-stack, controlled by an IT safety system and central screen (Figure 16.4). Syringe pumps are used in Europe much more that in the United States.

Elastomeric pumps are infusion devices that inflate with drug and have a flow restrictor that allows a specific infusion rate depending on drug viscosity (Figure 16.5). They have positive pressure so they are safer than gravity infusions but have not accurate infusion rate. They are disposable, single use and largely more expensive than an infusion set.

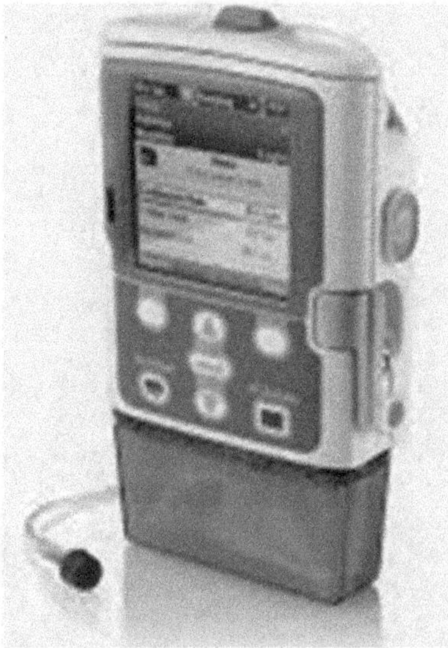

Figure 16.6 Ambulatory pump (SIMMS DELTEC)

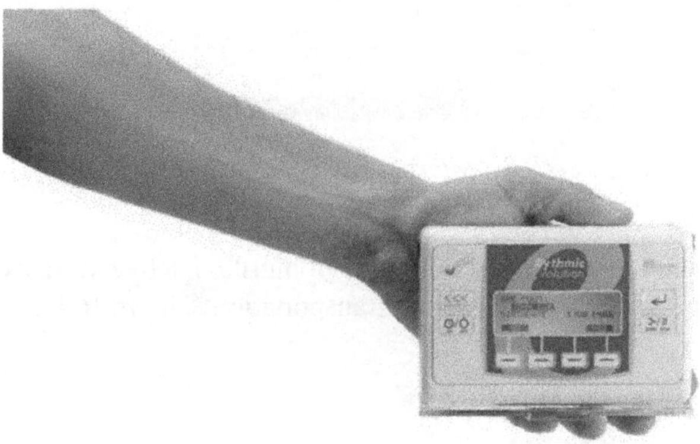

Figure 16.7 Ambulatory pump (Micrel Medical Devices)

Ambulatory infusion pump manufacturers are Deltec/Smiths Medical (Figure 16.6), Caesarea Medical Electronics Ltd, Israel/CareFusion/BD, Micrel Medical Devices (Figure 16.7) and Q-Core/Hospira/Pfizer. These pumps have longer battery autonomy and limited maximum infusion rate to usually 600 mL h^{-1} compared to 1,200 mL h^{-1} of LVPs. Two of them Micrel and Smiths have integrated drug reservoir compartment for true ambulatory use; the others being mostly pole mounted. For volumes larger than 200 mL or pre-filled reservoirs, a lockbox is enclosing drug for analgesia, prohibiting self-administration from patient.

Figure 16.8 Micrel Medical Devices rucksack

For parenteral nutrition needing litters of nutrition delivery, rucksacks are used carrying pumps and liquids, so easing transportation (Figure 16.8).

16.3 Infusion 'smart pumps'

Medication safety is a big concern since deaths from wrong infusion management are reported, so 5R:

- Right medication
- To right patient
- With right protocol
- At right time
- Through right delivery route

5R safety is being a standard in most developed countries over the world and so-called smart pumps are having the means to assure medication safety with 5R.

To do so, a nurse scans a patient ID and drug reservoir label by bar code scanner as a bedside personal computer (PC) or pump accessory; the drug is recognized by name and concentration from an embedded drug library, from patient name the pump downloads protocol from Hospital electronic health management systems, as well as time to infuse and delivery route. Nurse checks that all 5R are OK! and hospital system allows infusion.

16.4 Problems of infusion pumps at point of care

Safety of a single infusion is relatively assured with 5R, as scanning of a single drug is safe. But multiple infusions safety at POC is still to be found, since a nurse under pressure being in a hurry, can scan an adjacent, different drug than the one connected to the pump, just because of a mess of upstream tubing from pumps below to reservoirs on top makes difficult the connection verification. Medication errors are killing hundreds of patients in the United States (where official reports exist) and worldwide. Bar-code scanning and drug libraries have reduced medication errors but have not eliminated them. This is true especially in piggyback infusions in which pumps, using a Y connection upstream and two reservoirs at different height, in which the highest is only infusing by hydrostatic pressure difference, and infusion rates are different, the larger volume normally at higher rate than smaller drug volume. Errors are reported when nurses exchange the height without exchanging infusion rates.

Bedside and most of the times even ambulatory infusion pumps are pole mounted, and needing mains connection for longer infusions. This makes gravity infusions a by-far preferred infusion method both at hospital and home care, besides complications.

All peristaltic pumps have a ‘pulsed’ infusion causing sometimes a reflow, that is blood aspiration in the infusion catheter, due to momentary negative pressure of the pumping mechanism. This is caused by the geometry of the last finger when it disengages and aspirates from the output, whereas the first has just been engaged. Some proposals have been made and some few pumps have reduced pulse.

Syringe pumps are a pump category with linear not pulsed infusion, since a screw is pushing a syringe continuously. So they are preferred for drugs with small half-life in the body that need an accurate even at small time window (small fluctuation) flow. Syringe pumps have another advantage, they have no upstream tubing, so the drug you see written on the syringe is the one you are infusing, and so in cases of eight syringe pumps usually used in an operation, errors as described above for upstream mess, are not happening. But so many pumps are bulky and take a lot of floor space that is a problem in operating rooms.

16.5 Micrel Medical Devices patented innovation solving problems at point of care

Micrel Medical Devices has a line of ambulatory infusion pumps, called Rythmic, with integrated drug lock case, for 100, 200 and 500 mL, associated with an infusion and therapy monitoring system web based, called MicrelCare, so a therapist can get therapy feedback from any web interface, a unique feature in industry [1].

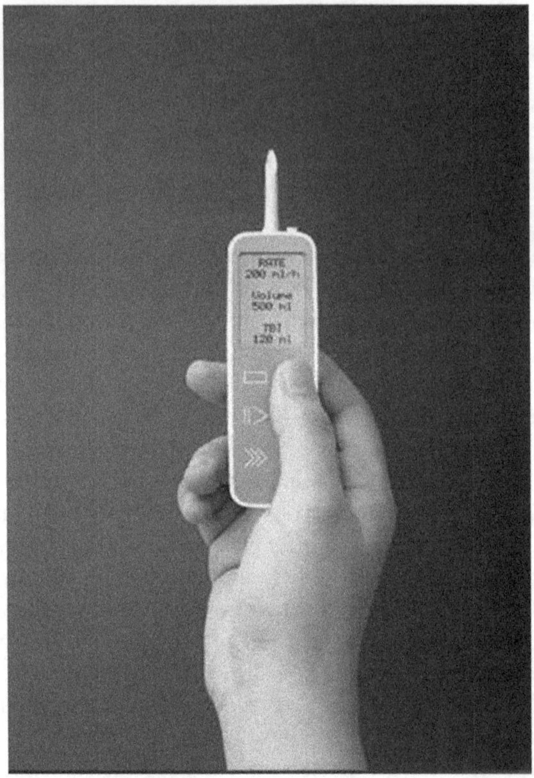

Figure 16.9 Micrel Medical Devices miniature LVP handy single finger operation

This feature is very much appreciated at home care in which long distances make communication more difficult.

So Rythmic pumps as an Internet of Things network are connecting with Global System for Mobile communications (GSM)/General Packet Radio Service (data over cellular network carrier) and Wi-Fi to a cloud-based server/data centre continuously and send infusion status data, but also patient feedback like bolus button press or answers to questions on the pump, each question set specific for a therapy. All this information is organized on a web page specific for each therapy, a page per patient and an alarm monitor for all therapists' patients. Parenteral and enteral nutrition daily infusion per liquid type is organized for easy view over a week, or month; therefore, multiple infusions are shown for a patient as care continuum. Also pressure in the catheter is recorded and can predict possible catheter occlusion, to clean it and prevent severe complications. Pain level for palliative care is recorded and displayed, and questions regarding motor blockage or pain in peripheral nerve block are shown for corrective therapeutic actions.

Micrel Rythmic pumps have Radio Frequency Identification (RFID) tag reading capability, so they can recognize a smart drug label on the reservoir loaded in integrated drug compartment, avoiding medication errors as pump displays drug

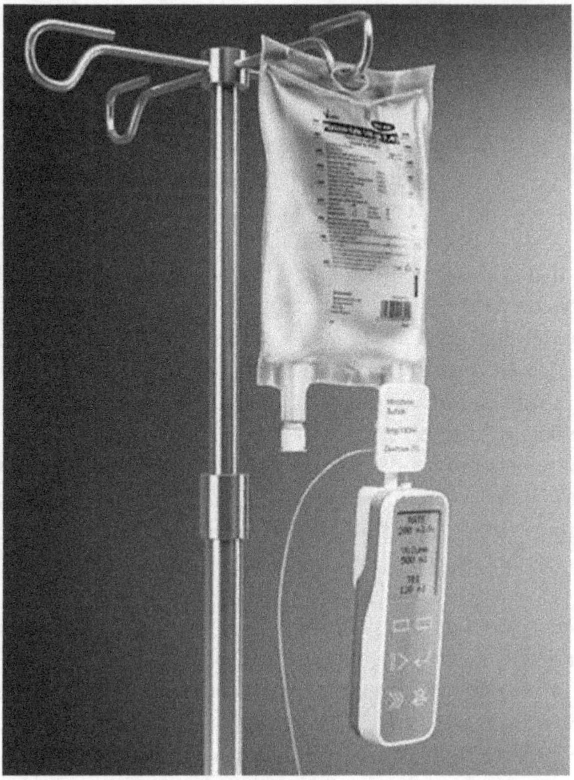

Figure 16.10 Micrel Medical Devices miniature LVP hanging from the spike

name and concentration and adapts accordingly to drug library limits. Pumps also have location tracking through GSM and Wi-Fi being displayed on Google maps at which location they are, so patient identification at home is automatic and infusion protocol is downloaded from Electronic Prescription on MicrelCare through interoperability standards from any other system.

A new type of pump (Figure 16.9) is going to be launched by 2018, which uses a single-use cartridge containing rotary peristaltic pumping mechanism [2], having a number of advantages.

The mechanism being rotary and all plastic is small and minimal weight, so the pump practically is a controller with the motor, battery and display, at unbelievable small size and weight overall, in which cartridge with spike is snap fitted at its back side. The pump/controller looks like a small TV remote control, size and form makes it handy to operate with a single finger as shown Figure 16.9. The technology of the cartridge is so high, and the power consumption is so low that a battery charge can infuse all liquids needed by a patient for his stay at hospital, usually less than 7 L.

The weight of the system being so low in the order of 100 g allows it to be hanged from the spike (Figure 16.10) whose standards need 1.5 kg force to disengage from a reservoir tube. Being hanged from an integrated into cartridge spike, the pump does not need a pole clamp, so nurse caries pump in her pocket and plugs it directly on the reservoir just as any gravity set and with just one charge manages all patient's infusions during his hospital stay.

The safety and usability of the system is unparalleled [3], as the pump can read bar-codes or RFID labels as a classic bar-code reader being handy as a TV remote control. So nurse can scan a drug label very close to the spike connection, so there is very limited possibility of mixed spike/drug association. There is a 100 per cent safety solution that is described below for piggyback infusions and RFID labeled reservoirs always read during infusion for single reservoir infusions. The nurse also scans patient's ID label, and through its integral Wi-Fi connection it sends the information to MicrelCare server, from which it receives all 5R to be checked on the pump, and latest drug library protocol limits. This is a pump with connectivity checking 5R, whereas in today's systems it is a PC that checks 5R.

The system is best for piggyback infusions, for which safety measures include label readers associated with flow controllers, so in case a drug reservoir is changing position from low to high, flow sensor will make pump to change rate for the actually flowing drug, and pump knows which drug from the label reader on the connection.

The pump also has a pulseless flow [4], and also high accuracy matching flow linearity and accuracy of best syringe pumps, even at 1,500 mL h^{-1} rate without reflow. This allows a single pump in a Hospital do almost all pumping functions, replacing LVPs, ambulatory pumps, elastomeric pumps and even bolus infusers.

The pump has a small display for more basic functions but internally can infuse very complex protocols, for this and for multiple pump use, an external larger display can be used pole mounted, which can programme and monitor many pumps with easiness.

References

[1] US patents US8551038B2.
[2] US patents US9468715B2.
[3] US patent applications US20160051750A1.
[4] US patent applications US20160123320A1.

Index